SECOND EDITION

IMAGE ANALYSIS, CLASSIFICATION, and CHANGE DETECTION in REMOTE SENSING

with Algorithms for ENVI/IDL

SECOND EDITION

IMAGE ANALYSIS, CLASSIFICATION, and CHANGE DETECTION in REMOTE SENSING

with Algorithms for ENVI/IDL

Morton J. Canty

CRC Press
Taylor & Francis Group
Boca Raton London New York

CRC Press is an imprint of the
Taylor & Francis Group, an **informa** business

CRC Press
Taylor & Francis Group
6000 Broken Sound Parkway NW, Suite 300
Boca Raton, FL 33487-2742

Printed in the United States of America on acid-free paper
10 9 8 7 6 5 4 3 2 1

International Standard Book Number: 978-1-4200-8713-0 (Hardback)

Library of Congress Cataloging-in-Publication Data

Canty, Morton John.
Image analysis, classification, and change detection in remote sensing : with algorithms for ENVI/IDL / author, Morton J. Canty. -- 2nd ed.
p. cm.
Includes bibliographical references and index.
ISBN 978-1-4200-8713-0 (hardcover : alk. paper)
1. Remote sensing--Mathematics. 2. Image analysis--Mathematics. 3. Image analysis--Data processing. I. Title.

G70.4.C36 2010
526.9'82--dc22 2009042041

Visit the Taylor & Francis Web site at
http://www.taylorandfrancis.com

and the CRC Press Web site at
http://www.crcpress.com

Contents

Preface to the Second Edition

Shortly after the manuscript for the first edition of this book went to the publisher, ENVI 4.3 appeared along with, among other new features, a support vector machine (SVM) classifier. Although my decision not to include the SVM in the original text was a conscious one (I balked at the thought of writing my own IDL implementation), this event did point to a rather glaring omission in a book purporting to be partly about land-use/land-cover classification. So, almost immediately, I began to dream of a revised second edition and to pester CRC Press for a contract. This was happily forthcoming and the present edition now has a fairly long section on supervised classification with SVMs.

The SVM is just one example of so-called kernel methods for nonlinear data analysis, and I decided to make kernelization one of the themes of the revised edition. The treatment begins with a dual formulation for ridge regression in Chapter 2 and continues through kernel principal components analysis in Chapters 3 and 4, SVMs in Chapter 6, kernel K-means clustering in Chapter 8, and nonlinear change detection in Chapter 9. Other new topics include entropy and mutual information (Chapter 1), adaptive boosting (Chapter 7), and image segmentation (Chapter 8). In order to accommodate the extended material on supervised classification, the discussion is now spread over both Chapters 6 and 7. The exercises at the end of each chapter have been extended and reworked and, as for the first edition, a solutions manual has been provided.

I have written several additional IDL extensions to ENVI to accompany the new themes, which can be downloaded, together with updated versions of previous programs, from the Internet. In order to accelerate some of the more computationally intensive routines for users with access to CUDA (parallel processing on NVIDIA graphics processors), code is included that can make use of the IDL bindings to CUDA provided by Tech-X Corporation in their GPULib product:

http://gpulib.txcorp.com

Notwithstanding the revisions, this edition remains a monograph on pixel-oriented analysis of intermediate resolution remote-sensing imagery with an emphasis on the development and programming of statistically motivated, data-driven algorithms. Important topics such as object-based feature analysis (for high-resolution imagery), or the physics of the radiation/surface interaction (e.g., in connection with hyperspectral sensing) are only touched upon briefly, and the vast field of radar remote sensing is left out completely. Nevertheless, I hope that the in-depth focus on the topics covered will continue to be of use both to practitioners as well as to teachers.

I would like to express my appreciation to Peter Reinartz and the German Aerospace Center for permission to use the traffic scene images in Chapter 9 and to NASA's Land Processes Distributed Active Archive Center for free and uncomplicated access to archived ASTER imagery. Thanks also go to Peter Messmer and Michael Galloy, Tech-X Corporation, for their prompt responses to my many cries for help with GPULib. I am especially grateful to my colleagues Harry Vereecken and Allan Nielsen, the former for generously providing me with the environment and resources needed to complete this book, the latter for the continuing inspiration of our friendship and long-time collaboration.

Morton John Canty

Preface to the First Edition

This textbook had its beginnings as a set of notes to accompany seminars and lectures conducted at the Geographical Institute of Bonn University and at its associated Center for Remote Sensing of Land Cover. Lecture notes typically continue to be refined and polished over the years until the question inevitably poses itself: "Why not have them published?" The answer of course is "By all means, if they contribute something new and useful."

So what is "new and useful" here? This is a book about remote sensing image analysis with a distinctly mathematical-algorithmic-computer-oriented flavor, intended for graduate level teaching and with, to borrow from the remote sensing jargon, a rather restricted FOV. It does not attempt to match the wider *fields of view* of existing texts on the subject, such as Schowengerdt (1997); Richards and Jia (2006); Jensen (2005) and others. However the topics that are covered are dealt with in considerable depth, and I believe that this coverage fills an important gap. Many aspects of the analysis of remote sensing data are quite technical and tend to be intimidating to students with moderate mathematical backgrounds. At the same time one often witnesses a desire on the part of students to apply advanced methods and to modify them to fit their particular research problems. Fulfilling the latter wish, in particular, requires more than superficial understanding of the material.

The focus of the book is on pixel-oriented analysis of visual/infrared Earth observation satellite imagery. Among the topics that get the most attention are the discrete wavelet transform, image fusion, supervised classification with neural networks, clustering algorithms and statistical change detection methods. The first two chapters introduce the mathematical and statistical tools necessary in order to follow later developments. Chapters 3 and 4 deal with spatial/spectral transformations, convolutions and filtering of multispectral image arrays. Chapter 5 treats image enhancement and some of the preprocessing steps that precede classification and change detection. Chapters 6 and 7 are concerned, respectively, with supervised and unsupervised land cover classification. The last chapter is about change detection with heavy emphasis on the use of canonical correlation analysis. Each of the 8 chapters concludes with exercises, some of which are small programming projects, intended to illustrate or justify the foregoing development. Solutions to the exercises are included in a separate booklet. Appendix A provides some additional mathematical/statistical background and Appendix B develops two efficient training algorithms for neural networks. Finally, Appendix C describes the installation and use of the many computer programs introduced in the course of the book.

I've made considerable effort to maintain a consistent, clear mathematical style throughout. Although the developments in the text are admittedly uncompromising, there is nothing that, given a little perseverance, cannot be followed by a reader who has grasped the elementary matrix algebra and statistical concepts explained in the first two chapters. If the student has ambitions to write his or her own image analysis programs, then he or she must be prepared to "get the maths right" beforehand. There are, heaven knows, enough pitfalls to worry about thereafter.

All of the illustrations and applications in the text are programmed in RSI's ENVI/IDL. The software is available for download at the publisher's website:

http://www.crcpress.com/e_products/downloads/default.asp

Given the plethora of image analysis and geographic information system (GIS) software systems on the market or available under open source license, one might think that the choice of computer environment would have been difficult. It wasn't. IDL is an extremely powerful, array- and graphics-oriented, universal programming language with a versatile interface (ENVI) for importing and analyzing remote sensing data—a peerless combination for my purposes. Extending the ENVI interface in IDL in order to implement new methods and algorithms of arbitrary sophistication is both easy and fun.

So, apart from some exposure to elementary calculus (and the aforesaid perseverance), the only other prerequisites for the book are a little familiarity with the ENVI environment and the basic knowledge of IDL imparted by such excellent introductions as Fanning (2000) or Gumley (2002). For everyday problems with IDL at any level from "newbie" on upward, help and solace are available at the newsgroup

comp.lang.idl-pvwave

frequented by some of the friendliest and most competent gurus on the net.

I would like to express my thanks to Rudolf Avenhaus and Allan Nielsen for their many comments and suggestions for improvement of the manuscript and to CRC Press for competent assistance in its preparation. Part of the software documented in the text was developed within the Global Monitoring for Security and Stability (GMOSS) network of excellence funded by the European Commission.

Morton John Canty

1

Images, Arrays, and Matrices

There are many satellite-based sensors currently in orbit that are used routinely for the observation of the Earth. Representative of these we describe briefly the ASTER system (Abrams et al., 1999); see Jensen (2005) for an overview of remote sensing platforms.

The Advanced Spaceborne Thermal Emission and Reflectance Radiometer (ASTER) was launched in December 1999 on the *Terra* spacecraft. It is being used to obtain detailed maps of land surface temperature, reflectance, and elevation, and consists of sensors to measure radiance in three spectral intervals:

- VNIR: Visible and near-infrared bands 1, 2, 3N, and 3B, in the spectral region between 0.52 and 0.86 μm (four arrays of charge-coupled detectors [CCDs] in pushbroom scanning mode).
- SWIR: Short-wavelength infrared bands 4–9 in the region between 1.60 and 2.43 μm (six cooled PtSi–Si Schottky barrier arrays, pushbroom scanning).
- TIR: Thermal infrared bands 10–14 covering a spectral range from 8.13 to 11.65 μm (cooled HgCdTe detector arrays, whiskbroom scanning).

The altitude of the spacecraft is 705 km. Ground resolutions or, more precisely, the across- and in-track *ground sample distances* (GSDs), i.e., the detector widths projected through the system optics onto the Earth's surface, are 15m (VNIR), 30m (SWIR), and 90m (TIR). The telescope associated with the 3B sensors is back-looking at an angle of 27.6° to provide, together with the 3N sensors, along-track stereo image pairs. In addition, the VNIR camera can be rotated from straight down (nadir) to ±24° across-track. The SWIR and TIR instrument mirrors can be pointed to ±8.5° across-track. Like most platforms in this ground resolution category, the orbit is near polar, sun-synchronous. Quantization levels are 8 bits for VNIR and SWIR bands and 12 bits for TIR bands. The sensor systems have an average duty cycle of 8% per orbit (about 650 scenes per day, each $60\times60\,\text{km}^2$ in area) with revisit times between 4 and 16 days.

Figure 1.1 shows a spatial/spectral subset of an ASTER scene. The image is a UTM (Universal Transverse Mercator) projection oriented along the satellite path (rotated approximately 16.4° from north) and orthorectified using a digital terrain model generated from the stereo bands.

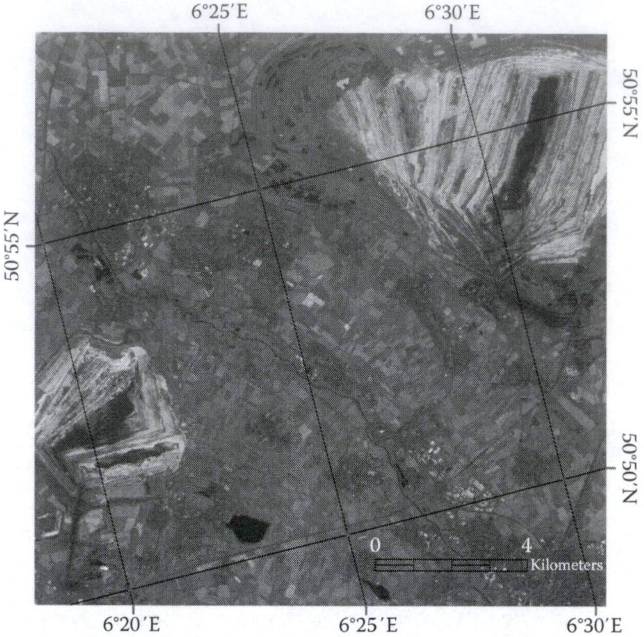

FIGURE 1.1
(See color insert following page 114.) ASTER color composite image (1000 × 1000 pixels) of VNIR bands 1 (blue), 2 (green), and 3N (red) over the town of Jülich in Germany, acquired on May 1, 2007. The bright areas are open cast coal mines.

1.1 Multispectral Satellite Images

A multispectral image like that shown in Figure 1.1 may be represented as a three-dimensional array of grayscale values or pixel intensities

$$g_k(i,j), \quad 1 \le i \le c,\ 1 \le j \le r,\ 1 \le k \le N,$$

where
c is the number of pixel columns (also called *samples*)
r is the number of pixel rows (or *lines*)

The index k denotes the spectral band, of which there are N in all. For data at an early processing stage a pixel may be stored as a *digital number* (DN), often in a single byte, so that $0 \le g_k \le 255$. This is the case for the ASTER VNIR and SWIR bands at the processing level L1A (unprocessed reconstructed instrument data), whereas the L1A TIR data are quantized to 12 bits (as unsigned integers) and thus stored as digital numbers from 0 to

$2^{12}-1=4095$. Processed image data may, of course, be stored in byte, integer, or floating point format and can have negative or even complex values.

The grayscale values in the various bands encode measurements of radiance, $L_{\Delta\lambda}(x,y)$, in a wavelength band, $\Delta\lambda$, due to sunlight reflected from some point (x,y) on the Earth's surface, or due to thermal emission from that surface, and focused by the instrument's optical system along the array of sensors. Ignoring all absorption and scattering effects of the intervening atmosphere, the at-sensor radiance available for measurement from reflected sunlight from a horizontal, *Lambertian* surface, i.e., a surface which scatters reflected radiation uniformly in all directions, is given by

$$L_{\Delta\lambda}(x,y) = E_{\Delta\lambda} \cdot \cos\theta_z \cdot R_{\Delta\lambda}(x,y)/\pi. \tag{1.1}$$

The units are $\mathrm{w}/(\mathrm{m}^2 \cdot \mathrm{sr} \cdot \mu\mathrm{m})$, $E_{\Delta\lambda}$ is the average spectral solar irradiance in the spectral band $\Delta\lambda$, θ_z is the solar zenith angle, $R_{\Delta\lambda}(x,y)$ is the surface reflectance at coordinates (x,y), a number between 0 and 1, and π accounts for the upper hemisphere of solid angle. The conversion between DN and the at-sensor radiance is determined by the sensor calibration as measured (and maintained) by the satellite image provider. For example, for ASTER VNIR and SWIR L1A data,

$$L_{\Delta\lambda}(x,y) = A \cdot DN/G + D. \tag{1.2}$$

The quantities A (linear coefficient), G (gain), and D (offset) are tabulated for each of the detectors in the arrays and included with each acquisition. Atmospheric scattering and absorption models may be used to deduce at-surface radiance, surface temperature, and emissivity or surface reflectance from the observed radiance at the sensor. Reflectance and emissivity are directly related to the physical properties of the surface being imaged. See Schowengerdt (1997) for a thorough discussion of atmospheric effects and their correction.

Various conventions are used for storing the image array $g_k(i,j)$ in computer memory or other storage media. In the *band interleaved by pixel* (BIP) format, for example, a two-channel, 3×3 pixel image would be stored as

$$\begin{matrix} g_1(1,1) & g_2(1,1) & g_1(2,1) & g_2(2,1) & g_1(3,1) & g_2(3,1) \\ g_1(1,2) & g_2(1,2) & g_1(2,2) & g_2(2,2) & g_1(3,2) & g_2(3,2) \\ g_1(1,3) & g_2(1,3) & g_1(2,3) & g_2(2,3) & g_1(3,3) & g_2(3,3), \end{matrix}$$

whereas in the *band interleaved by line* (BIL) format, it would be stored as

$$\begin{matrix} g_1(1,1) & g_1(2,1) & g_1(3,1) & g_2(1,1) & g_2(2,1) & g_2(3,1) \\ g_1(1,2) & g_1(2,2) & g_1(3,2) & g_2(1,2) & g_2(2,2) & g_2(3,2) \\ g_1(1,3) & g_1(2,3) & g_1(3,3) & g_2(1,3) & g_2(2,3) & g_2(3,3), \end{matrix}$$

Listing 1.1
Reading and displaying an ENVI image in IDL.

```
PRO EX1_1

envi_select, title='Choose␣multispectral␣image', $
             fid=fid, dims=dims,pos=pos
IF (fid EQ -1) THEN BEGIN
   PRINT, 'cancelled'
   RETURN
ENDIF

envi_file_query, fid, fname=fname

num_cols = dims[2]-dims[1]+1
num_rows = dims[4]-dims[3]+1
num_bands = n_elements(pos)

; BIP array
image = fltarr(num_bands,num_cols,num_rows)

FOR i=0,num_bands-1 DO image[i,*,*] = $
   envi_get_data(fid=fid,dims=dims,pos=pos[i])

window, 11, xsize=num_cols, ysize=num_rows, title=fname
tvscl, image[0,*,*]

END
```

and in the *band sequential* (BSQ) format as

$$
\begin{array}{ccc}
g_1(1,1) & g_1(2,1) & g_1(3,1) \\
g_1(1,2) & g_1(2,2) & g_1(3,2) \\
g_1(1,3) & g_1(2,3) & g_1(3,3) \\
g_2(1,1) & g_2(2,1) & g_2(3,1) \\
g_2(1,2) & g_2(2,2) & g_2(3,2) \\
g_2(1,3) & g_2(2,3) & g_2(3,3).
\end{array}
$$

In the computer language IDL, the so-called *column major indexing* is used for arrays (Gumley, 2002), and the elements in an array are numbered from zero. This means that if a grayscale image g is assigned to an IDL array variable `G`, then the intensity value $g(i,j)$ is addressed as `G[i-1,j-1]`. An N-band multispectral image is stored in the BIP format as an $N \times c \times r$ array in IDL, in the BIL format as a $c \times N \times r$, and in the BSQ format as a $c \times r \times N$ array. So for example in the BIP format, the value $g_k(i,j)$ is stored at `G[k-1,i-1,j-1]`.

Auxiliary information, such as image acquisition parameters and georeferencing, is normally included with the image data on the same file, and the format may or may not make use of compression algorithms. Examples are the GeoTIFF* file format used, for instance, by Space Imaging Inc. for distributing Carterra(c) imagery and which includes lossless compression, the HDF-EOS (Hierarchical Data Format-Earth Observing System) files in which ASTER images are distributed, and the PIX format employed by PCI Geomatics with its image processing software, in which auxiliary information is in plain ASCII and the image data are not compressed. ENVI uses a simple "flat binary" file structure with an additional ASCII header file.

The IDL program shown in Listing 1.1 makes use of the ENVI procedures `ENI_SELECT`, `ENVI_FILE_QUERY`, and the ENVI function `ENVI_GET_DATA()` to read a multispectral image into an IDL array with the BIP interleave format, and then display the first spectral band with the IDL procedures, `WINDOW` and `TVSCL`. In order to execute the program, ENVI should be up and running.

1.2 Algebra of Vectors and Matrices

It is very convenient, in fact essential, to use a vector representation for multispectral image pixels. In this book, a pixel will be represented in the form

$$g(i,j) = \begin{pmatrix} g_1(i,j) \\ \vdots \\ g_N(i,j) \end{pmatrix}, \tag{1.3}$$

which is understood to be a *column vector* of spectral intensities at the image position (i,j). It can be thought of as a point in N-dimensional Euclidean space, commonly referred to as *input space* or *feature space*.

Since we will be making extensive use of the vector notation of Equation 1.3, some of the basic properties of vectors, and of matrices which generalize them, will be reviewed here. One can illustrate—and the reader can easily verify—many of these properties for two-component vectors and 2×2 matrices. A list of frequently used mathematical symbols is given at the end of the book.

* GeoTIFF is an open source specification and refers to TIFF files that have geographic (or cartographic) data embedded as tags within the file. The geographic data can be used to position the image in the correct location and geometry on the screen of a geographic information display.

1.2.1 Elementary Properties

The *transpose* of the two-component column vector

$$x = \begin{pmatrix} x_1 \\ x_2 \end{pmatrix}, \tag{1.4}$$

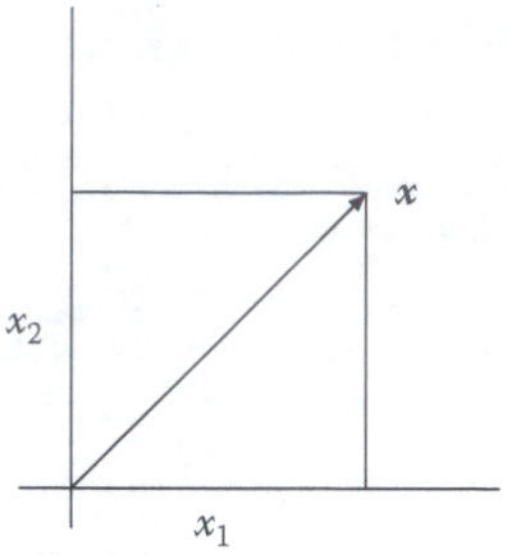

FIGURE 1.2
A vector with two components.

shown in Figure 1.2, is the *row vector*

$$\boldsymbol{x}^\top = (x_1, x_2). \tag{1.5}$$

The *sum* of two column vectors is given by

$$\boldsymbol{x} + \boldsymbol{y} = \begin{pmatrix} x_1 \\ x_2 \end{pmatrix} + \begin{pmatrix} y_1 \\ y_2 \end{pmatrix} = \begin{pmatrix} x_1 + y_1 \\ x_2 + y_2 \end{pmatrix}, \tag{1.6}$$

and their *inner product* by

$$\boldsymbol{x}^\top \boldsymbol{y} = (x_1, x_2) \begin{pmatrix} y_1 \\ y_2 \end{pmatrix} = x_1 y_1 + x_2 y_2. \tag{1.7}$$

The *length* or (2-)*norm* of the vector $\boldsymbol{x}$ is

$$\|\boldsymbol{x}\| = \sqrt{x_1^2 + x_2^2} = \sqrt{\boldsymbol{x}^\top \boldsymbol{x}}. \tag{1.8}$$

The programming language IDL is especially good at manipulating array structures like vectors and matrices. In the following dialog, a column vector is assigned to the variable X and then printed together with its transposed form:

```
IDL> X=[[1],[2]]
IDL> PRINT,X
        1
        2
IDL> PRINT,transpose(X)
        1        2
```

The inner product of $\boldsymbol{x}$ and $\boldsymbol{y}$ can be expressed in terms of the vector lengths and the angle θ between the two vectors:

$$\boldsymbol{x}^\top \boldsymbol{y} = \|\boldsymbol{x}\| \, \|\boldsymbol{y}\| \cos\theta, \tag{1.9}$$

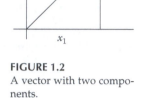

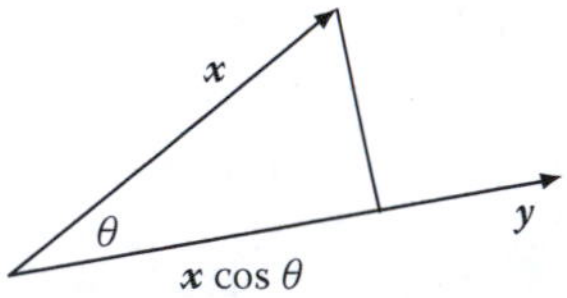

FIGURE 1.3
Illustrating the inner product.

see Figure 1.3 and Exercise 1. If $\theta = 90°$, the vectors are said to be *orthogonal*, in which case $\boldsymbol{x}^\top \boldsymbol{y} = 0$.

Any vector can be decomposed into orthogonal *unit vectors*, for example

$$x = \begin{pmatrix} x_1 \\ x_2 \end{pmatrix} = x_1 \begin{pmatrix} 1 \\ 0 \end{pmatrix} + x_2 \begin{pmatrix} 0 \\ 1 \end{pmatrix} = x_1 \boldsymbol{i} + x_2 \boldsymbol{j}, \tag{1.10}$$

where the symbols $\boldsymbol{i}$ and $\boldsymbol{j}$ denote vectors of unit length along the x and y directions, respectively.

A 2×2 *matrix* is written in the form

$$\boldsymbol{A} = \begin{pmatrix} a_{11} & a_{12} \\ a_{21} & a_{22} \end{pmatrix}. \tag{1.11}$$

Note that the indices of the elements *do not* correspond to the IDL convention. The first index of a matrix element indicates its row, the second its column. Thus IDL may be said to be array oriented but not matrix oriented. This can cause confusion and must be kept in mind when writing programs. *Depending on the context, we will refer to an $n \times m$ IDL array and an $m \times n$ matrix as one and the same object.*

When a matrix is multiplied with a vector, the result is another vector, e.g.,

$$\boldsymbol{A}x = \begin{pmatrix} a_{11} & a_{12} \\ a_{21} & a_{22} \end{pmatrix} \begin{pmatrix} x_1 \\ x_2 \end{pmatrix} = \begin{pmatrix} a_{11}x_1 + a_{12}x_2 \\ a_{21}x_1 + a_{22}x_2 \end{pmatrix} = x_1 \begin{pmatrix} a_{11} \\ a_{21} \end{pmatrix} + x_2 \begin{pmatrix} a_{11} \\ a_{22} \end{pmatrix}.$$

In general, for $\boldsymbol{A} = (\boldsymbol{a}_1, \boldsymbol{a}_2 \ldots \boldsymbol{a}_N)$, where the vectors $\boldsymbol{a}_i$ are the columns of $\boldsymbol{A}$,

$$\boldsymbol{A}x = x_1\boldsymbol{a}_1 + x_2\boldsymbol{a}_2 + \cdots + x_N\boldsymbol{a}_N. \tag{1.12}$$

The IDL operator for matrix and vector multiplication is `##`, for example

```
IDL> A=[[1,2],[3,4]]
IDL> PRINT,A
          1          2
          3          4
IDL> PRINT,A##X
              5
             11
```

The product of two 2×2 matrices is given by

$$\boldsymbol{AB} = \begin{pmatrix} a_{11} & a_{12} \\ a_{21} & a_{22} \end{pmatrix} \begin{pmatrix} b_{11} & b_{12} \\ b_{21} & b_{22} \end{pmatrix} = \begin{pmatrix} a_{11}b_{11} + a_{12}b_{21} & a_{11}b_{12} + a_{12}b_{22} \\ a_{21}b_{11} + a_{22}b_{21} & a_{21}b_{12} + a_{22}b_{22} \end{pmatrix},$$

and is another matrix. More generally, the matrix product $\boldsymbol{AB}$ is allowed whenever $\boldsymbol{A}$ has the same number of columns as $\boldsymbol{B}$ has rows. So if $\boldsymbol{A}$ has a dimension $\ell \times m$ and $\boldsymbol{B}$ has a dimension $m \times n$, then $\boldsymbol{AB}$ is $\ell \times n$ with elements

$$(\boldsymbol{AB})_{ij} = \sum_{k=1}^{m} a_{ik} b_{kj} \quad i = 1 \ldots \ell,\ j = 1 \ldots n. \tag{1.13}$$

Matrix multiplication is not *commutative*, i.e., $\boldsymbol{AB} \neq \boldsymbol{BA}$ in general. However, it is *associative*:

$$(\boldsymbol{AB})\boldsymbol{C} = \boldsymbol{A}(\boldsymbol{BC}) \tag{1.14}$$

so we can write, for example,

$$(\boldsymbol{AB})\boldsymbol{C} = \boldsymbol{A}(\boldsymbol{BC}) = \boldsymbol{ABC} \tag{1.15}$$

without ambiguity. The *outer product* of two vectors of equal length, written $\boldsymbol{xy}^\top$, is a matrix, e.g.,

$$\boldsymbol{xy}^\top = \begin{pmatrix} x_1 \\ x_2 \end{pmatrix} (y_1, y_2) = \begin{pmatrix} x_1 & 0 \\ x_2 & 0 \end{pmatrix} \begin{pmatrix} y_1 & y_2 \\ 0 & 0 \end{pmatrix} = \begin{pmatrix} x_1 y_1 & x_1 y_2 \\ x_2 y_1 & x_2 y_2 \end{pmatrix}. \tag{1.16}$$

Matrices, like vectors, have a transposed form, obtained by interchanging their rows and columns:

$$\boldsymbol{A}^\top = \begin{pmatrix} a_{11} & a_{21} \\ a_{12} & a_{22} \end{pmatrix}. \tag{1.17}$$

Transposition has the properties

$$\begin{aligned} (\boldsymbol{A} + \boldsymbol{B})^\top &= \boldsymbol{A}^\top + \boldsymbol{B}^\top \\ (\boldsymbol{AB})^\top &= \boldsymbol{B}^\top \boldsymbol{A}^\top. \end{aligned} \tag{1.18}$$

1.2.2 Square Matrices

A *square* matrix, $\boldsymbol{A}$, has equal numbers of rows and columns. The *determinant* of a $p \times p$ square matrix is written as $|\boldsymbol{A}|$ and is defined as

$$|\boldsymbol{A}| = \sum_{(j_1 \ldots j_p)} (-1)^{f(j_1 \ldots j_p)} a_{1j_1} a_{2j_2} \cdots a_{pj_p}. \tag{1.19}$$

The sum is over all permutations $(j_1 \ldots j_p)$ of the integers $(1 \ldots p)$ and $f(j_1 \ldots j_p)$ is the number of transpositions (interchanges of two integers) required to change $(1 \ldots p)$ into $(j_1 \ldots j_p)$. The determinant of a 2×2 matrix, for example, is given by

$$|A| = a_{11}a_{22} - a_{12}a_{21}.$$

The determinant has the properties

$$\begin{aligned} |AB| &= |A||B| \\ |A^\top| &= |A|. \end{aligned} \tag{1.20}$$

The *identity matrix* is a square matrix with ones along its diagonal and zeroes everywhere else. For example

$$I = \begin{pmatrix} 1 & 0 \\ 0 & 1 \end{pmatrix},$$

and, for any A,

$$IA = AI = A. \tag{1.21}$$

The *matrix inverse* A^{-1} of a square matrix A is defined in terms of the identity matrix by the requirements

$$A^{-1}A = AA^{-1} = I. \tag{1.22}$$

For example, it is easy to verify that a 2×2 matrix has the inverse:

$$A^{-1} = \frac{1}{|A|}\begin{pmatrix} a_{22} & -a_{12} \\ -a_{21} & a_{11} \end{pmatrix}.$$

In IDL, continuing the previous dialog,

```
IDL> PRINT, determ(A)
      -2.00000
IDL> PRINT, invert(A)
      -2.00000       1.00000
       1.50000     -0.500000
IDL> PRINT, A##invert(A)
       1.00000      0.000000
      0.000000       1.00000
```

The matrix inverse has the properties

$$\begin{aligned} (AB)^{-1} &= B^{-1}A^{-1} \\ (A^{-1})^\top &= (A^\top)^{-1}. \end{aligned} \tag{1.23}$$

If the transpose of a square matrix is its inverse, i.e., if

$$A^\top A = I, \tag{1.24}$$

then it is referred to as an *orthonormal matrix*.

A system of n linear equations of the form

$$y_i = \sum_{j=1}^{n} a_j x_j(i), \quad i = 1 \dots n, \tag{1.25}$$

can be written in matrix notation as

$$y = Aa, \tag{1.26}$$

where

$y = (y_1 \dots y_n)^\top$
$a = (a_1 \dots a_n)^\top$
$A_{ij} = x_j(i)$

Provided A is nonsingular (see below), the solution for the parameter vector a is given by

$$a = A^{-1} y. \tag{1.27}$$

The *trace* of a square matrix is the sum of its diagonal elements, e.g., for a 2×2 matrix,

$$\mathrm{tr}(A) = a_{11} + a_{22}. \tag{1.28}$$

The trace has the properties

$$\begin{aligned} \mathrm{tr}(A + B) &= \mathrm{tr}A + \mathrm{tr}B \\ \mathrm{tr}(AB) &= \mathrm{tr}(BA). \end{aligned} \tag{1.29}$$

1.2.3 Singular Matrices

If $|A| = 0$, then A has no inverse and is said to be a *singular matrix*. If A is nonsingular, then the equation

$$Ax = 0 \tag{1.30}$$

has only the so-called *trivial solution* $x = 0$. To see this, multiply from the left with A^{-1}. Then $A^{-1}Ax = Ix = x = 0$.

If $\boldsymbol{A}$ is singular, Equation 1.30 has at least one *nontrivial solution* $\boldsymbol{x} \neq \boldsymbol{0}$. This, again, is easy to see for a 2×2 matrix. Suppose $|\boldsymbol{A}| = 0$. Writing Equation 1.30 out fully,

$$a_{11}x_1 + a_{12}x_2 = 0$$

$$a_{21}x_1 + a_{22}x_2 = 0.$$

To get a nontrivial solution, assume without loss of generality that $a_{12} \neq 0$. Just choose $x_1 = 1$. Then the above two equations imply that

$$x_2 = -\frac{a_{11}}{a_{12}} \quad \text{and} \quad a_{21} - a_{22}\frac{a_{11}}{a_{12}} = 0.$$

The latter equality is satisfied because $|\boldsymbol{A}| = a_{11}a_{22} - a_{12}a_{21} = 0$.

1.2.4 Symmetric, Positive Definite Matrices

The *variance–covariance matrix* (often simply called covariance matrix), which we shall discuss in Chapter 2 and which plays a central role in digital image analysis, is both *symmetric* and *positive definite*.

Definition 1.1: A square matrix is said to be *symmetric* if $\boldsymbol{A}^\top = \boldsymbol{A}$. The $p \times p$ matrix $\boldsymbol{A}$ is *positive definite* if

$$\boldsymbol{x}^\top \boldsymbol{A}\boldsymbol{x} > 0 \tag{1.31}$$

for all p-dimensional vectors $\boldsymbol{x} \neq \boldsymbol{0}$.

The expression $\boldsymbol{x}^\top \boldsymbol{A}\boldsymbol{x}$ in the above definition is called a *quadratic form*. If $\boldsymbol{x}^\top \boldsymbol{A}\boldsymbol{x} \geq 0$ for all $\boldsymbol{x}$ then $\boldsymbol{A}$ is *positive semidefinite*.

Listing 1.2 shows how to extract the covariance matrix from an ENVI image. In line 21 the image is read into an array `G`, in which each pixel vector constitutes a row of the array. We will refer to such an array as a *data design matrix* or simply *data matrix*, see Section 2.3.1. Many of our IDL programs will involve manipulations of the data matrix. Here, it is passed to the IDL function `CORRELATE()` (with keyword `COVARIANCE` set) in line 22, which returns the covariance matrix. The result for the three VNIR bands 1, 2, and 3N of the Jülich image in Figure 1.1 is a symmetric 3×3 matrix:

```
ENVI> EX1_2
'Covariance␣matrix␣for␣image␣E:\satellite␣images\juelich'
       87.117627        92.394480       -10.518867
       92.394480        115.83875       -77.420886
      -10.518867       -77.420886        374.61514
```

We will see how to show that this matrix is positive definite in Section 1.3.

Listing 1.2
Determining the covariance matrix of an image.

```
PRO EX1_2

envi_select, title='Choose multispectral image', $
             fid=fid, dims=dims,pos=pos
IF (fid EQ -1) THEN BEGIN
   PRINT, 'cancelled'
   RETURN
ENDIF

envi_file_query, fid, fname=fname

num_cols = dims[2]-dims[1]+1
num_rows = dims[4]-dims[3]+1
num_bands = n_elements(pos)
num_pixels = num_cols*num_rows

; array with pixel vectors as rows
G = fltarr(num_bands,num_pixels)

FOR i=0,num_bands-1 DO $
   G[i,*]= envi_get_data(fid=fid,dims=dims,pos=pos[i])
C = correlate(G,/covariance,/double)

PRINT, 'Covariance matrix for image '+fname
PRINT, C

END
```

1.2.5 Linear Dependence and Vector Spaces

Vectors are said to be *linearly dependent* when any one can be expressed as a linear combination of the others. Here is a formal definition:

Definition 1.2: A set S of vectors $\boldsymbol{x}_1 \ldots \boldsymbol{x}_r$ is said to be *linearly dependent* if there exist scalars $c_1 \ldots c_r$, not all of which are zero, such that $\sum_{i=1}^{r} c_i \boldsymbol{x}_i = \boldsymbol{0}$. Otherwise they are *linearly independent*.

A matrix is said to have *rank* r if the maximum number of linearly independent columns is r. If the $p \times p$ matrix $\boldsymbol{A} = (\boldsymbol{a}_1 \ldots \boldsymbol{a}_p)$, where $\boldsymbol{a}_i$ is its ith column, is nonsingular, then it has full rank p. If this were not the case, then there must exist a set of scalars $c_1 \dots c_p$, not all of which are zero, for which

$$c_1 \boldsymbol{a}_1 + \cdots + c_p \boldsymbol{a}_p = \boldsymbol{A}\boldsymbol{c} = \boldsymbol{0}.$$

In other words there would be a nontrivial solution to $\boldsymbol{A}\boldsymbol{c} = \boldsymbol{0}$, contradicting the fact that $\boldsymbol{A}$ is nonsingular.

The set S in Definition 1.2 is said to constitute a *basis* for a *vector space* V, comprising all vectors that can be expressed as a linear combination of the vectors in S. The number r of vectors in the basis is called the *dimension* of V. The vector space, V, is also an *inner product space* by virtue of the inner product definition (Equation 1.7). Inner product spaces are elaborated upon in Appendix A.

1.3 Eigenvalues and Eigenvectors

In image analysis, it is frequently necessary to solve an *eigenvalue problem*. In the simplest case, and the one which will concern us primarily, the eigenvalue problem consists in finding *eigenvectors* $\boldsymbol{x}$ and *eigenvalues* λ that satisfy the matrix equation

$$\boldsymbol{A}\boldsymbol{x} = \lambda \boldsymbol{x}, \tag{1.32}$$

where $\boldsymbol{A}$ is both symmetric and positive definite. Equation 1.32 can be written equivalently as

$$(\boldsymbol{A} - \lambda \boldsymbol{I})\boldsymbol{x} = \boldsymbol{0}, \tag{1.33}$$

so for a nontrivial solution we must have

$$|\boldsymbol{A} - \lambda \boldsymbol{I}| = 0. \tag{1.34}$$

This is known as the *characteristic equation* for the matrix $\boldsymbol{A}$. For instance, in the case of a 2×2 matrix eigenvalue problem,

$$\begin{pmatrix} a_{11} & a_{12} \\ a_{21} & a_{22} \end{pmatrix} \begin{pmatrix} x_1 \\ x_2 \end{pmatrix} = \lambda \begin{pmatrix} x_1 \\ x_2 \end{pmatrix}, \tag{1.35}$$

the characteristic equation is

$$(a_{11} - \lambda)(a_{22} - \lambda) - a_{12}^2 = 0,$$

which is a quadratic equation in λ with solutions

$$\begin{aligned}\lambda^{(1)} &= \frac{1}{2}\left(a_{11} + a_{22} + \sqrt{(a_{11} + a_{22})^2 - 4(a_{11}a_{22} - a_{12}^2)}\right)\\ \lambda^{(2)} &= \frac{1}{2}\left(a_{11} + a_{22} - \sqrt{(a_{11} + a_{22})^2 - 4(a_{11}a_{22} - a_{12}^2)}\right).\end{aligned} \tag{1.36}$$

Thus there are two eigenvalues and, correspondingly, two eigenvectors $\boldsymbol{x}^{(1)}$ and $\boldsymbol{x}^{(2)}$.* The eigenvectors can be obtained by first substituting $\lambda^{(1)}$ and then $\lambda^{(2)}$ into Equation 1.35 and solving for x_1 and x_2 each time. It is easy to show (Exercise 8) that the eigenvectors are orthogonal:

$$(\boldsymbol{x}^{(1)})^\top \boldsymbol{x}^{(2)} = 0. \tag{1.37}$$

Moreover, since the left- and right-hand sides of Equation 1.35 can be multiplied by any constant, the eigenvectors can be always chosen to have unit length, $\|\boldsymbol{x}^{(1)}\| = \|\boldsymbol{x}^{(2)}\| = 1$. The matrix formed by two such eigenvectors, e.g.,

$$\boldsymbol{U} = (\boldsymbol{x}^{(1)}, \boldsymbol{x}^{(2)}) = \begin{pmatrix} x_1^{(1)} & x_1^{(2)} \\ x_2^{(1)} & x_2^{(2)} \end{pmatrix}, \tag{1.38}$$

is said to *diagonalize* the matrix $\boldsymbol{A}$. That is, if $\boldsymbol{A}$ is multiplied from the left by $\boldsymbol{U}^\top$ and from the right by $\boldsymbol{U}$, the result is a diagonal matrix with the eigenvalues along the diagonal:

$$\boldsymbol{U}^\top \boldsymbol{A} \boldsymbol{U} = \boldsymbol{\Lambda} = \begin{pmatrix} \lambda^{(1)} & 0 \\ 0 & \lambda^{(2)} \end{pmatrix}, \tag{1.39}$$

as can easily be verified. Note that $\boldsymbol{U}$ is an orthonormal matrix: $\boldsymbol{U}^\top \boldsymbol{U} = \boldsymbol{I}$ and therefore Equation 1.39 can be written as

$$\boldsymbol{A}\boldsymbol{U} = \boldsymbol{U}\boldsymbol{\Lambda}. \tag{1.40}$$

All of the above statements generalize in a straightforward fashion to square matrices of any size.

Suppose that $\boldsymbol{A}$ is an $N \times N$ symmetric matrix. Then its eigenvectors $\boldsymbol{u}^{(j)}$, $j = 1 \dots N$, are orthogonal, and any N-component vector $\boldsymbol{x}$ can be expressed

* The special case of degeneracy, which arises when the two eigenvalues are equal, will be ignored here.

Listing 1.3

Eigenvalues and eigenvectors of the covariance matrix.

```
PRO EX1_3

envi_select, title='Choose␣multispectral␣image', $
             fid=fid, dims=dims,pos=pos
IF (fid EQ -1) THEN BEGIN
   PRINT, 'cancelled'
   RETURN
ENDIF

envi_file_query, fid, fname=fname

num_cols = dims[2]-dims[1]+1
num_rows = dims[4]-dims[3]+1
num_bands = n_elements(pos)
num_pixels = num_cols*num_rows

; data matrix
G = fltarr(num_bands,num_pixels)

FOR i=0,num_bands-1 DO $
 G[i,*]=envi_get_data(fid=fid,dims=dims,pos=pos[i])

C = correlate(G,/covariance)

PRINT, 'Eigenvalues'
PRINT, eigenql(C, eigenvectors=U, /double)
PRINT, 'Eigenvectors␣(as␣rows)'
PRINT, U

PRINT, 'Orthogonality'
PRINT, U##transpose(U)

END
```

as a linear combination of them,

$$x = \sum_{j=1}^{N} \beta_j \boldsymbol{u}^{(j)}. \tag{1.41}$$

Multiplying from the right with $\boldsymbol{u}^{(i)\top}$,

$$\boldsymbol{u}^{(i)\top}\boldsymbol{x}\ (= \boldsymbol{x}^\top \boldsymbol{u}^{(i)}) = \sum_{j=1}^{N} \beta_j \boldsymbol{u}^{(i)\top}\boldsymbol{u}^{(j)} = \beta_i, \tag{1.42}$$

so that we have

$$x^\top A x = x^\top \sum_i \beta_i \lambda_i u^{(i)} = \sum_i \beta_i^2 \lambda_i, \tag{1.43}$$

where λ_i, $i = 1 \ldots N$, are the eigenvalues of A. We can conclude, from the definition of positive definite matrices given in Section 1.2.4, that A is positive definite if and only if all of its eigenvalues λ_i, $i = 1 \ldots N$, are positive.

The eigenvalue problem can be solved in IDL with the built-in function `EIGENQL(A)`. This function returns an array of the eigenvalues of the symmetric matrix A and, optionally, the orthonormal matrix $U^\top$ in the keyword `EIGENVECTORS`, see the program in Listing 1.3. Here is the output of the program with the VNIR bands 1, 2, and 3N of the Jülich image:

```
ENVI> Ex1_3
Eigenvalues
        934.08442       398.98373        4.0842497
Eigenvectors (as rows)
      -0.64092380     -0.71641836       0.27561100
      -0.22689155     -0.16619583      -0.95963492
       0.73330547     -0.67758666     -0.056030408
Orthogonality
        1.0000000      0.00000000    5.0306981e-017
       0.00000000       1.0000000    4.1633363e-017
   5.0306981e-017  4.1633363e-017        1.0000000
```

Notice that, due to rounding errors, $U^\top U$ has finite—but very small—nondiagonal elements. The eigenvalues are all positive, so the covariance matrix calculated by the program is positive definite.

1.4 Singular Value Decomposition

Rewriting Equation 1.40, we get a special form of *singular value decomposition* (SVD) for symmetric matrices:

$$A = U \Lambda U^\top. \tag{1.44}$$

This says that any symmetric matrix A can be factored into the product of an orthonormal matrix U times a diagonal matrix Λ times the transpose of U. This is also called the *eigen-decomposition* or *spectral decomposition* of A, and can be written alternatively as a sum over outer products of the eigenvectors (Exercise 9),

$$A = \sum_{i=1}^{N} \lambda_i \boldsymbol{u}^{(i)} \boldsymbol{u}^{(i)\top}. \tag{1.45}$$

For the general form for singular value decomposition of nonsymmetric or non-square matrices, see Press et al. (2002).

SVD is a powerful tool for the solution of systems of linear equations, and is often used when a solution cannot be determined by other numerical algorithms. To invert a symmetric matrix, $\boldsymbol{A}$, we simply write

$$\boldsymbol{A}^{-1} = \boldsymbol{U}\boldsymbol{\Lambda}^{-1}\boldsymbol{U}^\top, \tag{1.46}$$

since $\boldsymbol{U}^{-1} = \boldsymbol{U}^\top$. The matrix $\boldsymbol{\Lambda}^{-1}$ is the diagonal matrix whose diagonal elements are the inverses of the diagonal elements of $\boldsymbol{\Lambda}$. If $\boldsymbol{A}$ is singular (has no inverse), clearly at least one of its eigenvalues is zero.

The matrix $\boldsymbol{A}$ is said to be *ill-conditioned* or *nearly singular* if one or more of the diagonal elements of $\boldsymbol{\Lambda}$ is close to zero. SVD detects this situation effectively, since the factorization in Equation 1.44 is always possible, even if $\boldsymbol{A}$ is truly singular. The IDL procedure for singular value decomposition is SVDC. For example, the program

```
PRO svd_test
   b = [1,2,3]
; an almost singular matrix
   A = transpose(b)##b+randomu(seed,3,3)*0.001
; a symmetric almost singular matrix
   A = A + transpose(A)
   PRINT,'det(A)␣=␣',determ(A)
; singular value decomposition A = U L V^T
   svdc,A,L,U,V,/double
   PRINT,'L␣='
   PRINT, diag_matrix(L)
   PRINT,'U␣=␣V␣='
   PRINT, U
END
```

generates the output

```
% Compiled module: SVD_TEST.
det(A) = -4.21913e-006
L =
       28.002635      0.00000000      0.00000000
      0.00000000    0.0010748116      0.00000000
      0.00000000      0.00000000   0.00014015930
U = V =
     -0.26730887      0.88174905     -0.38867026
     -0.53453525      0.19991515      0.82116137
     -0.80175934     -0.42726168     -0.41788685
```

indicating that A is ill-conditioned (two of the three diagonal elements of Λ are close to zero).

1.5 Vector Derivatives

In order to maximize some desirable property of a multispectral image, such as signal to noise ratio or spread in intensity, we often need to take derivatives with respect to vectors. A *vector partial derivative* operator is written in the form $\frac{\partial}{\partial x}$. For instance, in two dimensions,

$$\frac{\partial}{\partial x} = \begin{pmatrix}1\\0\end{pmatrix}\frac{\partial}{\partial x_1} + \begin{pmatrix}0\\1\end{pmatrix}\frac{\partial}{\partial x_2} = i\frac{\partial}{\partial x_1} + j\frac{\partial}{\partial x_2}.$$

Such an operator (also called a *gradient*) arranges the partial derivatives of any scalar function of the vector x with respect to each of its components into a column vector, e.g.,

$$\frac{\partial f(x)}{\partial x} = \begin{pmatrix}\frac{\partial f(x)}{\partial x_1}\\ \frac{\partial f(x)}{\partial x_2}\end{pmatrix}. \tag{1.47}$$

Many of the operations with vector derivatives correspond exactly to operations with ordinary scalar derivatives (and can all be verified easily by writing out the expressions component-by-component):

$$\frac{\partial}{\partial x}(x^\top y) = y \quad \text{analogous to} \quad \frac{\partial}{\partial x}xy = y$$

$$\frac{\partial}{\partial x}(x^\top x) = 2x \quad \text{analogous to} \quad \frac{\partial}{\partial x}x^2 = 2x.$$

For quadratic forms we have

$$\begin{aligned}\frac{\partial}{\partial x}(x^\top A y) &= Ay\\ \frac{\partial}{\partial y}(x^\top A y) &= A^\top x\end{aligned} \tag{1.48}$$

and

$$\frac{\partial}{\partial x}(x^\top A x) = Ax + A^\top x. \tag{1.49}$$

Note that, if A is a symmetric matrix, the last equation can be written as

$$\frac{\partial}{\partial \boldsymbol{x}}(\boldsymbol{x}^\top \boldsymbol{A}\boldsymbol{x}) = 2\boldsymbol{A}\boldsymbol{x}. \tag{1.50}$$

1.6 Finding Minima and Maxima

Suppose x^* is a stationary point of the function $f(x)$, which means that its first derivative vanishes at that point:

$$\frac{d}{dx}f(x^*) = \frac{d}{dx}f(x)\bigg|_{x=x^*} = 0, \tag{1.51}$$

see Figure 1.4. Then $f(x^*)$ is a local minimum if the second derivative at x^* is positive,

$$\frac{d^2}{dx^2}f(x^*) > 0.$$

This becomes obvious if $f(x)$ is expanded in a *Taylor series* about x^*,

$$f(x) = f(x^*) + (x - x^*)\frac{d}{dx}f(x^*) + \frac{1}{2}(x - x^*)^2\frac{d^2}{dx^2}f(x^*) + \cdots . \tag{1.52}$$

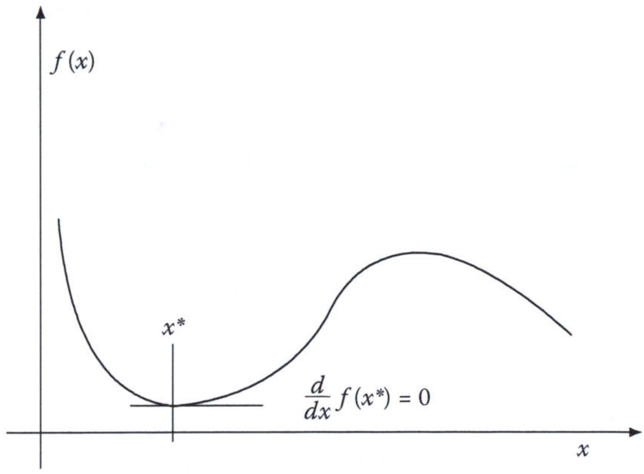

FIGURE 1.4
A function of one variable with a minimum at x^*.

The second term is zero, so for $|x-x^*|$ sufficiently small, $f(x)$ is approximately a quadratic function

$$f(x) \approx f(x^*) + \frac{1}{2}(x - x^*)^2 \frac{d^2}{dx^2} f(x^*) \tag{1.53}$$

with a minimum at x^* when the second derivative is positive, a maximum when it is negative, and a point of inflection when it is zero.

The situation is similar for scalar functions of a vector. Now the Taylor expansion is

$$f(\boldsymbol{x}) = f(\boldsymbol{x}^*) + (\boldsymbol{x} - \boldsymbol{x}^*)^\top \frac{\partial f(\boldsymbol{x}^*)}{\partial \boldsymbol{x}} + \frac{1}{2}(\boldsymbol{x} - \boldsymbol{x}^*)^\top \boldsymbol{H}(\boldsymbol{x} - \boldsymbol{x}^*) + \cdots, \tag{1.54}$$

where $\boldsymbol{H}$ is called the *Hessian matrix*. Its elements are given by

$$(\boldsymbol{H})_{ij} = \frac{\partial^2 f(\boldsymbol{x}^*)}{\partial x_i \partial x_j}, \tag{1.55}$$

and it is a symmetric matrix.* At the stationary point the vector derivative with respect to $\boldsymbol{x}$ vanishes,

$$\frac{\partial f(\boldsymbol{x}^*)}{\partial \boldsymbol{x}} = \boldsymbol{0},$$

so we get the second-order approximation

$$f(\boldsymbol{x}) \approx f(\boldsymbol{x}^*) + \frac{1}{2}(\boldsymbol{x} - \boldsymbol{x}^*)^\top \boldsymbol{H}(\boldsymbol{x} - \boldsymbol{x}^*). \tag{1.56}$$

Now the condition for a local minimum is clearly that the Hessian matrix be positive definite at the point $\boldsymbol{x}^*$. Note that the Hessian can be expressed in terms of an outer product of vector derivatives as

$$\frac{\partial}{\partial \boldsymbol{x}} \frac{\partial f(\boldsymbol{x})}{\partial \boldsymbol{x}^\top} = \frac{\partial^2 f(\boldsymbol{x})}{\partial \boldsymbol{x} \partial \boldsymbol{x}^\top}. \tag{1.57}$$

Suppose that we wish to find the position $\boldsymbol{x}^*$ of a minimum (or maximum) of a scalar function $f(\boldsymbol{x})$. If there are no constraints, then we solve the set

* Because the order of partial differentiation does not matter.

of equations

$$\frac{\partial f(x)}{\partial x_i} = 0, \quad i = 1, 2 \ldots$$

or, in terms of our notation for vector derivatives,

$$\frac{\partial f(\boldsymbol{x})}{\partial \boldsymbol{x}} = \boldsymbol{0}, \tag{1.58}$$

and examine the Hessian matrix at the solution point. However, suppose that $\boldsymbol{x}$ is *constrained* by the condition

$$h(\boldsymbol{x}) = 0. \tag{1.59}$$

For example, in two dimensions we might have the constraint

$$h(\boldsymbol{x}) = x_1^2 + x_2^2 - 1 = 0,$$

which requires $\boldsymbol{x}$ to lie on a circle of radius 1. An obvious procedure would be to solve the above equation for x_1, say, and then substitute the result into Equation 1.58 to determine x_2. This method is not always practical, however. It may be the case, for example, that the constraint Equation 1.59 cannot be solved analytically.

A more convenient and generally applicable way to find an extremum of $f(\boldsymbol{x})$ subject to $h(\boldsymbol{x}) = 0$ is to determine an *unconstrained* minimum or maximum of the expression

$$L(\boldsymbol{x}) = f(\boldsymbol{x}) + \lambda h(\boldsymbol{x}). \tag{1.60}$$

This is called a *Lagrange function* and λ is a *Lagrange multiplier*. The Lagrange multiplier is treated as though it were an additional variable. To find the extremum, we solve the set of equations

$$\begin{aligned} \frac{\partial}{\partial \boldsymbol{x}}(f(\boldsymbol{x}) + \lambda h(\boldsymbol{x})) &= \boldsymbol{0}, \\ \frac{\partial}{\partial \lambda}(f(\boldsymbol{x}) + \lambda h(\boldsymbol{x})) &= 0 \end{aligned} \tag{1.61}$$

for $\boldsymbol{x}$ and λ. To see this, note that a minimum or maximum need not generally occur at a point $\boldsymbol{x}^*$ for which

$$\frac{\partial}{\partial \boldsymbol{x}} f(\boldsymbol{x}^*) = \boldsymbol{0}, \tag{1.62}$$

because of the presence of the constraint. This possibility is taken into account in the first of Equations 1.61, which at an extremum reads

$$\frac{\partial}{\partial \boldsymbol{x}} f(\boldsymbol{x}^*) = -\lambda \frac{\partial}{\partial \boldsymbol{x}} h(\boldsymbol{x}^*). \tag{1.63}$$

It implies that, if $\lambda \neq 0$, any small change in $\boldsymbol{x}$ away from $\boldsymbol{x}^*$ causing a change in $f(\boldsymbol{x}^*)$, i.e., any change in $\boldsymbol{x}$ which is not orthogonal to the gradient of $f(\boldsymbol{x}^*)$, would be accompanied by a proportional change in $h(\boldsymbol{x}^*)$. This would necessarily violate the constraint, so $\boldsymbol{x}^*$ is an extremum. The second equation is just the constraint $h(\boldsymbol{x}^*) = 0$ itself. For more detailed justifications of this procedure, see Bishop (1995, Appendix C), Milman (1999, Chapter 14), or Cristianini and Shawe-Taylor (2000, Chapter 5).

As a first example illustrating the Lagrange method, let $f(\boldsymbol{x}) = ax_1^2 + bx_2^2$ and $h(\boldsymbol{x}) = x_1 + x_2 - 1$. Then we get the three equations

$$\frac{\partial}{\partial x_1}(f(\boldsymbol{x}) + \lambda h(\boldsymbol{x})) = 2ax_1 + \lambda = 0$$

$$\frac{\partial}{\partial x_2}(f(\boldsymbol{x}) + \lambda h(\boldsymbol{x})) = 2bx_2 + \lambda = 0$$

$$\frac{\partial}{\partial \lambda}(f(\boldsymbol{x}) + \lambda h(\boldsymbol{x})) = x_1 + x_2 - 1 = 0.$$

The solution for $\boldsymbol{x}$ is

$$x_1 = \frac{b}{a+b}, \quad x_2 = \frac{a}{a+b}.$$

A second and very important example is as follows. Let us find the maximum of

$$f(\boldsymbol{x}) = \boldsymbol{x}^\top \boldsymbol{C} \boldsymbol{x}, \tag{1.64}$$

where $\boldsymbol{C}$ is symmetric positive definite, subject to the constraint

$$h(\boldsymbol{x}) = \boldsymbol{x}^\top \boldsymbol{x} - 1 = 0. \tag{1.65}$$

The Lagrange function is

$$L(\boldsymbol{x}) = \boldsymbol{x}^\top \boldsymbol{C} \boldsymbol{x} - \lambda(\boldsymbol{x}^\top \boldsymbol{x} - 1). \tag{1.66}$$

Listing 1.4
Principal components analysis.

```
PRO EX1_4

envi_select, title='Choose␣multispectral␣image', $
             fid=fid, dims=dims,pos=pos
IF (fid EQ -1) THEN BEGIN
   PRINT, 'cancelled'
   RETURN
ENDIF
envi_file_query, fid, fname=fname
num_cols = dims[2]-dims[1]+1
num_rows = dims[4]-dims[3]+1
num_bands = n_elements(pos)
num_pixels = num_cols*num_rows

; data matrix
G = fltarr(num_bands,num_pixels)

; read in image and subtract means
FOR i=0,num_bands-1 DO BEGIN
  temp =  envi_get_data(fid=fid,dims=dims,pos=pos[i])
  G[i,*] = temp-mean(temp)
END

; principal components transformation
C = correlate(G,/covariance,/double)
void = eigenql(C, eigenvectors=U, /double)
PCs = G##transpose(U)

; BSQ array
image = fltarr(num_cols,num_rows,num_bands)

FOR i = 0, num_bands-1 DO image[*,*,i] = $
   reform(PCs[i,*],num_cols,num_rows)

; map tie point
map_info = envi_get_map_info(fid=fid)
envi_convert_file_coordinates, fid, $
   dims[1], dims[3], e, n, /to_map
map_info.mc = [0D,0D,e,n]

envi_enter_data, image, map_info = map_info

END
```

The maximum of L occurs at a value of $\boldsymbol{x}$ for which

$$\frac{\partial L}{\partial \boldsymbol{x}} = 2\boldsymbol{C}\boldsymbol{x} - 2\lambda\boldsymbol{x} = 0, \tag{1.67}$$

which is the eigenvalue problem

$$\boldsymbol{C}\boldsymbol{x} = \lambda\boldsymbol{x}. \tag{1.68}$$

The vector that maximizes Equation 1.66 is the eigenvector corresponding to the maximum eigenvalue of Equation 1.68.

If $\boldsymbol{C}$ is the covariance matrix of a multispectral image, then, as we shall see in Chapter 3, the above maximization corresponds to a *principal components analysis* (PCA). Listing 1.4 gives a rudimentary procedure for doing PCA on an ENVI image. After solving the eigenvalue problem, Equation 1.68, in line 26, the principal components are calculated by projecting the original image bands along the eigenvectors (line 27). Then the transformed data matrix is rearranged in the BSQ format and returned to ENVI with the `ENVI_ENTER_DATA` procedure, line 41. Note that a map tie-point is recalculated with the `ENVI_CONVERT_FILE_ COORDINATES` procedure (lines 36–39) in case the user happens to choose a spatial subset of the original image. The first principal component of VNIR bands 1, 2, and 3N of the Jülich image determined with this program is displayed in Figure 1.5.

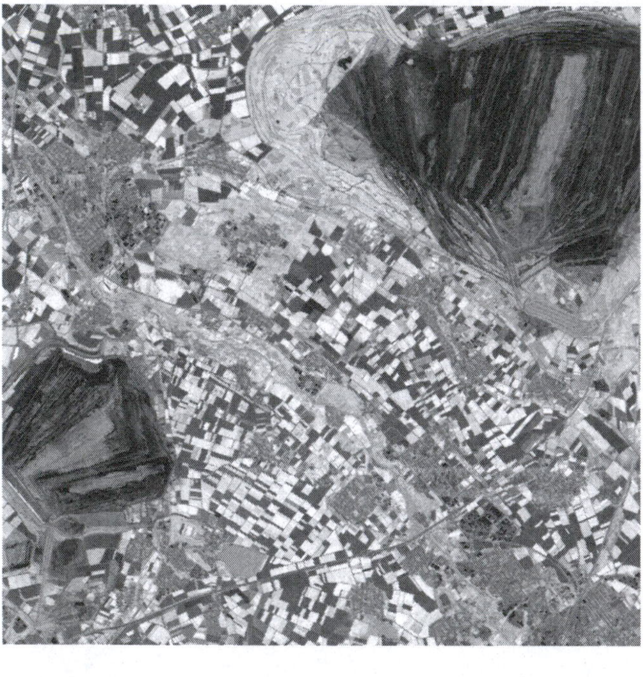

FIGURE 1.5
First principal component of the VNIR spectral bands for the Jülich scene.

1.7 Exercises

1. Demonstrate that the definition

$$x^\top y = \|x\| \, \|y\| \cos\theta$$

is equivalent to

$$x^\top y = x_1 y_1 + x_2 y_2.$$

Hint: Use the trigonometric identity $\cos(\alpha - \beta) = \cos\alpha \cos\beta + \sin\alpha \sin\beta$.

2. Show that the outer product of two two-dimensional vectors is a singular matrix.
3. Verify the matrix identity

$$(AB)^\top = B^\top A^\top$$

in IDL.

4. Show that three two-dimensional vectors representing three points, all lying on the same line, are linearly dependent.
5. Show that the determinant of a symmetric 2×2 matrix is given by the product of its eigenvalues.
6. Prove that the inverse of a symmetric nonsingular matrix is symmetric.
7. Prove that the eigenvectors of A^{-1} are the same as those of A, but with reciprocal eigenvalues.
8. Prove that the eigenvectors or a 2×2 symmetric matrix are orthogonal.
9. Demonstrate the equivalence of Equations 1.44 and 1.45 for a symmetric 2×2 matrix.
10. Prove from Equation 1.45 that the trace of a symmetric matrix is the sum of its eigenvalues.
11. Differentiate the function

$$\frac{1}{x^\top A y}$$

with respect to y.

12. Calculate the eigenvectors of the (nonsymmetric!) matrix

$$\begin{pmatrix} 1 & 2 & 3 \\ 4 & 5 & 6 \\ 7 & 8 & 9 \end{pmatrix}$$

with IDL.

13. Plot the function $f(\boldsymbol{x}) = x_1^2 - x_2^2$ with IDL. Find its minima and maxima subject to the constraint $h(\boldsymbol{x}) = x_1^2 + x_2^2 - 1 = 0$.
14. Modify the program in Listing 1.4 to make use of the IDL function `PCOMP()`.

2

Image Statistics

In a multispectral image, a given pixel intensity or gray-scale value, $g(i,j)$, derived from the measured radiance at a satellite sensor, is never exactly reproducible. It is the outcome of a complex measurement influenced by instrument noise, atmospheric conditions, changing illumination, and so forth. It may be assumed, however, that there is an underlying random process with an associated probability distribution that restricts the possible outcomes in some way. Each time we make an observation, we are sampling from that probability distribution or, put another way, we are observing a different possible *realization* of the random process. In this chapter, some basic statistical concepts for multispectral images viewed as random processes will be introduced.

2.1 Random Variables

A *random variable* can be used to represent a quantity, in the present context a grayscale value, which changes in an unpredictable way each time it is observed. In order to make a precise definition, let us consider some random experiment that has a set, Ω, of possible *outcomes*. This set is referred to as the *sample space* for the experiment. Subsets of Ω are called *events*. An event will be said to have *occurred* whenever the outcome of the random experiment is contained within it.

To make this clearer, consider the random experiment consisting of the throw of two dice. The sample space is the set of 36 possible outcomes

$$\Omega = \{(1,1),(1,2),(2,1)\ldots(6,6)\}.$$

An event is then, for example, that the sum of the points is 7. It is the subset

$$\{(1,6),(2,5),(3,4),(4,3),(5,2),(6,1)\}$$

of the sample space. If for instance $(3,4)$ is thrown, then the event has occurred.

Definition 2.1: A *random variable* $G : \Omega \mapsto \mathbb{R}$ is a function which maps all outcomes onto the set $\mathbb{R}$ of real numbers such that the set

$$\{\omega \in \Omega \mid G(\omega) \leq g\}$$

is an event, that is, a subset of Ω. This subset is usually abbreviated as $\{G \leq g\}$.

Thus, for the throw of two dice, the sum of points, S, is a random variable, since it maps all outcomes onto real numbers,

$$S(1,1) = 2, \; S(1,2) = S(2,1) = 3, \; \ldots \; S(6,6) = 12,$$

and sets such as

$$\{S \leq 4\} = \{(1,1),(1,2),(2,1),(1,3),(3,1),(2,2)\}$$

are subsets of the sample space. The set $\{S \leq 1\}$ is the empty set, whereas $\{S \leq 12\} = \Omega$, the entire sample space.

On the basis of the *probabilities* for the individual outcomes, we can associate a function, $P(g)$, with the random variable, G, as follows:

$$P(g) = \Pr(G \leq g).$$

This is the probability of observing the event that the random variable, G, takes on a value less than or equal to g. The probability of an event may be thought of as the relative frequency with which it occurs in n repetitions of a random experiment in the limit as $n \to \infty$. (For the complete, axiomatic definition, see Freund (1992).) In the dice example, the probability of throwing a four or less is

$$P(4) = \Pr(S \leq 4) = 6/36 = 1/6,$$

for instance.

Definition 2.2: Given the random variable G, then

$$P(g) = \Pr(G \leq g), \quad -\infty < g < \infty \tag{2.1}$$

is called its *distribution function*.

2.1.1 Discrete Random Variables

When, as in the case of the dice throw, a random variable, G, is discrete and takes on values $g_1 < g_2 < g_3 < \ldots$, then the probabilities of the separate outcomes

$$p(g_i) = \Pr(G = g_i) = P(g_i) - P(g_{i-1}), \quad i = 1, 2, \ldots$$

are said to constitute the *mass function* for the random variable. This is best illustrated with a practical example. As we shall see in Chapter 7, the evaluation of a land cover classification model involves repeated trials with a finite number of independent test observations, keeping track of the number of times the model fails to predict the correct land cover category. This will lead us to the consideration of a discrete random variable having the so-called *binomial distribution*. We can derive its mass function as follows.

Let θ be the probability of failure in a single trial. The probability of getting y misclassifications (and hence $n - y$ correct classifications) in n trials *in a specific sequence* is

$$\theta^y (1 - \theta)^{n-y}.$$

In this expression, there is a factor θ for each of the y misclassifications and a factor $(1 - \theta)$ for each of the $n - y$ correct classifications. Taking the product is justified by the assumption that the trials are independent of each other. The number of such sequences is just the number of ways of selecting y trials from n possible ones. This is given by the *binomial coefficient*

$$\binom{n}{y} = \frac{n!}{(n-y)!\, y!}, \tag{2.2}$$

so that the probability for y misclassifications in n trials is

$$\binom{n}{y} \theta^y (1 - \theta)^{n-y}.$$

Accordingly, a discrete random variable, Y, is said to be *binomially distributed* with parameters n and θ if its mass function is given by

$$p_{n,\theta}(y) = \begin{cases} \binom{n}{y} \theta^y (1 - \theta)^{n-y} & \text{for } y = 0, 1, 2 \ldots n \\ 0 & \text{otherwise.} \end{cases} \tag{2.3}$$

Note that the values of $p_{n,\theta}(y)$ are the terms in the binomial expansion of

$$[\theta + (1 - \theta)]^n = 1^n = 1,$$

so the sum over the probabilities equals 1, as it should.

2.1.2 Continuous Random Variables

In the case of continuous random variables, which we are in effect dealing with when we speak of pixel intensities, the distribution function is not expressed in terms of the discrete probabilities of a mass function, but rather in terms of a *probability density function*, $p(g)$. The quantity $p(g)dg$ is interpreted as the probability that the associated random variable, G, lies within the infinitesimal interval, $[g, g + dg]$. Now it is the integral over all such intervals which must yield 1:

$$\int_{-\infty}^{\infty} p(g)dg = 1. \tag{2.4}$$

The distribution function, $P(g)$ (Equation 2.1), is written in terms of the density function as

$$P(g) = \int_{-\infty}^{g} p(t)dt. \tag{2.5}$$

Conversely, differentiating with respect to g, we get

$$p(g) = \frac{d}{dg}P(g). \tag{2.6}$$

The distribution function has the limiting values

$$P(-\infty) = 0, \quad P(\infty) = \int_{-\infty}^{\infty} p(t)dt = 1.$$

For many practical applications, it is sufficient to characterize a distribution function by a small number of its *moments*. The most important of these are the *mean* or *expected value* and the *variance*. The mean of a continuous random variable, G, is usually written $\langle G\rangle$ or $E(G)$.* The mean is defined in terms of the density function according to

$$\langle G\rangle = \int_{-\infty}^{\infty} g \cdot p(g)dg. \tag{2.7}$$

The mean has the important property that, for two random variables G_1 and G_2 and real numbers a_0, a_1, and a_2,

$$\langle a_0 + a_1G_1 + a_2G_2\rangle = a_0 + a_1\langle G_1\rangle + a_2\langle G_2\rangle, \tag{2.8}$$

a fact which follows directly from Equation 2.7.

* We prefer to use the former.

The variance of G, written $\mathrm{var}(G)$, describes how widely the realizations scatter around the mean. It is defined as

$$\mathrm{var}(G) = \left\langle (G - \langle G \rangle)^2 \right\rangle, \tag{2.9}$$

that is, as the mean of the random variable $(G - \langle G \rangle)^2$. In terms of the density function, the variance is given by

$$\mathrm{var}(G) = \int_{-\infty}^{\infty} (g - \langle G \rangle)^2 p(g) dg. \tag{2.10}$$

For example, consider a *uniformly distributed* random variable, X, with density function

$$p(x) = \begin{cases} 1 & \text{if } 0 \le x \le 1 \\ 0 & \text{otherwise.} \end{cases}$$

We calculate the moments to be

$$\langle X \rangle = \int_0^1 x \cdot 1 \, dx = 1/2$$

$$\mathrm{var}(X) = \int_0^1 (x - 1/2)^2 \cdot 1 \, dx = 1/12.$$

For discrete random variables, the integrals in Equations 2.7 and 2.10 are replaced by summations over the allowed values of G and the density is replaced by the mass function.

Two very useful identities follow from the definition of variance (Exercise 1):

$$\begin{aligned} \mathrm{var}(G) &= \langle G^2 \rangle - \langle G \rangle^2 \\ \mathrm{var}(a_0 + a_1 G) &= a_1^2 \, \mathrm{var}(G). \end{aligned} \tag{2.11}$$

The following theorem can often be used to determine the probability density of a function of some random variable whose density is known. For a proof, see Freund (1992).

THEOREM 2.1

Let $p_x(x)$ be the density function for random variable X and

$$y = u(x) \tag{2.12}$$

a monotonic function of x for all values of x for which $p_x(x) \neq 0$. For these x values Equation 2.12 can be solved for x to give $x = w(y)$. Then the density function of the random variable $Y = u(X)$ is given by

$$p_y(y) = p_x(w(y)) \left| \frac{dx}{dy} \right|. \tag{2.13}$$

As an example of the application of Theorem 2.1, suppose that a random variable X has the exponential distribution with density function

$$p_x(x) = \begin{cases} e^{-x} & \text{for } x > 0 \\ 0 & \text{otherwise,} \end{cases}$$

and we wish to determine the probability density of the random variable $Y = \sqrt{X}$. The monotonic function, $y = u(x) = \sqrt{x}$, can be inverted to give

$$x = w(y) = y^2.$$

Thus

$$\left| \frac{dx}{dy} \right| = |2y|, \quad p_x(w(y)) = e^{-y^2},$$

and we obtain

$$p_y(y) = 2ye^{-y^2}, \quad y > 0.$$

2.1.3 Normal Distribution

It is very often the case that random variables are well described by the *normal* or *Gaussian density function*

$$p(g) = \frac{1}{\sqrt{2\pi}\sigma} \exp\left(-\frac{1}{2\sigma^2}(g-\mu)^2 \right), \tag{2.14}$$

where $-\infty < \mu < \infty$ and $\sigma^2 > 0$. In that case, it follows from Equations 2.7 and 2.10 that

$$\langle G \rangle = \mu, \quad \text{var}(G) = \sigma^2$$

(see Exercise 2). This is commonly abbreviated by writing

$$G \sim \mathcal{N}(\mu, \sigma^2).$$

If G is normally distributed, then the *standardized* random variable, $(G-\mu)/\sigma$, has the *standard normal distribution*, $\Phi(g)$, with zero mean and unit variance

$$\Phi(g) = \frac{1}{\sqrt{2\pi}} \int_{-\infty}^{g} \exp(-t^2/2)dt = \int_{-\infty}^{g} \phi(t)dt, \tag{2.15}$$

where the *standard normal density*, $\phi(t)$, is given by

$$\phi(t) = \frac{1}{\sqrt{2\pi}} \exp(-t^2/2). \tag{2.16}$$

Since it is not possible to express the normal distribution function, $\Phi(g)$, in terms of simple analytical functions, it is tabulated. From the symmetry of the density function, it follows that

$$\Phi(-g) = 1 - \Phi(g), \tag{2.17}$$

so it is sufficient to give tables only for $g \geq 0$. Note that

$$P(g) = \Pr(G \leq g) = \Pr\left(\frac{G-\mu}{\sigma} \leq \frac{g-\mu}{\sigma}\right) = \Phi\left(\frac{g-\mu}{\sigma}\right), \tag{2.18}$$

so that values for any normally distributed random variable can be read from the table.

The ubiquity of the normal distribution is often attributed to the *central limit theorem* of statistics.

THEOREM 2.2 (Central Limit Theorem)

If random variables $G_1, G_2 \ldots G_n$ have equal distributions with mean μ and variance σ^2, then the random variable

$$\frac{1}{\sigma\sqrt{n}} \sum_{i=1}^{n} (G_i - \mu)$$

is standard normally distributed in the limit $n \to \infty$.

A proof is given in Freund (1992). This theorem explains why random measurement errors in general can be assumed to be normally distributed. The overall error is made up of a large number of small contributions, and the theorem implies that their sum will tend to a normal distribution. As an illustration, the IDL code in Listing 2.1 calculates 10,000 sums of 12 random

Listing 2.1
Illustrating the central limit theorem.

```
PRO ex2_1

thisDevice =!D.Name
set_plot, 'PS'
Device,Filename='fig2_1.eps',xsize=6,ysize=4,$
   /inches,/encapsulated
PLOT,histogram(total(randomu(seed,10000,12),2), $
      nbins=12)
device,/close_file
set_plot, thisDevice

END
```

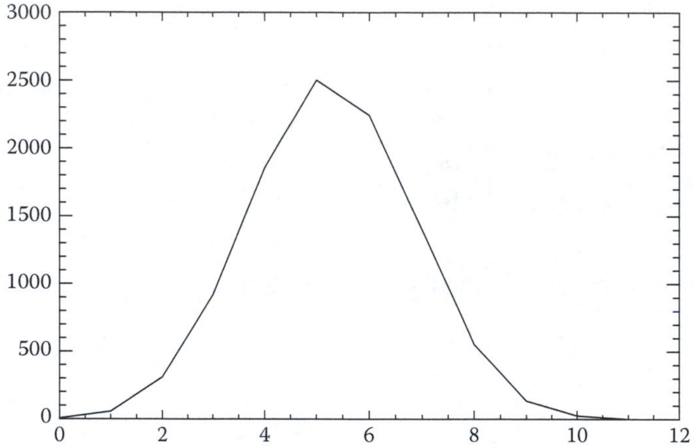

FIGURE 2.1
Histogram of sums of 12 uniformly distributed random numbers.

numbers uniformly distributed on the interval [0, 1] and plots their histogram to a PostScript file (see Figure 2.1). Note the use of the IDL function `TOTAL(A,2)`, that sums a two-dimensional array `A` along its second dimension, that is, along its columns. The histogram closely approximates a normal distribution.

2.2 Random Vectors

The idea of a distribution function may be extended to more than one random variable. For convenience, we consider only two continuous random variables in the following discussion, but the generalization to any number of continuous or discrete random variables is straightforward.

Let $\mathbf{G} = (G_1, G_2)^\top$ be a *random vector*, that is, a vector the components of which are random variables. The *joint distribution function* of $\mathbf{G}$ is defined by

$$P(\mathbf{g}) = P(g_1, g_2) = \Pr(G_1 \leq g_1 \text{ and } G_2 \leq g_2) \tag{2.19}$$

or, in terms of the *joint density function* $p(g_1, g_2)$,

$$P(g_1, g_2) = \int_{-\infty}^{g_1} \int_{-\infty}^{g_2} p(g_1, g_2) dg_1 dg_2 \tag{2.20}$$

and, conversely,

$$p(g_1, g_1) = \frac{\partial^2}{\partial g_1 \partial g_2} P(g_1, g_2). \tag{2.21}$$

The *marginal distribution function* for G_1 is given by

$$P_1(g_1) = P(g_1, \infty) = \int_{-\infty}^{g_1} \int_{-\infty}^{\infty} p(g_1, g_2) dg_1 dg_2, \tag{2.22}$$

and similarly for G_2. The *marginal density* is defined as

$$p_1(g_1) = \int_{-\infty}^{\infty} p(g_1, g_2) dg_2 \tag{2.23}$$

with a similar expression for $p_2(g_2)$.

The mean of the random vector, $\mathbf{G}$, is the vector of mean values of G_1 and G_2,

$$\langle \mathbf{G} \rangle = \begin{pmatrix} \langle G_1 \rangle \\ \langle G_2 \rangle \end{pmatrix},$$

where the components are calculated with Equation 2.7 using the corresponding marginal densities.

Definition 2.3: Two random variables G_1 and G_2 are said to be *independent* when their joint distribution is the product of their marginal distributions,

$$P(g_1, g_2) = P_1(g_1)P_2(g_2)$$

or, equivalently, when their joint density is the product of their marginal densities,

$$p(g_1, g_2) = p_1(g_1)p_2(g_2).$$

Thus we have, for the mean of the product of two independent random variables,

$$\begin{aligned}\langle G_1 G_2 \rangle &= \int_{-\infty}^{\infty}\int_{-\infty}^{\infty} g_1 g_2 p(g_1, g_2) dg_1 dg_2 \\ &= \int_{-\infty}^{\infty} g_1 p_1(g_1) dg_1 \int_{-\infty}^{\infty} g_2 p_2(g_2) dg_2 = \langle G_1 \rangle \langle G_2 \rangle.\end{aligned}$$

In particular, if G_1 and G_2 have the same distribution function with mean $\langle G \rangle$, then

$$\langle G_1 G_2 \rangle = \langle G \rangle^2. \tag{2.24}$$

An important property of the normal distribution is *additivity*. We state again without proof (but see Exercise 6).

THEOREM 2.3

If the random variables G_1 and G_2 are independent and normally distributed, then the linear combination

$$\boldsymbol{a}^\top \boldsymbol{G} = a_1 G_1 + a_2 G_2$$

is normally distributed with moments

$$\mu = \boldsymbol{a}^\top \boldsymbol{\mu}, \quad \sigma^2 = a_1^2 \sigma_1^2 + a_2^2 \sigma_2^2. \tag{2.25}$$

The *covariance* of random variables G_1 and G_2 is a measure of how their realizations are dependent upon each other and is defined to be the mean of the random variable $(G_1 - \langle G_1\rangle)(G_2 - \langle G_2\rangle)$, that is,

$$\text{cov}(G_1, G_2) = \Big\langle (G_1 - \langle G_1\rangle)(G_2 - \langle G_2\rangle)\Big\rangle. \tag{2.26}$$

Their *correlation* is defined by

$$\rho_{12} = \frac{\text{cov}(G_1, G_2)}{\sqrt{\text{var}(G_1)\text{var}(G_2)}}. \tag{2.27}$$

The correlation is unitless and restricted to values $-1 \le \rho_{12} \le 1$. If $|\rho_{12}| = 1$, then G_1 and G_2 are *linearly dependent*. Two simple consequences of the definition of covariance are

$$\begin{aligned} \text{cov}(G_1, G_2) &= \langle G_1 G_2\rangle - \langle G_1\rangle\langle G_2\rangle \\ \text{cov}(G_1, G_1) &= \text{var}(G_1). \end{aligned} \tag{2.28}$$

A convenient way of representing the variances and covariances of the components of a random vector is in terms of the *covariance matrix*. Let $\boldsymbol{a} = (a_1, a_2)^\top$ be any constant vector. Then the variance of the random variable $\boldsymbol{a}^\top \boldsymbol{G} = a_1 G_1 + a_2 G_2$ is, according to the preceding definitions,

$$\begin{aligned} \text{var}(\boldsymbol{a}^\top \boldsymbol{G}) &= \text{cov}(a_1 G_1 + a_2 G_2, a_1 G_1 + a_2 G_2) \\ &= a_1^2 \text{var}(G_1) + a_1 a_2 \text{cov}(G_1, G_2) + a_1 a_2 \text{cov}(G_2, G_1) + a_2^2 \text{var}(G_2) \\ &= (a_1, a_2) \begin{pmatrix} \text{var}(G_1) & \text{cov}(G_1, G_2) \\ \text{cov}(G_2, G_1) & \text{var}(G_2) \end{pmatrix} \begin{pmatrix} a_1 \\ a_2 \end{pmatrix}. \end{aligned}$$

The matrix in the above equation is the covariance matrix,* usually denoted by the symbol $\boldsymbol{\Sigma}$,

$$\boldsymbol{\Sigma} = \begin{pmatrix} \text{var}(G_1) & \text{cov}(G_1, G_2) \\ \text{cov}(G_2, G_1) & \text{var}(G_2) \end{pmatrix}. \tag{2.29}$$

Therefore, we have

$$\text{var}(\boldsymbol{a}^\top \boldsymbol{G}) = \boldsymbol{a}^\top \boldsymbol{\Sigma} \boldsymbol{a}. \tag{2.30}$$

Note that, since $\text{cov}(G_1, G_2) = \text{cov}(G_2, G_1)$, $\boldsymbol{\Sigma}$ is a symmetric matrix. Moreover, since $\boldsymbol{a}$ is arbitrary and the variance of any random variable is positive, $\boldsymbol{\Sigma}$ is also positive definite (see Definition 1.1).

* More correctly, the *variance–covariance matrix*, but for the sake of brevity we will use the shorter terminology throughout.

The covariance matrix can be written as an outer product,

$$\boldsymbol{\Sigma} = \langle (G - \langle G \rangle)(G - \langle G \rangle)^\top \rangle = \langle GG^\top \rangle - \langle G \rangle \langle G \rangle^\top, \tag{2.31}$$

as is easily verified. Indeed, if $\langle G \rangle = \mathbf{0}$, we can write simply

$$\boldsymbol{\Sigma} = \langle GG^\top \rangle. \tag{2.32}$$

The *correlation matrix* C is similar to the covariance matrix, except that each matrix element, $(C)_{ij}$, is divided by $\sqrt{\text{var}(G_i)\text{var}(G_j)}$ as in Equation 2.27,

$$C = \begin{pmatrix} 1 & \rho_{12} \\ \rho_{21} & 1 \end{pmatrix} = \begin{pmatrix} 1 & \frac{\text{cov}(G_1,G_2)}{\sqrt{\text{var}(G_1)\text{var}(G_2)}} \\ \frac{\text{cov}(G_2,G_1)}{\sqrt{\text{var}(G_1)\text{var}(G_2)}} & 1 \end{pmatrix}, \tag{2.33}$$

where $\rho_{12} = \rho_{21}$ is the correlation of G_1 and G_2.

Analogously to continuous random variables, the continuous random vector $G = (G_1, G_2)^\top$ is often assumed to be described by a *bivariate normal density function*, $p(g)$, given by

$$p(g) = \frac{1}{(2\pi)^{N/2}\sqrt{|\boldsymbol{\Sigma}|}} \exp\left(-\frac{1}{2}(g - \boldsymbol{\mu})^\top \boldsymbol{\Sigma}^{-1}(g - \boldsymbol{\mu})\right), \tag{2.34}$$

with $N=2$. For $N > 2$, the same definition applies and we speak of a *multivariate* normal density function. The mean vector of G is $\langle G \rangle = \boldsymbol{\mu}$ and the covariance matrix is $\boldsymbol{\Sigma}$, which is indicated by writing

$$G \sim \mathcal{N}(\boldsymbol{\mu}, \boldsymbol{\Sigma}).$$

Finally, we have the following important result (see again Freund (1992)).

THEOREM 2.4

If two random variables G_1 and G_2 have a bivariate normal distribution, they are independent if and only if $\rho_{12} = 0$, that is, if and only if they are uncorrelated.

In general, a zero correlation does not imply that two random variables are independent. However, this theorem says that it does if the variables are normally distributed.

2.3 Parameter Estimation

Having introduced distribution functions for random variables and vectors, the question arises as to how to estimate the parameters that characterize those distributions, most importantly their means, variances, and covariance matrices, from observations.

2.3.1 Sampling a Distribution

Consider a multispectral image and a specific land cover category within it. We might choose n pixels belonging to that category and use them to estimate the moments of the underlying distribution. That distribution will be determined not only by measurement noise, atmospheric disturbances, etc., but also by the spread in reflectances characterizing the land cover category itself. For example, Figure 2.2 shows the 3N band of the ASTER image of Figure 1.1, with region of interest (ROI) masks marking areas of mixed forest. The histogram of the observations under the masks is given in Figure 2.3 and is, roughly speaking, a normal distribution. The masked observations

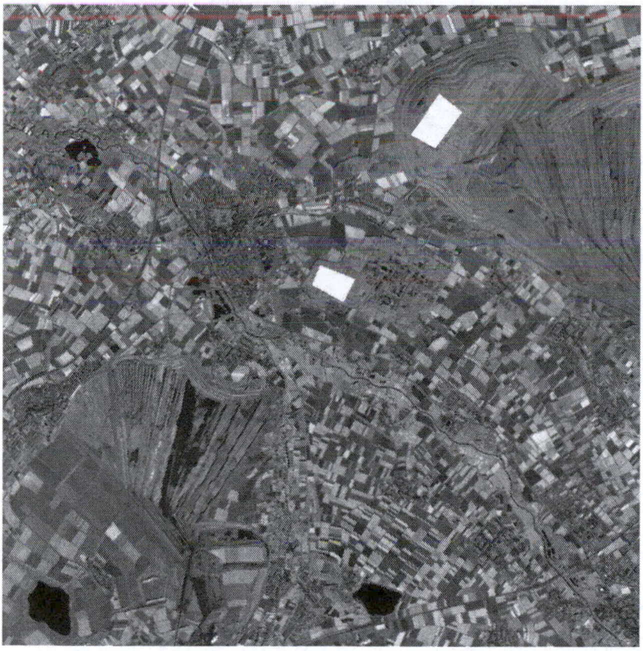

FIGURE 2.2
Spectral band 3N of the ASTER image over Jülich with ROIs covering areas of mixed forest (set with ENVI's ROI tool).

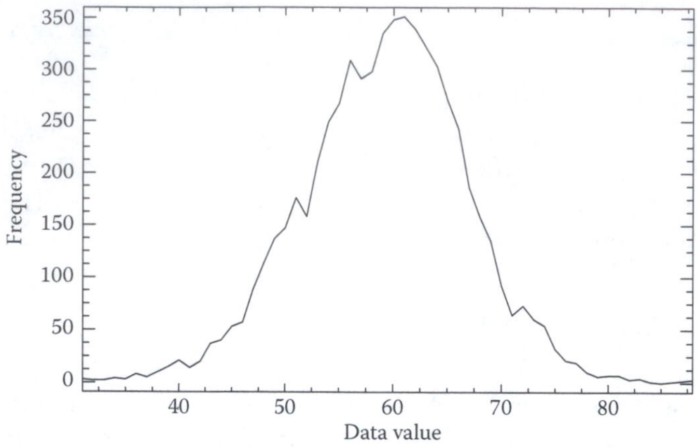

FIGURE 2.3
Histogram of the pixels in Figure 2.2 under the ROIs.

might thus be used to calculate an approximate mean and variance for a random variable describing mixed forest land cover.

More formally, let $\mathbf{G}(1), \mathbf{G}(2) \ldots \mathbf{G}(m)$ be independent random vectors that all have the same joint distribution function, $P(\mathbf{g})$, with mean, $\langle \mathbf{G} \rangle$, and covariance matrix, $\mathbf{\Sigma}$. These random vectors are referred to as a *sample of the distribution* and are said to be *independent and identically distributed* (i.i.d.). Any function of them is called a *sample function* and is itself a random variable, vector, or matrix. The pixel intensities contributing to Figure 2.3 are a particular realization of some—in this instance scalar—sample of the distribution corresponding to the land cover category mixed forest. For our present purposes, the sample functions of interest are those that can be used to estimate the vector mean and covariance matrix of a distribution $P(\mathbf{g})$. These are the *sample mean**

$$\bar{\mathbf{G}} = \frac{1}{m} \sum_{\nu=1}^{m} \mathbf{G}(\nu) \tag{2.35}$$

and the *sample covariance matrix*

$$\mathbf{S} = \frac{1}{m-1} \sum_{\nu=1}^{m} (\mathbf{G}(\nu) - \bar{\mathbf{G}})(\mathbf{G}(\nu) - \bar{\mathbf{G}})^\top. \tag{2.36}$$

* We prefer to use the Greek letter ν to index random vectors and their realizations throughout this book.

These two sample functions are also called *unbiased estimators* because their expected values are equal to the corresponding moments of $P(g)$, that is,

$$\langle \bar{G} \rangle = \langle G \rangle \tag{2.37}$$

and

$$\langle S \rangle = \mathbf{\Sigma}. \tag{2.38}$$

The first result, Equation 2.37, follows immediately:

$$\langle \bar{G} \rangle = \frac{1}{m} \sum_{\nu} \langle G(\nu) \rangle = \frac{1}{m} m \langle G \rangle = \langle G \rangle.$$

The reader is asked to demonstrate Equation 2.38 for the matrix element $(S)_{11}$ in Exercise 8.

Suppose that we have made independent observations $g(\nu)$ of $G(\nu)$, $\nu = 1 \ldots m$. These may be arranged into an $m \times N$ matrix $\mathcal{G}$ in which each N-component observation vector forms a row, that is,

$$\mathcal{G} = \begin{pmatrix} g(1)^\top \\ g(2)^\top \\ \vdots \\ g(m)^\top \end{pmatrix}, \tag{2.39}$$

which, as first introduced in Section 1.2.4, we call the *data matrix*. The calligraphic font is chosen to avoid confusion with the random variable G. If the m observations have zero mean, then the data matrix is said to be *column centered*, in which case an unbiased estimate, $\mathcal{C}$, for the covariance matrix, $\mathbf{\Sigma}$, is given by

$$\mathcal{C} = \frac{1}{m-1} \mathcal{G}^\top \mathcal{G}. \tag{2.40}$$

Note that the order of transposition is reversed relative to Equations 2.32 and 2.36 because the observations are stored as rows. The rules of matrix multiplication (Equation 1.13) take care of the sum over all observations, so it only remains to divide by $m - 1$ to obtain the desired estimate of $\mathbf{S}$, Equation 2.36. In terms of a column-centered data matrix stored in the IDL array `G`,

```
C = transpose(G)##G/(n_elements(G[0,*])-1)
```

Of course, it is simpler to use the built-in function `correlate()`:

```
C = correlate(G,/covariance)
```

See Listing 1.4.

2.3.2 Interval Estimation

The estimators in Equations 2.35 and 2.36 are referred to as *point estimators,* since their realizations involve real numbers. The estimated value of any distribution parameter will of course differ from the true value. Generally, one prefers to quote an interval within which, to some specified degree of confidence, the true value will lie.

To give an important example, consider an $\mathcal{N}(\mu, \sigma^2)$-distributed random sample $G(1) \dots G(m)$. The sample mean

$$\bar{G} = \frac{1}{m} \sum_{\nu=1}^{m} G(\nu)$$

is, according to Theorem 2.3,* also normally distributed and has a mean value μ (see Equation 2.37). With the second of Equations 2.11, we can write

$$\text{var}(\bar{G}) = \frac{1}{m^2} \sum_{\nu} \text{var}(G(\nu)) = \frac{1}{m^2} m \,\text{var}(G) = \frac{1}{m} \sigma^2, \tag{2.41}$$

so that

$$\bar{G} \sim \mathcal{N}(\mu, \sigma^2/m).$$

The variance in the sample mean thus decreases inversely with sample size m. Moreover, $(\bar{G} - \mu)/(\sigma/\sqrt{m})$ is standard normally distributed. Therefore, for any $s > 0$, and with Equation 2.17,

$$\Pr\left(-s < \frac{\bar{G} - \mu}{\sigma/\sqrt{m}} \le s\right) = \Phi(s) - \Phi(-s) = 2\Phi(s) - 1.$$

This may be written in the form

$$\Pr\left(\bar{G} - s\frac{\sigma}{\sqrt{m}} \le \mu < \bar{G} + s\frac{\sigma}{\sqrt{m}}\right) = 2\Phi(s) - 1.$$

Thus, we can say that the probability that the *random interval*

$$\left(\bar{G} - s\frac{\sigma}{\sqrt{m}}, \bar{G} + s\frac{\sigma}{\sqrt{m}}\right) \tag{2.42}$$

covers the unknown mean value, μ, is $2\Phi(s) - 1$. Once a realization of the random interval has been obtained and reported, that is, by plugging a realization of $\bar{G}$ into Equation 2.42, then μ either lies within it or it does not.

* Generalized to a linear combination of n random variables.

Therefore, one can no longer properly speak of probabilities. Instead, a *degree of confidence* for the reported interval is conventionally given and expressed in terms of a (usually small) quantity α defined by

$$1 - \alpha = 2\Phi(s) - 1.$$

This determines s in Equation 2.42 according to

$$\Phi(s) = 1 - \alpha/2. \tag{2.43}$$

For given α, s can be read from the table of the normal distribution. One then says that the interval *covers the unknown parameter* μ *with confidence* $1 - \alpha$.

2.3.3 Provisional Means

In Listing 1.2, an IDL program was given that estimated the covariance matrix of an entire image by sampling *all* of its pixels. As we will see in subsequent chapters, this operation is required for many useful transformations of multispectral images. For large datasets, however, the use of the IDL function `CORRELATE()` may become impractical since it requires that the image array be stored completely in memory. An alternative is to use an iterative algorithm and take advantage of the ENVI *tiling facility*, reading small portions of the image at a time. The procedure we describe here is referred to as the *method of provisional means*, and we give it in a form which includes the possibility of weighting each sample.

Let $\boldsymbol{g}(\nu) = (g_1(\nu), g_2(\nu) \dots g_N(\nu))^\top$ denote the νth sample from an N-band multispectral image with some distribution function $P(\boldsymbol{g})$. Set $\nu = 0$ and define the following quantities and their initial values:

$$\bar{g}_k(\nu = 0) = 0, \quad k = 1 \dots N$$
$$c_{k\ell}(\nu = 0) = 0, \quad k, \ell = 1 \dots N.$$

The $(\nu + 1)$st observation is given weight $w_{\nu+1}$ and we define the update constant

$$r_{\nu+1} = \frac{w_{\nu+1}}{\sum_{\nu'=1}^{\nu+1} w_{\nu'}}.$$

Each new observation $\boldsymbol{g}(\nu + 1)$ leads to the following updates:

$$\bar{g}_k(\nu + 1) = \bar{g}_k(\nu) + (g_k(\nu + 1) - \bar{g}_k(\nu))r_{\nu+1}$$
$$c_{k\ell}(\nu + 1) = c_{k\ell}(\nu) + w_{\nu+1}(g_k(\nu + 1) - \bar{g}_k(\nu))(g_\ell(\nu + 1) - \bar{g}_\ell(\nu)(1 - r_{\nu+1}).$$

Then, after m observations, $\bar{\boldsymbol{g}}(m) = (\bar{g}_1(m) \dots \bar{g}_N(m))^\top$ is a realization of the (weighted) sample mean, Equation 2.35, and $c_{k\ell}(m)$ is a realization of the (k, ℓ)th element of the (weighted) sample covariance matrix, Equation 2.36.

To see that this prescription gives the desired result, consider the case in which $w_\nu = 1$ for all ν. The first two mean values are

$$\bar{g}_k(1) = 0 + (g_k(1) - 0) \cdot 1 = g_k(1)$$
$$\bar{g}_k(2) = g_k(1) + (g_k(2) - g_k(1)) \cdot \frac{1}{2} = \frac{g_k(1) + g_k(2)}{2}$$

as expected. The first two cross products are

$$\begin{aligned}
c_{k\ell}(1) &= 0 + (g_k(1) - 0)(g_\ell(1) - 0)(1 - 1) = 0 \\
c_{k\ell}(2) &= 0 + (g_k(2) - g_k(1))(g_\ell(2) - g_\ell(1))(1 - 1/2) \\
&= \frac{1}{2}(g_k(2) - g_k(1))(g_\ell(2) - g_\ell(1)) \\
&= \frac{1}{2-1} \sum_{\nu=1}^{2} \left(g_k(\nu) - \frac{g_k(1) + g_k(2)}{2} \right) \left(g_\ell(\nu) - \frac{g_\ell(1) + g_\ell(2)}{2} \right)
\end{aligned}$$

again as expected.

An IDL program for the provisional means method is given in Listing 2.2. It is implemented for convenience as an IDL *object class* COVPM (see Fanning (2000) for an introduction to object-oriented programming in IDL). In order to avoid an expensive IDL FOR-loop over the pixels, an external library function written in C is called with the CALL_EXTERNAL() function to process an input array. Listing 2.3 is a modification of Listing 1.2 to calculate the covariance matrix of a multispectral image using the object class and spectral tiling. The result for the VNIR bands of the Jülich image is, apart from some cumulative rounding errors, as before:

```
'Covariance matrix for image E:\satellite images\juelich'
       87.117540        92.394387       -10.518856
       92.394387        115.83864       -77.420809
      -10.518856       -77.420809        374.61476
```

2.4 Hypothesis Testing and Sample Distribution Functions

A *statistical hypothesis* is a conjecture about the distributions of one or more random variables. It might, for instance, be an assertion about the mean of a distribution, or about the equivalence of the variances of two different distributions. One distinguishes between *simple* hypotheses, for which the

Listing 2.2
An object class for the method of provisional means.

```
FUNCTION COVPM::Init, p
   self.mn = ptr_new(dblarr(p))
   self.cov= ptr_new(dblarr(p,p))
   self.sw = 0.0000001
   RETURN, 1
END

PRO COVPM::Cleanup
   Ptr_Free, self.mn
   Ptr_Free, self.cov
END

PRO COVPM::Update, Xs, weights = Ws
   sz = size(Xs)
   NN = sz[1]
   n = sz[2]
   IF n_elements(Ws) EQ 0 THEN Ws = fltarr(sz[2])+1.0
   sw  =  self.sw
   mn  = *self.mn
   cov = *self.cov
   result = call_external('prov_means', 'prov_means', $
       float(Xs), float(Ws), NN, n, sw, mn, cov, $
       value = [0,0,1,1,0,0,0], /i_value,  /cdecl)
   self.sw   = sw
   *self.mn  = mn
   *self.cov = cov
END

FUNCTION COVPM::Covariance
   c = *self.cov/self.sw
   d = diag_matrix(diag_matrix(c))
   RETURN, c+transpose(c)-d
END

FUNCTION COVPM::Means
   RETURN, *self.mn
END

PRO COVPM__Define
class = {COVPM, $
         sw: 0.0, $      ;current sum of weights
         mn: ptr_new(), $;current mean
         cov: ptr_new()} ;current sum of cross products
END
```

Listing 2.3
Calculation of the covariance matrix of an image with the provisional means method and ENVI tiling.

```
PRO EX2_2

envi_select, title='Choose␣multispectral␣image', $
             fid=fid, dims=dims,pos=pos
IF (fid EQ -1) THEN BEGIN
   PRINT, 'cancelled'
   RETURN
ENDIF

envi_file_query,fid,fname=fname,interleave=interleave
IF interleave NE 2 THEN BEGIN
   PRINT, 'not␣BIP␣format'
   RETURN
ENDIF

num_cols = dims[2]-dims[1]+1
num_bands = n_elements(pos)

covpm = Obj_New("COVPM",num_bands)

tile_id = envi_init_tile(fid,pos,num_tiles=num_tiles, $
       interleave=2,xs=dims[1],xe=dims[2], $
       ys=dims[3],ye=dims[4])

; spectral tiling
FOR tile_index=0L,num_tiles-1 DO $
   covpm->update, envi_get_tile(tile_id,tile_index)

PRINT, 'Covariance␣matrix␣for␣image␣'+fname
PRINT, covpm->covariance()

Obj_Destroy, covpm
envi_tile_done, tile_id

END
```

distributions are completely specified, for example, *the mean of a normal distribution with variance* σ^2 *is* $\mu = 0$, and *composite* hypotheses, for which this is not the case, for example, *the mean is* $\mu \geq 0$.

In order to test such assertions on the basis of samples of the distributions involved, it is also necessary to formulate *alternative* hypotheses. To distinguish these from the original assertions, the latter are traditionally called *null* hypotheses. Thus we might be interested in testing the simple null hypothesis $\mu = 0$ against the composite alternative hypothesis $\mu \neq 0$. An appropriate sample function for deciding whether or not to reject the null hypothesis in favor of its alternative is referred to as a *test statistic*. An appropriate *test procedure* will partition the possible realizations of the test statistic into two subsets: an acceptance region for the null hypothesis and a rejection region. The latter is customarily referred to as the *critical region*.

Definition 2.4: Referring to the null hypothesis as H_0, there are two kinds of errors which can arise from any test procedure:

1. H_0 may be rejected when in fact it is true. This is called an *error of the first kind* and the probability that it will occur is denoted α.
2. H_0 may be accepted when in fact it is false, which is called an *error of the second kind* with probability of occurrence β.

The probability of obtaining a value of the test statistic within the critical region when H_0 is true is thus α. The probability α is also referred to as the *level of significance* of the test. It is generally the case that the lower the value of α, the higher is the probability β of making an error of the second kind. Traditionally, significance levels of 0.01 or 0.05 are used. Such values are obviously arbitrary, and for exploratory data analysis it is common to avoid specifying them altogether. Instead the *P-value* for the test is stated.

Definition 2.5: Given the observed value of a test statistic, the *P-value* is the lowest level of significance at which the null hypothesis could have been rejected.

High *P*-values provide evidence in favor of accepting the null hypothesis, without actually forcing one to commit to a decision.

The theory of statistical testing specifies methods for determining the most appropriate test statistic for a given null hypothesis and its alternative.* To illustrate, consider a sample $G(1), G(2) \ldots G(m)$ from a normal distribution with mean μ and variance σ^2. The sample mean, as we have seen, is distributed as $\bar{G} \sim \mathcal{N}(\mu, \sigma^2/m)$. Tests for hypotheses regarding μ are easily formulated in terms of the test statistic $\bar{G}$ and the standard normal

---

* Fundamental to the theory is the *Neyman–Pearson Lemma*, which gives for simple hypotheses a prescription for finding the test procedure that maximizes the probability $1 - \beta$ of rejecting the null hypothesis when it is false for a fixed level of significance α (see Freund [1992] for an excellent discussion).

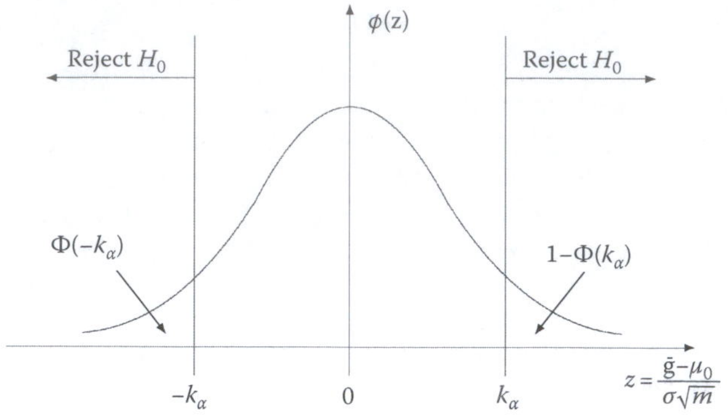

FIGURE 2.4
Critical region for rejecting the hypothesis $\mu = \mu_0$.

distribution $\Phi(x)$. For example, the test at significance level α for the simple hypothesis

$$H_0\colon \mu = \mu_0$$

against the alternative composite hypothesis

$$H_1\colon \mu \neq \mu_0$$

has critical region

$$\left|\frac{\bar{g} - \mu_0}{\sigma/\sqrt{m}}\right| \geq k_\alpha,$$

where k_α is determined by

$$\Phi(-k_\alpha) + 1 - \Phi(k_\alpha) = \alpha,$$

or

$$1 - \Phi(k_\alpha) = \alpha/2$$

(see Figure 2.4).

Suppose however that the variance σ^2 is unknown and that we wish nevertheless to make a statement about μ in terms of some realization of an appropriate statistic. To treat this and similar problems it is first necessary to define some additional distribution functions.

2.4.1 Chi-Square Distribution

If the random variables $G(\nu)$, $\nu = 1 \ldots m$, are independent and standard normally distributed (i.e., with mean 0 and variance 1) then the random variable

$$Z = \sum_{\nu=1}^{m} G(\nu)^2$$

is *chi-square distributed with m degrees of freedom*. The chi-square density function is given by (Exercise 4)

$$p_{\chi^2;m}(z) = \begin{cases} c_m z^{(m-2)/2} e^{-z/2} & \text{for } z > 0 \\ 0 & \text{otherwise} \end{cases} \tag{2.44}$$

and the distribution function by

$$P_{\chi^2;m}(z) = \int_0^z p_{\chi^2;m}(t)dt. \tag{2.45}$$

The normalization factor, c_m, is determined by the requirement

$$\int_0^\infty p_{\chi^2;nm}(z)dz = 1.$$

One can use the IDL function `CHISQR_PDF()` to calculate the chi-square distribution function or `CHISQR_CVF()` to calculate its percentiles (values of z for given $P_{\chi^2;m}(z)$).

2.4.2 Student-*t* Distribution

If the random variables $G(\nu)$, $\nu = 0 \ldots m$, are independent and standard normally distributed, then the random variable

$$T = \frac{G(0)}{\sqrt{\frac{1}{m}(G(1)^2 + \cdots + G(m)^2)}} \tag{2.46}$$

is said to be *Student-t distributed with m degrees of freedom*. The corresponding density is given by

$$p_{t;m}(t) = d_m \left(1 + \frac{t^2}{m}\right)^{-(m+1)/2}, \quad -\infty < t < \infty, \tag{2.47}$$

with d_m again being a normalization factor. The Student-*t* distribution converges to the standard normal distribution for $m \to \infty$. The IDL functions `T_PDF()` and `T_CVF()` may be used to calculate it, and respectively its percentiles.

The Student-t distribution is used to make statements regarding the mean when the variance is unknown. Thus if $G(\nu)$, $\nu = 1 \dots m$, is a sample from a normal distribution with mean μ and unknown variance, and

$$\bar{G} = \frac{1}{m}\sum_{\nu=1}^{m} G(\nu), \quad S = \frac{1}{m-1}\sum_{\nu}^{n}(G(\nu) - \bar{G})^2,$$

then the random variable

$$T = \frac{\bar{G} - \mu}{\sqrt{S/m}} \tag{2.48}$$

is Student-t distributed with $m - 1$ degrees of freedom. It can be used to test hypotheses regarding μ, for example, that $\mu = 0$.

If $G(\nu)$, $\nu = 1 \dots n$, and $H(\nu)$, $\nu = 1 \dots m$, are samples from normal distributions with equal variance, then the random variable

$$T_d = \frac{\bar{G} - \bar{H} - (\mu_G - \mu_H)}{\sqrt{\left(\frac{1}{n} + \frac{1}{m}\right)\frac{(n-1)S_G + (m-1)S_H}{n+m-2}}} \tag{2.49}$$

is Student-t distributed with $n + m - 2$ degrees of freedom. It may be used to test the hypothesis $\mu_G = \mu_H$, for example.

2.4.3 *F*-Distribution

Given independent and standard normally distributed random variables $G(\nu)$, $\nu = 1 \dots n$, and $H(\nu)$, $\nu = 1 \dots m$, the random variable

$$F = \frac{\frac{1}{m}(H(1)^2 + \cdots + H(m)^2)}{\frac{1}{n}(G(1)^2 + \cdots + G(n)^2)} \tag{2.50}$$

is *F-distributed with m and n degrees of freedom*. Its density function is

$$p_{f;m,n}(f) = \begin{cases} c_{mn} f^{(m-2)/2}\left(1 + \frac{m}{n}f\right)^{-(m+n)/2} & \text{for } f > 0 \\ 0 & \text{otherwise,} \end{cases} \tag{2.51}$$

with normalization factor c_{mn}. Note that F is the ratio of two chi-square distributed random variables. One can compute the F-distribution function and its percentiles with `F_PDF()` and `F_CVF()` in IDL.

The F-distribution can be used for hypothesis tests regarding two variances. If S_G and S_H are sample variances for samples of size n and m, respectively, drawn from normally distributed populations with variances σ_G^2 and σ_H^2, then

$$F = \frac{\sigma_H^2 S_G}{\sigma_G^2 S_H} \tag{2.52}$$

is a random variable having an F-distribution with $n-1$ and $m-1$ degrees of freedom. It is thus suitable for calculating the critical region for rejection of the null hypothesis that $\sigma_G^2 = \sigma_H^2$.

2.5 Conditional Probabilities, Bayes' Theorem, and Classification

If A and B are two events, that is, two subsets of a sample space Ω, such that the probability of A and B occurring simultaneously is $\Pr(A, B)$, and if $\Pr(B) \neq 0$, then the *conditional probability of A occurring given that B occurs* is defined to be

$$\Pr(A \mid B) = \frac{\Pr(A, B)}{\Pr(B)}. \tag{2.53}$$

We have the following (Freund, 1992).

THEOREM 2.5 (Theorem of Total Probability)

If $A_1, A_2 \ldots A_m$ are disjoint events associated with some random experiment, and if their union is the set of all possible events, then for any event B

$$\Pr(B) = \sum_{i=1}^{m} \Pr(B \mid A_i)\Pr(A_i) = \sum_{i=1}^{m} \Pr(B, A_i). \tag{2.54}$$

It should be noted that both Equations 2.53 and 2.54 have their counterparts for probability density functions:

$$p(x, y) = p(x \mid y)p(y) \tag{2.55}$$

$$p(x) = \int_{-\infty}^{\infty} p(x \mid y)p(y)dy = \int_{-\infty}^{\infty} p(x, y)dy. \tag{2.56}$$

Bayes' theorem* follows directly from the definition of conditional probability and is the basic starting point for inference problems using probability theory as logic.

THEOREM 2.6 (Bayes' Theorem)

If $A_1, A_2 \dots A_m$ are disjoint events associated with some random experiment, their union is the set of all possible events, and if $\Pr(A_i) \neq 0$ for $i = 1 \dots m$, then for any event B for which $\Pr(B) \neq 0$

$$\Pr(A_k \mid B) = \frac{\Pr(B \mid A_k)\Pr(A_k)}{\sum_{i=1}^{m} \Pr(B \mid A_i)\Pr(A_i)}. \tag{2.57}$$

Proof

From the definition of conditional probability, Equation 2.53,

$$\Pr(A_k \mid B)\Pr(B) = \Pr(A_k, B) = \Pr(B \mid A_k)\Pr(A_k),$$

and therefore

$$\Pr(A_k \mid B) = \frac{\Pr(B \mid A_k)\Pr(A_k)}{\Pr(B)} = \frac{\Pr(B \mid A_k)\Pr(A_k)}{\sum_{i=1}^{m} \Pr(B \mid A_i)\Pr(A_i)}$$

from the theorem of total probability. □

We will use Bayes' theorem primarily in the following form: Let g be a realization of a random vector associated with a distribution of pixel intensities, and let $\{k \mid k = 1 \dots K\}$ be a set of possible *class labels* (e.g., land cover categories) for all of the pixels. Then the *a posteriori* conditional probability for class k, *given the observation* g, may be written using Bayes' theorem as

$$\Pr(k \mid g) = \frac{\Pr(g \mid k)\Pr(k)}{\Pr(g)}, \tag{2.58}$$

* Named after Rev. Thomas Bayes, an eighteenth century hobby mathematician who derived a special case.

where $\Pr(k)$ is the *a priori* probability for class k, $\Pr(g \mid k)$ is the conditional probability of observing the value g if it belongs to class k, and

$$\Pr(g) = \sum_{k=1}^{K} \Pr(g \mid k)\Pr(k) \tag{2.59}$$

is the total probability for g.

One can formulate the problem of *classification* of multispectral imagery as the process of determining posterior conditional probabilities for all of the classes. This is accomplished by specifying prior probabilities $\Pr(k)$ (if prior information exists), modeling the class-specific probabilities $\Pr(g \mid k)$, and then applying Bayes' theorem. Thus for continuous variables we might choose a class-specific density function as our model:

$$\Pr(g \mid k) \sim p(g \mid \theta_k),$$

where θ_k is a set of parameters for the density function describing the kth class (for a multivariate normal distribution for instance, just the mean vector $\boldsymbol{\mu}_k$ and covariance matrix $\boldsymbol{\Sigma}_k$). In this case Equation 2.58 takes the form

$$\Pr(k \mid g) = \frac{p(g \mid \theta_k)\Pr(k)}{p(g)}, \tag{2.60}$$

where $p(g)$ is the unconditional density function for g,

$$p(g) = \sum_{k=1}^{K} p(g \mid \theta_k)\Pr(k), \tag{2.61}$$

which is constant when g is given. As we will see in Chapter 6, under reasonable assumptions the observation g should be assigned to the class k which maximizes $\Pr(k \mid g)$.

In order to find that maximum, we require estimates of the parameters θ_k, $k = 1 \ldots K$. If we have access to measured values (realizations) that are known to be in class k, $g(\nu)$, $\nu = 1 \ldots m_k$, say, then we can form the product of probability densities

$$\prod_{\nu=1}^{m_k} p(g(\nu) \mid \theta_k),$$

which is called a *likelihood function*, and take its logarithm

$$\mathcal{L}(\theta_k) = \sum_{\nu=1}^{m_k} \log p(g(\nu) \mid \theta_k), \tag{2.62}$$

which is the *log-likelihood*. Taking products is justified when the $g(\nu)$ are realizations of independent random vectors. The parameter set θ_k which maximizes the likelihood function or its logarithm, that is, which gives the largest value for all of the realizations, is called the *maximum likelihood estimate* of θ_k. For normally distributed random vectors, maximum likelihood parameter estimators for $\theta_k = \{\boldsymbol{\mu}_k, \boldsymbol{\Sigma}_k\}$ turn out to correspond to the unbiased estimators, Equations 2.35 and 2.36.

To illustrate this for the class mean, write out Equation 2.62 for the multivariate normal distribution, Equation 2.34:

$$\mathcal{L}(\boldsymbol{\mu}_k, \boldsymbol{\Sigma}_k) = m_k \frac{N}{2} \log 2\pi - m_k \frac{1}{2} \log |\boldsymbol{\Sigma}_k| - \frac{1}{2} \sum_{\nu=1}^{m_k} (g(\nu) - \boldsymbol{\mu}_k)^\top \boldsymbol{\Sigma}_k^{-1} (g(\nu) - \boldsymbol{\mu}_k).$$

To maximize with respect to $\boldsymbol{\mu}_k$, we set

$$\frac{\partial \mathcal{L}(\boldsymbol{\mu}_k, \boldsymbol{\Sigma}_k)}{\partial \boldsymbol{\mu}_k} = \sum_{\nu=1}^{m_k} \boldsymbol{\Sigma}_k^{-1} (g(\nu) - \boldsymbol{\mu}_k) = \mathbf{0},$$

giving

$$\boldsymbol{\mu}_k \approx \mathbf{m}_k = \frac{1}{m_k} \sum_{i=1}^{m_k} g(\nu),$$

which is the realization of the sample mean, Equation 2.35. In a similar way (see, e.g., Duda et al., 2001) one can show that

$$\boldsymbol{\Sigma}_k \approx \mathcal{C}_k = \frac{1}{m_k} \sum_{\nu=1}^{m_k} (g(\nu) - \boldsymbol{\mu}_k)(g(\nu) - \boldsymbol{\mu}_k)^\top,$$

the realization of the sample covariance matrix, Equation 2.36 (except that the denominator $m - 1$ is replaced by m, a fact which usually can be ignored).

The observations, of course, must be chosen from the appropriate class k in each case. For *supervised classification* (Chapters 6 and 7), there exists a set of training data with known class labels. Therefore maximum likelihood estimates can be obtained and posterior probability distributions for the classes can be calculated from Equation 2.60 and then used to generalize to all of the image data. In the case of *unsupervised classification* the class memberships are not initially known. How this conundrum is solved will be discussed in Chapter 8.

2.6 Ordinary Linear Regression

Many applications in remote sensing, such as radiometric calibration of digital images, involve fitting a set of data points to a straight line

$$y(x) = a + bx. \tag{2.63}$$

We review here the standard procedure for determining the parameters a and b and their uncertainties, namely *ordinary linear regression*, in which it is assumed that the dependent variable, y, has a random error, but that the independent variable x is exact. The procedure is then generalized to more than one independent variable, and the concepts of regularization and duality are introduced.

Linear regression can also be carried out sequentially by updating the best fit after each new observation. This topic, as well as *orthogonal linear regression*, where it is assumed that both variables have random errors associated with them, are discussed in Appendix A.

2.6.1 One Independent Variable

Suppose that the dataset consists of m pairs $\{x(\nu), y(\nu) \mid \nu = 1 \ldots m\}$. An appropriate statistical model for linear regression is

$$Y(\nu) = a + bx(\nu) + R(\nu), \quad \nu = 1 \ldots m, \tag{2.64}$$

where

$Y(\nu)$ is a random variable representing the νth measurement of the dependent variable

$R(\nu)$, referred to as the *residual error*, is a random variable representing the measurement uncertainty

The $x(\nu)$ are exact. We will assume that the individual measurements are uncorrelated and that they all have the same variance:

$$\mathrm{cov}(R(\nu), R(\nu')) = \begin{cases} \sigma^2 & \text{for } \nu = \nu' \\ 0 & \text{otherwise.} \end{cases} \tag{2.65}$$

The realizations of $R(\nu)$ are thus $y(\nu) - a - bx(\nu)$, $\nu = 1 \ldots m$, from which we define a least squares *goodness of fit* function

$$z(a, b) = \sum_{\nu=1}^{m} \left(\frac{y(\nu) - a - bx(\nu)}{\sigma} \right)^2. \tag{2.66}$$

If the residuals $R(\nu)$ are normally distributed, then we recognize Equation 2.66 as a realization of a chi-square distributed random variable. For the "best" values of a and b, Equation 2.66 is in fact chi-square distributed with $m-2$ degrees of freedom (Press et al., 2002). The best values for the parameters a and b are obtained by minimizing $z(a, b)$, that is, by solving the equations

$$\frac{\partial z}{\partial a} = \frac{\partial z}{\partial b} = 0$$

for a and b. The solution is (Exercise 10)

$$\hat{b} = \frac{s_{xy}}{s_{xx}}, \quad \hat{a} = \bar{y} - \hat{b}\bar{x}, \tag{2.67}$$

where

$$\begin{aligned} s_{xy} &= \frac{1}{m}\sum_{\nu=1}^{m}(x(\nu)-\bar{x})(y(\nu)-\bar{y}) \\ s_{xx} &= \frac{1}{m}\sum_{\nu=1}^{m}(x(\nu)-\bar{x})^2 \\ \bar{x} &= \frac{1}{m}\sum_{\nu=1}^{m}x(\nu), \quad \bar{y} = \frac{1}{m}\sum_{\nu=1}^{m}y(\nu). \end{aligned} \tag{2.68}$$

The uncertainties in the estimates $\hat{a}$ and $\hat{b}$ are given by (Exercise 11)

$$\begin{aligned} \sigma_{\hat{a}}^2 &= \frac{\sigma^2 \sum x(\nu)^2}{m\sum x(\nu)^2 - (\sum x(\nu))^2} \\ \sigma_{\hat{b}}^2 &= \frac{\sigma^2}{\sum x(\nu)^2 - (\sum x(\nu))^2}. \end{aligned} \tag{2.69}$$

The goodness of fit can be determined by substituting the estimates $\hat{a}$ and $\hat{b}$ into Equation 2.66 which, as we have said, will then be chi-square distributed with $m-2$ degrees of freedom. The probability of finding a value $z = z(\hat{a}, \hat{b})$ *or higher* by chance is

$$Q = 1 - P_{\chi^2;m-2}(z),$$

where $P_{\chi^2;m-2}(z)$ is given by Equation 2.44. If $Q < 0.001$, one would typically reject the fit as being unsatisfactory.

If σ^2 is not known *a priori*, then it can be estimated by

$$\hat{\sigma}^2 = \frac{1}{m-2}\sum_{\nu=1}^{m}(y(\nu) - \hat{a} - \hat{b}x(\nu))^2, \tag{2.70}$$

in which case the goodness of fit procedure cannot be applied, since we *assume* the fit to be good in order to estimate σ^2 with Equation 2.70.

2.6.2 More Than One Independent Variable

The statistical model of Section 2.6.1 may be written more generally in the form

$$Y(\nu) = w_o + \sum_{i=1}^{N} w_i x_i(\nu) + R(\nu), \quad \nu = 1 \dots m, \tag{2.71}$$

relating m measurements of the N independent variables $x_1 \dots x_N$ to a measured dependent variable Y via the parameters $w_0, w_1 \dots w_N$. Equivalently, in vector notation we can write

$$Y(\nu) = \boldsymbol{w}^\top \boldsymbol{x}(\nu) + R(\nu), \quad \nu = 1 \dots m, \tag{2.72}$$

where

$\boldsymbol{x} = (x_0 = 1, x_1 \dots x_N)^\top$

$\boldsymbol{w} = (w_0, w_1 \dots w_N)^\top$

The random variables $R(\nu)$ again represent the measurement uncertainty in the realizations $y(\nu)$ of $Y(\nu)$. We again assume that they are uncorrelated and normally distributed with zero mean and variance σ^2, whereas the values $\boldsymbol{x}(\nu)$ are, as before, assumed to be exact. Now we wish to determine the best value for parameter vector $\boldsymbol{w}$.

Introducing the $m \times (N+1)$ data matrix

$$\mathcal{X} = \begin{pmatrix} \boldsymbol{x}(1)^\top \\ \vdots \\ \boldsymbol{x}(m)^\top \end{pmatrix},$$

we can express Equation 2.71 or 2.72 in the form

$$\boldsymbol{Y} = \mathcal{X}\boldsymbol{w} + \boldsymbol{R}, \tag{2.73}$$

where $\boldsymbol{Y} = (Y(1)\dots Y(m))^\top$, $\boldsymbol{R} = (R(1)\dots R(m))^\top$ and, by assumption,

$$\boldsymbol{\Sigma}_R = \langle \boldsymbol{R}\boldsymbol{R}^\top \rangle = \sigma^2 \boldsymbol{I}.$$

The identity matrix $\boldsymbol{I}$ is $m \times m$. The goodness of fit function analog to Equation 2.66 is

$$z(\boldsymbol{w}) = \sum_{\nu=1}^{m} \left[\frac{y(\nu) - \boldsymbol{w}^\top \boldsymbol{x}(\nu)}{\sigma} \right]^2 = \frac{1}{\sigma^2}(\boldsymbol{y} - \boldsymbol{\mathcal{X}}\boldsymbol{w})^\top (\boldsymbol{y} - \boldsymbol{\mathcal{X}}\boldsymbol{w}). \tag{2.74}$$

This is minimized by solving the equations

$$\frac{\partial z(\boldsymbol{w})}{\partial w_k} = 0, \quad k = 0\dots N.$$

Using the rules for vector differentiation we obtain (Exercise 12)

$$\boldsymbol{\mathcal{X}}^\top \boldsymbol{y} = (\boldsymbol{\mathcal{X}}^\top \boldsymbol{\mathcal{X}})\boldsymbol{w}. \tag{2.75}$$

Equation 2.75 is referred to as the *normal equation*. The estimated parameters of the model are obtained by solving for $\boldsymbol{w}$,

$$\hat{\boldsymbol{w}} = (\boldsymbol{\mathcal{X}}^\top \boldsymbol{\mathcal{X}})^{-1} \boldsymbol{\mathcal{X}}^\top \boldsymbol{y} =: \boldsymbol{L}\boldsymbol{y}. \tag{2.76}$$

The matrix

$$\boldsymbol{L} = (\boldsymbol{\mathcal{X}}^\top \boldsymbol{\mathcal{X}})^{-1} \boldsymbol{\mathcal{X}}^\top \tag{2.77}$$

is called the *pseudoinverse* of the data matrix $\boldsymbol{\mathcal{X}}$.

In order to obtain the uncertainty in the estimate $\hat{\boldsymbol{w}}$, Equation 2.76, we can think of $\boldsymbol{w}$ as a random vector with mean value $\langle \boldsymbol{w} \rangle$. Its covariance matrix is then given by

$$\begin{aligned}
\boldsymbol{\Sigma}_w &= \langle (\boldsymbol{w} - \langle \boldsymbol{w} \rangle)(\boldsymbol{w} - \langle \boldsymbol{w} \rangle)^\top \rangle \\
&\approx \langle (\boldsymbol{w} - \hat{\boldsymbol{w}})(\boldsymbol{w} - \hat{\boldsymbol{w}})^\top \rangle \\
&= \langle (\boldsymbol{w} - \boldsymbol{L}\boldsymbol{y})(\boldsymbol{w} - \boldsymbol{L}\boldsymbol{y})^\top \rangle \\
&= \langle (\boldsymbol{w} - \boldsymbol{L}(\boldsymbol{\mathcal{X}}\boldsymbol{w} + \boldsymbol{r}))(\boldsymbol{w} - \boldsymbol{L}(\boldsymbol{\mathcal{X}}\boldsymbol{w} + \boldsymbol{r}))^\top \rangle,
\end{aligned}$$

where $\boldsymbol{r}$ is some realization of $\boldsymbol{R}$. But from Equation 2.77 we see that $\boldsymbol{L}\boldsymbol{\mathcal{X}} = \boldsymbol{I}$, so

$$\begin{aligned}\boldsymbol{\Sigma}_w &\approx \langle(-\boldsymbol{L}\boldsymbol{r})(-\boldsymbol{L}\boldsymbol{r})^\top\rangle = \boldsymbol{L}\langle\boldsymbol{r}\boldsymbol{r}^\top\rangle\boldsymbol{L}^\top \\ &\approx \boldsymbol{L}\langle\boldsymbol{R}\boldsymbol{R}^\top\rangle\boldsymbol{L}^\top \\ &= \sigma^2\boldsymbol{L}\boldsymbol{L}^\top.\end{aligned}$$

Again with Equation 2.77 we have finally

$$\boldsymbol{\Sigma}_w = \sigma^2(\boldsymbol{\mathcal{X}}^\top\boldsymbol{\mathcal{X}})^{-1}. \tag{2.78}$$

To check that this is indeed a generalization of ordinary linear regression on a single independent variable, identify the parameter vector $\boldsymbol{w}$ with the straight line parameters a and b, that is,

$$\boldsymbol{w} = \begin{pmatrix} w_0 \\ w_1 \end{pmatrix} = \begin{pmatrix} a \\ b \end{pmatrix}.$$

The matrix $\boldsymbol{\mathcal{X}}$ and vector $\boldsymbol{y}$ are similarly

$$\boldsymbol{\mathcal{X}} = \begin{pmatrix} 1 & x(1) \\ 1 & x(2) \\ \vdots & \vdots \\ 1 & x(m) \end{pmatrix}, \quad \boldsymbol{y} = \begin{pmatrix} y(1) \\ y(2) \\ \vdots \\ y(m) \end{pmatrix}.$$

Thus the best estimates for the parameters are

$$\hat{\boldsymbol{w}} = \begin{pmatrix} \hat{a} \\ \hat{b} \end{pmatrix} = (\boldsymbol{\mathcal{X}}^\top\boldsymbol{\mathcal{X}})^{-1}(\boldsymbol{\mathcal{X}}^\top\boldsymbol{y}).$$

Evaluating,

$$(\boldsymbol{\mathcal{X}}^\top\boldsymbol{\mathcal{X}})^{-1} = \begin{pmatrix} m & \sum x(\nu) \\ \sum x(\nu) & \sum x(\nu)^2 \end{pmatrix}^{-1} = \begin{pmatrix} m & m\bar{x} \\ m\bar{x} & \sum x(\nu)^2 \end{pmatrix}^{-1}.$$

Recalling the expression for the inverse of a 2×2 matrix in Chapter 1, we then have

$$(\boldsymbol{\mathcal{X}}^\top\boldsymbol{\mathcal{X}})^{-1} = \frac{1}{m\sum x(\nu)^2 + m^2\bar{x}^2}\begin{pmatrix} \sum x(\nu)^2 & -m\bar{x} \\ -m\bar{x} & m \end{pmatrix}.$$

Furthermore,

$$\boldsymbol{\mathcal{X}}^\top\boldsymbol{y} = \begin{pmatrix} m\bar{y} \\ \sum x(\nu)y(\nu) \end{pmatrix}.$$

Therefore the estimate for b is

$$\hat{b} = \frac{1}{m\sum x(\nu)^2 + m^2\bar{x}^2}(-m^2\bar{x}\bar{y} + m\sum x(\nu)y(\nu)) = \frac{-m\bar{x}\bar{y} + \sum x(\nu)y(\nu)}{m\sum x(\nu)^2 + m^2\bar{x}^2}. \tag{2.79}$$

From Equation 2.78 the uncertainty in b is given by σ^2 times the (2,2) element of $(\mathcal{X}^\top\mathcal{X})^{-1}$,

$$\sigma_b^2 = \sigma^2 \frac{m}{m\sum x(\nu)^2 + m^2\bar{x}^2}. \tag{2.80}$$

Equations 2.79 and 2.80 correspond to Equations 2.67 and 2.69, respectively.

2.6.3 Regularization, Duality, and the Gram Matrix

For *ill-conditioned* regression problems (e.g., large amount of noise, insufficient data, or $\mathcal{X}^\top\mathcal{X}$ nearly singular) the solution $\hat{w}$ in Equation 2.76 may be unreliable. A remedy is to restrict w in some way, the simplest one being to favor a small length or, equivalently a small squared norm $\|w\|^2$. In the modified goodness of fit function

$$z(w) = (y - \mathcal{X}w)^\top(y - \mathcal{X}w) + \lambda\|w\|^2, \tag{2.81}$$

where we have assumed $\sigma^2 = 1$ for simplicity, the parameter λ defines a trade-off between minimum residual error and minimum norm. Equating the vector derivative with respect to w with zero as before then leads to the normal equation

$$\mathcal{X}^\top y = (\mathcal{X}^\top\mathcal{X} + \lambda I_{N+1})w, \tag{2.82}$$

where the identity matrix I_{N+1} has dimensions $(N+1)\times(N+1)$. The least squares estimate for the parameter vector w is now

$$\hat{w} = (\mathcal{X}^\top\mathcal{X} + \lambda I_{N+1})^{-1}\mathcal{X}^\top y. \tag{2.83}$$

For $\lambda > 0$ the matrix $\mathcal{X}^\top\mathcal{X} + \lambda I_{N+1}$ can always be inverted. This procedure is known as *ridge regression*, and Equation 2.83 may be referred to as its *primal solution*. Regularization will be encountered in Chapter 9 in the context of change detection.

With a simple manipulation, Equation 2.83 can be put in the form

$$\hat{w} = \mathcal{X}^\top\alpha = \sum_{\nu=1}^{m}\alpha_\nu x(\nu), \tag{2.84}$$

where $\boldsymbol{\alpha}$ is given by

$$\boldsymbol{\alpha} = \frac{1}{\lambda}(y - \mathcal{X}\hat{w}) \tag{2.85}$$

(see Exercise 14). Equation 2.84 expresses the unknown parameter vector $\hat{w}$ as a linear combination of the observation vectors $x(\nu)$. It remains to find a suitable expression for the vector $\boldsymbol{\alpha}$. We can eliminate $\hat{w}$ from Equation 2.85 by substituting Equation 2.84 and solving for $\boldsymbol{\alpha}$,

$$\boldsymbol{\alpha} = (\mathcal{X}\mathcal{X}^\top + \lambda \boldsymbol{I}_m)^{-1}\boldsymbol{y}, \tag{2.86}$$

where $\boldsymbol{I}_m$ is the $m \times m$ identity matrix. Equations 2.84 and 2.86 taken together constitute the *dual solution* of the ridge regression problem, and the components of $\boldsymbol{\alpha}$ are called the *dual parameters*. Once they have been determined from Equation 2.86, the solution for the original parameter vector $\hat{w}$ is recovered from Equation 2.84. Note that in the primal solution, Equation 2.83, we are inverting an $(N+1) \times (N+1)$ matrix,

$$\mathcal{X}^\top \mathcal{X} + \lambda \boldsymbol{I}_{N+1}$$

whereas in the dual solution we must invert the (often much larger) $m \times m$ matrix

$$\mathcal{X}\mathcal{X}^\top + \lambda \boldsymbol{I}_m.$$

The matrix $\mathcal{X}\mathcal{X}^\top$ is called a *Gram matrix*. Its elements consist of all inner products of the observation vectors

$$x(\nu)^\top x(\nu'), \quad \nu, \nu' = 1 \ldots m,$$

and it is obviously a symmetric matrix. Moreover, it is positive semidefinite since, for any vector z with m components,

$$z^\top \mathcal{X}\mathcal{X}^\top z = \|\mathcal{X}^\top z\|^2 \geq 0.$$

The dual solution is interesting because the dual parameters are expressed entirely in terms of inner products of observation vectors $x(\nu)$. Moreover, predicting values of y from new observations x can be expressed purely in terms of inner products as well. Thus

$$y = \hat{w}^\top x = \left(\sum_{\nu=1}^{m} \alpha_\nu x(\nu)^\top\right) x = \sum_{\nu=1}^{m} \alpha_\nu \, x(\nu)^\top x. \tag{2.87}$$

Later we will see that the inner products can be substituted by so-called *kernel functions* which allow very elegant and powerful nonlinear generalizations of linear methods such as ridge regression.

2.7 Entropy and Information

Suppose we make an observation on a discrete random variable X with mass function

$$p\big(X = x(i)\big) = p(i), \quad i = 1 \ldots n.$$

Qualitatively speaking, the *amount of information* we receive on observing a particular realization $x(i)$ may be thought of as the "amount of surprise" associated with the result. The information content of the observation should be a monotonically decreasing function of the probability $p(i)$ for that observation: if the probability is unity, there is no surprise, if $p(i) \ll 1$, the surprise is large. The function chosen to express the information content of $x(i)$ is

$$h(x(i)) = -\log p(i). \tag{2.88}$$

This function is monotonically decreasing, and zero when $p(i) = 1$. It also has the desirable property that, for two independent observations $x(i), x(j)$,

$$\begin{aligned} h(x(i), x(j)) &= -\log p(X = x(i), X = x(j)) \\ &= -\log\left[p(X = x(i))p(X = x(j))\right] \\ &= -\log\left[p(i)p(j)\right] = h(x(i)) + h(x(j)), \end{aligned}$$

that is, information gained from two independent observations is additive.

The *average amount of information* that we expect to receive on observing the random variable X is called the *entropy* of X and is given by

$$H(X) = -\sum_{i=1}^{n} p(i) \log p(i). \tag{2.89}$$

The entropy can also be interpreted as the *average amount of information required to specify the random variable.*

The discrete distribution with maximum entropy can be determined by maximizing the Lagrange function

$$L\left(p(1) \ldots p(n)\right) = -\sum_i p(i) \log p(i) + \lambda \left(\sum_i p(i) - 1)\right).$$

Equating the derivatives to zero,

$$\frac{\partial L}{\partial p(i)} = -\log p(i) - 1 + \lambda = 0,$$

so $p(i)$ is independent of i. The condition $\sum_i p(i) = 1$ then requires that

$$p(i) = 1/n.$$

The Hessian matrix is easily seen to have diagonal elements given by

$$(\boldsymbol{H})_{ii} = \frac{\partial^2 L}{\partial p(i)^2} = -\frac{1}{p(i)},$$

and off-diagonal elements, zero. It is therefore negative definite, that is, for any $\boldsymbol{x} > \boldsymbol{0}$,

$$\boldsymbol{x}^\top \boldsymbol{H}\boldsymbol{x} = -\frac{1}{p_1}x_1^2 - \cdots - \frac{1}{p_n}x_n^2 < 0.$$

Thus the uniform distribution indeed maximizes the entropy.

If X is a continuous random variable with probability density function $p(x)$, then its entropy* is defined analogously to Equation 2.89 as

$$H(X) = -\int p(x)\log[p(x)]dx. \tag{2.90}$$

The continuous distribution function which has maximum entropy is the normal distribution (see Bishop 2006, Chapter 1), for a derivation of this fact.

If $p(x, y)$ is a joint density function for random variables X and Y, then the *conditional entropy* of Y given X is

$$\begin{aligned} H(Y \mid X) &= -\int p(x)\left(\int p(y \mid x)\log[p(y \mid x)]dy\right)dx \qquad (2.91)\\ &= -\iint p(x, y)\log[p(y \mid x)]dy\,dx. \end{aligned}$$

(For the second equality we have used Equation 2.55.) This is just the information $-\log[p(y \mid x)]$ gained on observing y given x, averaged over the joint probability for x and y. If Y is independent of X, then $p(y \mid x) = p(y)$ and $H(Y \mid X) = H(Y)$.

---

* More correctly, *differential entropy* (Bishop, 2006).

We can express the entropy $H(X, Y)$ associated with the random vector $(X, Y)^\top$ in terms of conditional entropy as follows:

$$\begin{aligned} H(X,Y) &= -\iint p(x,y)\log[p(x,y)]dx\,dy \\ &= -\iint p(x,y)ln[p(y \mid x)p(x)]dx\,dy \\ &= -\iint p(x,y)\log[p(y \mid x)]dx\,dy - \iint p(x,y)\log[p(x)]dx\,dy \\ &= H(Y \mid X) - \int \left(\int p(x,y)dy\right)\log[p(x)]dx \\ &= H(Y \mid X) - \int p(x)\log[p(x)]dx \end{aligned}$$

or

$$H(X,Y) = H(Y \mid X) + H(X). \tag{2.92}$$

2.7.1 Kullback–Leibler Divergence

Let $p(x)$ be some unknown density function for a random variable X, and let $q(x)$ represent an approximation of that density function. Then the information required to specify X when using $q(x)$ as an approximation for $p(x)$ is given by

$$-\int p(x)\log[q(x)]dx.$$

The *additional information* required relative to that for the correct density function is called the *Kullback–Leibler* (KL) *divergence* between density functions $p(x)$ and $q(x)$ and is given by

$$\begin{aligned} \mathrm{KL}(p,q) &= -\int p(x)\log[q(x)]dx - \left(-\int p(x)\log[p(x)]dx\right) \\ &= -\int p(x)\log\left[\frac{q(x)}{p(x)}\right]dx. \end{aligned} \tag{2.93}$$

The KL divergence can be shown to satisfy (Exercise 15)

$$\mathrm{KL}(p,q) > 0,\ p(x) \neq q(x), \quad \mathrm{KL}(p,p) = 0$$

and is thus a measure of the dissimilarity between $p(x)$ and $q(x)$.

2.7.2 Mutual Information

Consider two grayscale images represented by random variables X and Y. Their joint probability distribution is $p(x, y)$. If the images are completely independent, then

$$p(x, y) = p(x)p(y).$$

Thus the extent $I(X, Y)$ to which they are *not* independent can be measured by the KL divergence between $p(x, y)$ and $p(x)p(y)$:

$$I(X, Y) = \text{KL}(p(x, y), p(x)p(y)) = -\iint p(x, y) \log\left[\frac{p(x)p(y)}{p(x, y)}\right] dx\, dy, \tag{2.94}$$

which is called the *mutual information* between X and Y. Expanding,

$$\begin{aligned} I(X, Y) &= -\iint p(x, y)\big[\log[p(x)] + \log[p(y)] - \log[p(x, y)]\big] dx\, dy \\ &= H(X) + H(Y) + \iint p(x, y) \log[p(x \mid y)p(y)] dx\, dy \\ &= H(X) + H(Y) - H(X \mid Y) - H(Y), \end{aligned}$$

and thus

$$I(X, Y) = H(X) - H(X \mid Y). \tag{2.95}$$

Mutual information measures the degree of dependence between the two images, a value of zero indicating statistical independence. This is to be contrasted with correlation, where a value of zero implies statistical independence only for normally distributed quantities (see Theorem 2.4).

In practice the images are quantized, so that if p_1 and p_2 are their normalized histograms (i.e., $\sum_i p_1(i) = \sum_i p_2(i) = 1$) and p_{12} is the normalized two-dimensional histogram, $\sum_{ij} p_{12}(i, j) = 1$, then the mutual information is

$$\begin{aligned} I(1, 2) &= -\sum_{ij} p_{12}(i, j)\big(\log[p_1(i)] + \log[p_2(j)] - \log[p_{12}(i, j)]\big) \\ &= \sum_{ij} p_{12}(i, j) \log \frac{p_{12}(i, j)}{p_1(i)p_2(j)}. \end{aligned} \tag{2.96}$$

The IDL program in Listing 2.4 calculates the mutual information between two image bands.

2.8 Exercises

1. Derive Equation 2.11.
2. Write the multivariate normal probability density function, $p(g)$, for the case $\mathbf{\Sigma} = \sigma^2 \boldsymbol{I}$. Show that probability density function, $p(g)$, for a

Listing 2.4
Mutual information using IDL histogram functions.

```
FUNCTION MI, image1, image2
; returns the mutual information of
; two grayscale byte images
   p12 = hist_2d(image1,image2,min1=0,max1=255, $
                 min2=0,max2=255)
   p12 = float(p12)/total(p12)
   p1  = histogram(image1,min=0,max=255)
   p1  = float(p1)/total(p1)
   p2  = histogram(image2,min=0,max=255)
   p2  = float(p2)/total(p2)
   p1p2 = transpose(p2)##p1
   i = where(p1p2 GT 0 AND p12 GT 0)
   RETURN, total(p12[i]*alog(p12[i]/p1p2[i]))
END

PRO ex2_3
  envi_select, title='Choose␣first␣single␣band␣image',$
             fid=fid1, dims=dims,pos=pos1, /band_only
  IF (fid1 EQ -1) THEN RETURN
  envi_select,title='Choose␣second␣single␣band␣image',$
             fid=fid2, dims=dims,pos=pos2, /band_only
  IF (fid2 EQ -1) THEN RETURN
  image1 = $
   (envi_get_data(fid=fid1,dims=dims,pos=pos1))[*]
  image2 = $
   (envi_get_data(fid=fid2,dims=dims,pos=pos2))[*]
  PRINT, MI(bytscl(image1),bytscl(image2))
END
```

one-dimensional random variable G is a special case. Using the fact that $\int_{-\infty}^{\infty} p(g)dg = 1$, demonstrate that $\langle G\rangle = \mu$.

3. Let the random variable X be standard normally distributed with density function

$$\phi(x) = \frac{1}{\sqrt{2\pi}} \exp(-x^2/2), \quad -\infty < x < \infty.$$

Show that the random variable $|X|$ has the density function

$$p(x) = \begin{cases} 2\phi(x) & \text{for } x > 0 \\ 0 & \text{otherwise.} \end{cases}$$

4. Use Equation 2.13 and the result of Exercise 3 to show that the random variable $Y = X^2$, where X is standard normally distributed, has the chi-square density function, Equation 2.44, with $n = 1$ degree of freedom.
5. For constant vectors $\boldsymbol{a}$ and $\boldsymbol{b}$ and random vector $\boldsymbol{G}$ with covariance matrix $\boldsymbol{\Sigma}$, demonstrate that $\text{cov}(\boldsymbol{a}^\top \boldsymbol{G}, \boldsymbol{b}^\top \boldsymbol{G}) = \boldsymbol{a}^\top \boldsymbol{\Sigma} \boldsymbol{b}$.
6. If X_1 and X_2 are independent random variables, both standard normally distributed, show that $X_1 + X_2$ is normally distributed with mean 0 and variance 2. (*Hint*: Write down the joint density function $f(x_1, x_2)$ for X_1 and X_2. Then treat x_1 as fixed and apply Theorem 2.1.)
7. Write an IDL program to generate two normal distributions with the random number generator `RANDOMU()` and test them for equal means with the Student-t test (IDL function `TM_TEST()`) and for equal variance with the F-test (IDL function `FV_TEST()`).
8. Demonstrate (partially) that the sample covariance matrix, Equation 2.36, is an unbiased estimator by showing that $\langle (\mathbf{S})_{11} \rangle = (\boldsymbol{\Sigma})_{11}$.
9. In the *Monty Hall* game a contestant is asked to choose between one of three doors. Behind one of the doors is an automobile as prize for choosing the correct door. After the contestant has chosen, Monty Hall opens one of the other two doors to show that the automobile is not there. He then asks the contestant if she wishes to change her mind and switch from her original choice to the other unopened door. Use Bayes' theorem to prove that her correct answer is "yes."
10. Prove Equation 2.67 for the regression parameter estimates $\hat{a}$ and $\hat{b}$. Demonstrate that these values correspond to a minimum of the goodness of fit function, Equation 2.66, and not to a maximum.
11. Derive the uncertainty for a in Equation 2.69 from the formula for error propagation for uncorrelated errors

$$\sigma_a^2 = \sum_{i=1}^{N} \sigma^2 \left(\frac{\partial a}{\partial y(i)} \right)^2 .$$

12. Derive Equation 2.75 by applying the rules for vector differentiation to minimize the goodness of fit function, Equation 2.74.
13. Write an IDL program to calculate the regression coefficients of spectral band 2 on spectral band 1 of a multispectral image. (The built-in IDL function for ordinary linear regression is `REGRESS()`.)
14. Show that Equations 2.84 and 2.85 are equivalent to Equation 2.83.
15. A *convex function* $f(x)$ satisfies $f(\lambda a + (1 - \lambda)b) \leq \lambda f(a) + (1 - \lambda) f(b)$. *Jensen's inequality* states that, for any convex function $f(x)$, any

function $g(x)$, and any probability density $p(x)$,

$$\int f(g(x))p(x)dx \geq f\left(\int g(x)p(x)dx\right). \tag{2.97}$$

Use this to show that the KL divergence satisfies

$$\mathrm{KL}(p,q) > 0,\ p(x) \neq q(x), \quad \mathrm{KL}(p,p) = 0.$$

3

Transformations

Thus far we have thought of multispectral images as three-dimensional arrays of pixel intensities representing, more or less directly, measured radiances. In this chapter we consider other, more abstract representations which are useful in image interpretation and analysis and which will play an important role in later chapters.

The discrete Fourier and wavelet transforms that we treat in Sections 3.1 and 3.2 convert the pixel values in a given spectral band to linear combinations of orthogonal functions of spatial frequency and distance. They may therefore be classified as *spatial transformations*. The principal components, minimum noise fraction and maximum autocorrelation factor transformations (Sections 3.3 through 3.5), on the other hand, create at each pixel location new linear combinations of the pixel intensities from all of the spectral bands, and can properly be called *spectral transformations* (Schowengerdt, 1997).

3.1 Discrete Fourier Transform

Let the function $g(x)$ represent the radiance at a point x focused along a row of pushbroom-geometry sensors, and $g(j)$ be the corresponding pixel intensities stored in a row of a digital image. We can think of $g(j)$ approximately as a discrete sample* of the function $g(x)$, sampled c times at some sampling interval Δ, c being the number of columns in the image, that is,

$$g(j) = g(x = j\Delta), \quad j = 0 \ldots c-1.$$

For convenience the pixels are numbered from zero, a convention that will be adhered to in the remainder of the book. The interval Δ is the sensor width or, projected back to the Earth's surface, the across-track ground sample distance (GSD).

* More correctly, $g(j)$ is a result of convolutions of the spatial and spectral response functions of the detector with the focused signal (see Chapter 4).

The theory of Fourier analysis states that the function $g(x)$ can be expressed in the form

$$g(x) = \int_{-\infty}^{\infty} \hat{g}(f) e^{\mathbf{i}2\pi f x} df, \tag{3.1}$$

where $\hat{g}(f)$ is called the *Fourier transform* of $g(x)$. Equation 3.1 describes a continuous superposition of periodic complex functions of x,

$$e^{\mathbf{i}2\pi f x} = \cos(2\pi f x) + \mathbf{i}\sin(2\pi f x),$$

having frequency* f. In general, we require such a continuum of periodic functions $e^{\mathbf{i}2\pi f x}$ to represent $g(x)$ in this way. However if $g(x)$ is in fact itself periodic with period T, that is, if $g(x+T) = g(x)$, then the integral in Equation 3.1 can be replaced by an infinite sum of *discrete* periodic functions of frequency kf for $-\infty < k < \infty$, where f is the *fundamental frequency* $f = 1/T$:

$$g(x) = \sum_{k=-\infty}^{\infty} \hat{g}(k) e^{\mathbf{i}2\pi (kf) x}. \tag{3.2}$$

Consequently, if we think of the sampled series of pixels $g(j)$, $j = 0 \ldots c-1$, as also being periodic with period $T = c\Delta$, that is, repeating itself to infinity in both positive and negative directions, then we can replace x in Equation 3.2 by $j\Delta$ and express $g(j)$ in a similar way:

$$g(j) = g(j\Delta) = \sum_{k=-\infty}^{\infty} \hat{g}(k) e^{\mathbf{i}2\pi (kf) j\Delta} = \sum_{k=-\infty}^{\infty} \hat{g}(k) e^{\mathbf{i}2\pi kj/c}, \tag{3.3}$$

where in the last equality we have used $f\Delta = \Delta/T = 1/c$.

The limits in the summation in Equation 3.3 must, however, be truncated. This is due to the fact that there is a limit to the highest frequency, $k_{max}f$, that can be measured by sampling at the interval Δ. The limit is called the *Nyquist critical frequency*, f_N. It may be determined simply by observing that the minimum number of samples needed to describe a sine wave completely is two per period (e.g., at the maximum and minimum values). Therefore the shortest period measurable is 2Δ and the Nyquist frequency is

$$f_N = \frac{1}{2\Delta} = \frac{cf}{2}.$$

* Most often frequency is associated with inverse time (cycles per second). Here, of course we are speaking of *spatial frequency*, or cycles per meter.

Hence $k_{max} = c/2$. Taking this into account in Equation 3.3, we obtain

$$g(j) = \sum_{k=-c/2}^{c/2} \hat{g}(k)e^{\mathbf{i}2\pi kj/c}, \quad j = 0 \ldots c-1. \tag{3.4}$$

The effect of truncation depends upon the nature of the function $g(x)$ being sampled. According to the *sampling theorem* (see Press et al., 2002) $g(x)$ is completely determined by the samples $g(j)$ in Equation 3.4 if it is *bandwidth limited* to frequencies smaller than f_N, that is, provided that, in Equation 3.1, $\hat{g}(f) = 0$ for all $|f| \geq f_N$. If this is not the case, then any frequency component outside the interval $(-f_N, f_N)$ is spuriously moved into that range, a phenomenon referred to as *aliasing*.

To bring Equation 3.4 into a more convenient form we have to make a few simple manipulations. To begin with, note that the exponents in the first and last terms in the summation are equal, that is,

$$e^{\mathbf{i}2\pi(-c/2)j/c} = e^{-\mathbf{i}\pi j} = (-1)^j = e^{\mathbf{i}\pi j} = e^{\mathbf{i}2\pi(c/2)j/c},$$

so we can lump those two terms together and write Equation 3.4 equivalently as

$$g(j) = \sum_{k=-c/2}^{c/2-1} \hat{g}(k)e^{\mathbf{i}2\pi kj/c}, \quad j = 0 \ldots c-1.$$

Rearranging further,

$$\begin{aligned} g(j) &= \sum_{k=0}^{c/2-1} \hat{g}(k)e^{\pi 2\pi kj/c} + \sum_{k=-c/2}^{-1} \hat{g}(k)e^{\mathbf{i}2\pi kj/c} \\ &= \sum_{k=0}^{c/2-1} \hat{g}(k)e^{\mathbf{i}2\pi kj/c} + \sum_{k'=c/2}^{c-1} \hat{g}(k'-c)e^{\mathbf{i}2\pi(k'-c)j/c} \\ &= \sum_{k=0}^{c/2-1} \hat{g}(k)e^{\mathbf{i}2\pi kj/c} + \sum_{k'=c/2}^{c-1} \hat{g}(k'-c)e^{\mathbf{i}2\pi k'j/c}. \end{aligned}$$

Thus we have finally

$$g(j) = \sum_{k=0}^{c-1} \hat{g}(k)e^{\mathbf{i}2\pi kj/c}, \quad j = 0 \ldots c-1, \tag{3.5}$$

provided that we interpret $\hat{g}(k)$ as meaning $\hat{g}(k-c)$ when $k \geq c/2$.*

* This is the convention adopted in IDL in the fast Fourier transform function `FFT()`.

Equation 3.5 is a set of c equations in the c unknown frequency components $\hat{g}(k)$. Its solution is called the *discrete Fourier transform* and is given by

$$\hat{g}(k) = \frac{1}{c}\sum_{j=0}^{c-1} g(j)e^{-\mathrm{i}2\pi kj/c}, \quad k = 0\dots c-1. \tag{3.6}$$

This follows (Exercise 2) from the orthogonality property of the exponentials:

$$\sum_{j=0}^{c-1} e^{\mathrm{i}2\pi(k-k')j/c} = c\delta_{k,k'}, \tag{3.7}$$

where $\delta_{k,k'}$ is the *delta function*,

$$\delta_{k,k'} = \begin{cases} 1 & \text{if } k = k' \\ 0 & \text{otherwise.} \end{cases}$$

Equation 3.5 itself is the *discrete inverse Fourier transform*. We write

$$g(j) \Leftrightarrow \hat{g}(k),$$

to signify that $g(j)$ and $\hat{g}(k)$ constitute a *discrete Fourier transform pair*.

Determining the frequency components in Equation 3.6 from the original pixel intensities would appear to involve, in all, c^2 floating point multiplication operations. The *fast Fourier transform* (FFT) exploits the structure of the complex e-functions to reduce this to order $c\log c$, a very considerable saving in computation time for large arrays. For good explanations of the FFT algorithm, see Press et al. (2002) or Gonzalez and Woods (2002).

The discrete Fourier transform is easily generalized to two dimensions. Let $g(i,j)$, $i = 0, c-1$, $j = 0, r-1$, represent a grayscale image. Its discrete inverse Fourier transform is

$$g(i,j) = \sum_{k=0}^{c-1}\sum_{\ell=0}^{r-1} \hat{g}(k,\ell)e^{\mathrm{i}2\pi(ik/c+j\ell/r)} \tag{3.8}$$

and the discrete Fourier transform is

$$\hat{g}(k,\ell) = \frac{1}{cr}\sum_{i=0}^{c-1}\sum_{j=0}^{r-1} g(i,j)e^{-\mathrm{i}2\pi(ik/c+j\ell/r)}. \tag{3.9}$$

The frequency coefficients $\hat{g}(k,\ell)$ in Equations 3.8 and 3.9 are complex numbers. In order to represent an image in the frequency domain as a raster, one can calculate its *power spectrum*, which is defined as*

$$P(k,\ell) = |\hat{g}(k,\ell)|^2 = \hat{g}(k,\ell)\hat{g}^*(k,\ell). \tag{3.10}$$

Rather than displaying $P(k,\ell)$ directly, which, according to ENVI's display convention, would place zero frequency components $k = 0$, $\ell = 0$ in the upper left-hand corner, use can be made of the *translation property* (Exercise 4) of the Fourier transform:

$$g(i,j)e^{\mathbf{i}2\pi(k_0 i/c+\ell_0 j/r)} \Leftrightarrow \hat{g}(k-k_0,\ell-\ell_0). \tag{3.11}$$

In particular, for $k_0 = c/2$ and $\ell_0 = r/2$, we can write

$$e^{\mathbf{i}2\pi(k_0 i/c+\ell_0 j/r)} = e^{\mathbf{i}\pi(i+j)} = (-1)^{i+j}.$$

Therefore,

$$g(i,j)(-1)^{i+j} \Leftrightarrow \hat{g}(k-c/2,\ell-r/2),$$

so if we multiply an image by $(-1)^{i+j}$ before transforming, zero frequency will be at the center. This is illustrated in Listing 3.1, which performs an FFT of an image band using the IDL function `FFT()` and displays the logarithm of the power spectrum with zero frequency at the center, Figure 3.1. We shall return to discrete Fourier transforms in Chapter 4 when we discuss convolutions and filters.

3.2 Discrete Wavelet Transform

Unlike the Fourier transform, which represents an array of pixel intensities in terms of pure frequency functions, the wavelet transform expresses an image array in terms of functions which are restricted both in terms of frequency and spatial extent. In many image processing applications this turns out to be particularly efficient and useful. The traditional (and most intuitive) way of introducing wavelets is in terms of the Haar scaling function (Strang, 1989; Aboufadel and Schlicker, 1999; Gonzalez and Woods, 2002), and we will adopt this approach here as well, in particular following the development in Aboufadel and Schlicker (1999).

---

* The magnitude $|z|$ of a complex number $z = x + \mathbf{i}y$ is $\sqrt{x^2+y^2} = \sqrt{zz^*}$, where $z^* = x - \mathbf{i}y$ is the complex conjugate of z.

Listing 3.1
Displaying the power spectrum of an image band.

```
PRO EX3_1

envi_select, title='Choose␣multispectral␣band', $
                 fid=fid, dims=dims,pos=pos, /band_only
IF (fid EQ -1) THEN BEGIN
   PRINT, 'cancelled'
   RETURN
ENDIF

num_cols = dims[2]-dims[1]+1
num_rows = dims[4]-dims[3]+1

image = envi_get_data(fid=fid,dims=dims,pos=pos[0])

; arrays of i and j values
a = lindgen(num_cols,num_rows)
i = a MOD num_cols
j = a/num_cols

; shift Fourier transform to center
image = (-1)^(i+j)*image

envi_enter_data, alog((abs(FFT(image)))^2)

END
```

Fundamental to the definition of wavelet transforms is the concept of an *inner product* of real-valued functions and the associated *inner product space* (see also Appendix A).

Definition 3.1: If f and g are two real functions on the set of real numbers $\mathbb{R}$, then their *inner product* is given by

$$\langle f,g\rangle = \int_{-\infty}^{\infty} f(x)g(x)dx. \tag{3.12}$$

The *inner product space*, $L_2(\mathbb{R})$, is the collection of all functions $f : \mathbb{R} \mapsto \mathbb{R}$ with the property that

$$\langle f,f\rangle = \int_{-\infty}^{\infty} f(x)^2 dx. \tag{3.13}$$

is finite.

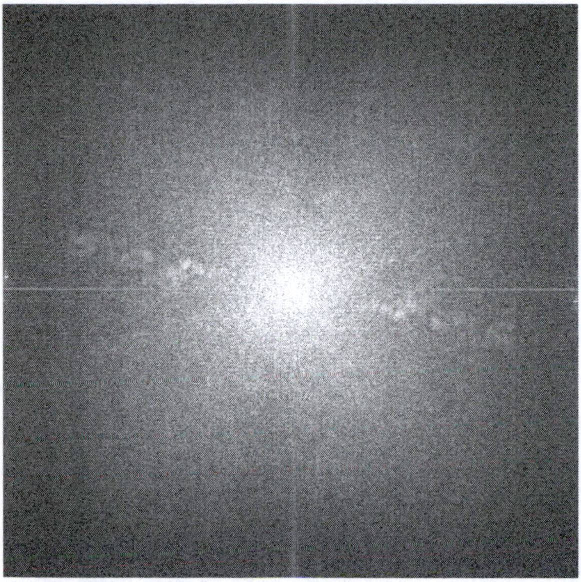

FIGURE 3.1
Logarithm of the power spectrum for the 3N band of the Jülich ASTER image.

3.2.1 Haar Wavelets

The *Haar scaling function* is the function

$$\phi(x) = \begin{cases} 1 & \text{if } 0 \le x \le 1 \\ 0 & \text{otherwise} \end{cases} \tag{3.14}$$

shown in Figure 3.2. We shall use it to represent pixel intensities.

The quantities $g(j)$, $j = 0, c-1$, representing a row of pixel intensities, can be thought of as a piecewise constant function of x. This is indicated in Figure 3.3. The abscissa has been normalized to the interval $[0, 1]$, so that j measures the distance in increments of $1/c$ along the pixel row, with the last pixel occupying the interval $\left[\frac{c-1}{c}, 1\right]$. We have called this piecewise constant function $\bar{g}(x)$ to distinguish it from $g(j)$. According to Definition 3.1 it is in $L_2(\mathbb{R})$.

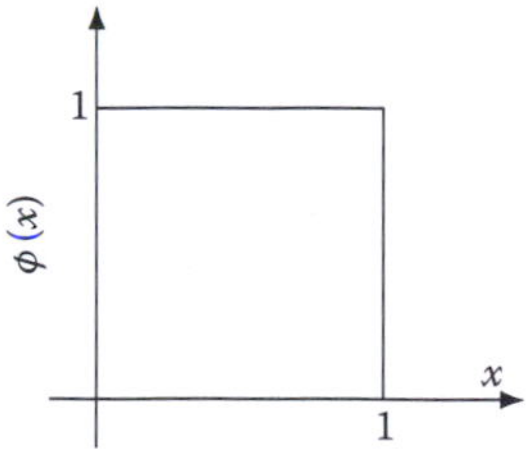

FIGURE 3.2
The Haar scaling function.

Now let V_n be the collection of *all* piecewise constant functions on the interval $[0, 1]$ that have possible discontinuities at the rational points $j \cdot 2^{-n}$, where j and n are nonnegative integers. If the number of pixels c in Figure 3.3 is a power of two, $c = 2^n$ say, then $\bar{g}(x)$ clearly is a function which belongs to

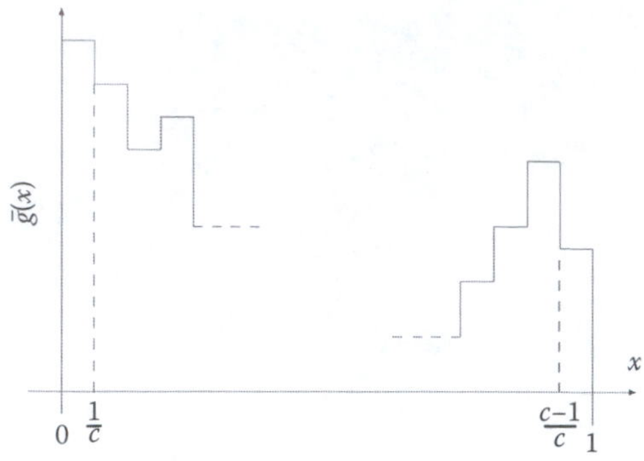

FIGURE 3.3
A row of c pixel intensities on the interval $[0, 1]$ as a piecewise constant function $\bar{g}(x) \in L_2(\mathbb{R})$. In the text it is assumed that $c = 2^n$ for some integer n.

V_n, that is, $\bar{g}(x) \in V_n$. The (possible) discontinuities occur at

$$x = 1 \cdot 2^{-n}, 2 \cdot 2^{-n} \ldots (c-1) \cdot 2^{-n}.$$

Certainly all members of V_n also belong to the function space $L_2(\mathbb{R})$, so that $V_n \subset L_2(\mathbb{R})$. Any function in V_n confined to the interval $[0, 1]$ in this way can be expressed as a linear combination of the *standard Haar basis functions*. These are scaled and shifted versions of the Haar scaling function of Figure 3.2 and comprise the set

$$C_n = \{\phi_{n,k}(x) = \phi(2^n x - k) \mid k = 0, 1 \ldots 2^n - 1\} \tag{3.15}$$

(see Figure 3.4).

Note that $\phi_{0,0}(x) = \phi(x)$. The index n corresponds to a compression or change of scale by a factor of 2^{-n}, whereas the index k shifts the basis function across the interval $[0, 1]$. The row of pixels in Figure 3.3 can be expanded in terms of the standard Haar basis trivially as

$$\bar{g}(x) = g(0)\phi_{n,0}(x) + g(1)\phi_{n,1}(x) + \cdots$$
$$+ g(c-1)\phi_{n,c-1}(x) = \sum_{j=0}^{c-1} g(j)\phi_{n,j}(x). \tag{3.16}$$

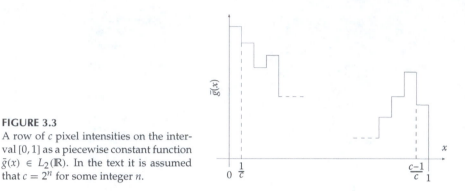

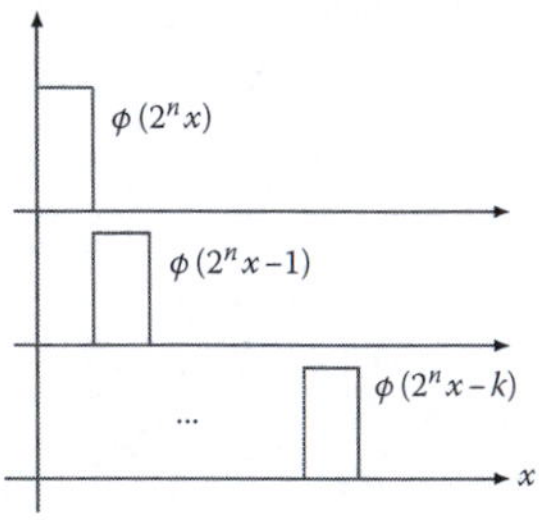

FIGURE 3.4
Basis functions C_n for space V_n.

The Haar basis functions are clearly orthogonal:

$$\langle \phi_{n,k}, \phi_{n,k'} \rangle = \int_0^1 \phi_{n,k}(x)\phi_{n,k'}(x)dx = \frac{1}{2^n}\delta_{k,k'}. \tag{3.17}$$

The expansion coefficients $g(j)$ are, from this orthogonality, given formally by

$$g(j) = \frac{\langle \bar{g}, \phi_{n,j} \rangle}{\langle \phi_{n,j}, \phi_{n,j} \rangle} = 2^n \langle \bar{g}, \phi_{n,j} \rangle. \tag{3.18}$$

We will now derive a new and more interesting orthogonal basis for V_n. Consider first of all the function spaces V_0 and V_1 with standard Haar bases $\{\phi_{0,0}(x)\}$ and $\{\phi_{1,0}(x), \phi_{1,1}(x)\}$, respectively. According to the orthogonal decomposition theorem (Theorem A.4) any function in V_1 can be expressed as a linear combination of the basis for V_0 plus some function in a *residual space* $V_0^{\perp}$ which is orthogonal to V_0 (any function in $V_0^{\perp}$ is orthogonal to any function in V_0). This is denoted formally by writing

$$V_1 = V_0 \oplus V_0^{\perp}. \tag{3.19}$$

For example, the basis function $\phi_{1,0}(x)$ of V_0 is also in V_1, and so can be written in the form

$$\phi_{1,0}(x) = \frac{\langle \phi_{1,0}, \phi_{0,0} \rangle}{\langle \phi_{0,0}, \phi_{0,0} \rangle} \phi_{0,0}(x) + r(x) = \frac{1}{2} \phi_{0,0}(x) + r(x).$$

The function $r(x)$ is in the residual space $V_0^{\perp}$. We see that, in this case,

$$r(x) = \phi_{1,0}(x) - \frac{1}{2} \phi_{0,0}(x) = \phi(2x) - \frac{1}{2} \phi(x).$$

But we can express $\phi(x)$ as $\phi(x) = \phi(2x) + \phi(2x - 1)$ so that

$$r(x) = \phi(2x) - \frac{1}{2}(\phi(2x) + \phi(2x-1)) = \frac{1}{2}(\phi(2x) - \phi(2x-1)) =: \frac{1}{2} \psi(x).$$

The function

$$\psi(x) = \phi(2x) - \phi(2x - 1) \tag{3.20}$$

is shown in Figure 3.5. It is orthogonal to $\phi(x)$ and is called the *Haar mother wavelet*. Thus an *alternative basis* for V_1 is

$$B_1 = \{\phi_{0,0}, \psi_{0,0}\},$$

where for consistency we have defined $\psi_{0,0}(x) = \psi(x)$.

This argument can be repeated (Exercise 6) for $V_2 = V_1 \oplus V_1^{\perp}$ to obtain the basis

$$B_2 = \{\phi_{0,0}, \psi_{0,0}, \psi_{1,0}, \psi_{1,1}\}$$

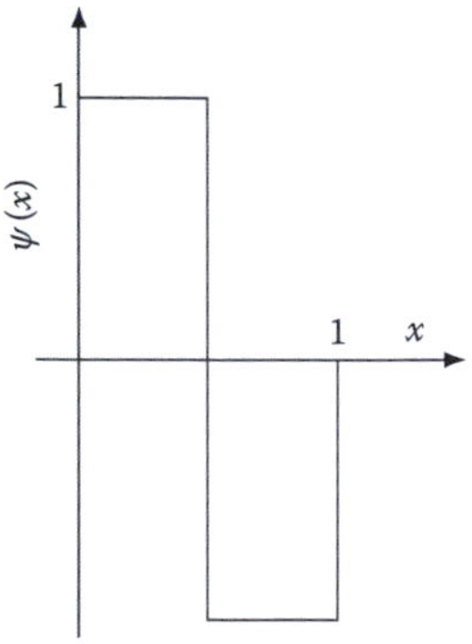

FIGURE 3.5
The Haar mother wavelet.

for V_2, where now $\{\psi_{1,0}, \psi_{1,1}\}$ is an orthogonal basis for $V_1^\perp$ given by

$$\psi_{1,0} = \psi(2x), \ \psi_{1,1} = \psi(2x-1).$$

Indeed, the argument can be continued indefinitely, so in general the *Haar wavelet basis* for V_n is

$$B_n = \{\phi_{0,0}, \psi_{0,0}, \psi_{1,0}, \psi_{1,1} \dots \psi_{n-1,0}, \psi_{n-1,1} \dots \psi_{n-1,2^n-1}\},$$

where $\{\psi_{m,k} = \psi(2^m x - k) \mid k = 0 \dots 2^m - 1\}$ is an orthogonal basis for $V_m^\perp$, and

$$V_n = V_{n-1} \oplus V_{n-1}^\perp = V_0 \oplus V_0^\perp \oplus \cdots \oplus V_{n-2}^\perp \oplus V_{n-1}^\perp.$$

In terms of this new basis, the function $\bar{g}(x)$ in Figure 3.3 can now be expressed as

$$\bar{g}(x) = \hat{g}(0)\phi_{0,0}(x) + \hat{g}(1)\psi_{0,0}(x) + \cdots + \hat{g}(c-1)\psi_{n-1,c-1}(x), \tag{3.21}$$

where $c = 2^n$. The expansion coefficients $\hat{g}(j)$ are called the *wavelet coefficients*. They are still to be determined.

In the case of the Haar wavelets, their determination turns out to be quite easy because there is a simple correspondence between the basis functions (ϕ, ψ) and the space of 2^n-component vectors (Strang, 1989). Consider for instance $n = 2$. Then the correspondence is

$$\phi_{0,0} = \begin{pmatrix}1\\1\\1\\1\end{pmatrix}, \ \phi_{1,0} = \begin{pmatrix}1\\1\\0\\0\end{pmatrix}, \ \phi_{1,1} = \begin{pmatrix}0\\0\\1\\1\end{pmatrix}, \ \phi_{2,0} = \begin{pmatrix}1\\0\\0\\0\end{pmatrix}, \ \dots$$

and

$$\psi_{0,0} = \begin{pmatrix}1\\1\\-1\\-1\end{pmatrix}, \ \psi_{1,0} = \begin{pmatrix}1\\-1\\0\\0\end{pmatrix}, \ \psi_{1,1} = \begin{pmatrix}0\\0\\1\\-1\end{pmatrix}.$$

Thus the orthogonal basis B_2 may be represented equivalently by the mutually orthogonal vectors

$$B_2 = \left\{ \begin{pmatrix}1\\1\\1\\1\end{pmatrix}, \begin{pmatrix}1\\1\\-1\\-1\end{pmatrix}, \begin{pmatrix}1\\-1\\0\\0\end{pmatrix}, \begin{pmatrix}0\\0\\1\\-1\end{pmatrix} \right\}.$$

This gives us a more convenient representation of the expansion in Equation 3.21, namely

$$\bar{g} = \boldsymbol{B}_n \hat{g}, \tag{3.22}$$

where
$\bar{g} = \big(g(0)\dots g(c-1)\big)^\top$ is a column vector of the original pixel intensities
$\hat{g} = \big(\hat{g}(0)\dots \hat{g}(c-1)\big)^\top$ is a column vector of the wavelet coefficients
$\boldsymbol{B}_n$ is a transformation matrix whose columns are the basis vectors of B_n

The wavelet coefficients for the pixel vector are then given by inverting the representation:

$$\hat{g} = \boldsymbol{B}_n^{-1} \bar{g}. \tag{3.23}$$

A full grayscale image is transformed by first applying Equation 3.23 to its columns and then to its rows. Wavelet coefficients for grayscale images thus obtained tend to have "simple statistics" (Gonzalez and Woods, 2002), for example, they might be approximately Gaussian with zero mean.

3.2.2 Image Compression

The fact that many of the wavelet coefficients are close to zero makes the wavelet transformation useful for image compression. This is illustrated in Listing 3.2, which is adapted from an example in the IDL Reference Guide. First the auxiliary functions `PSI_M()` and `PSI()` are used to generate the Haar wavelet transformation matrix $\boldsymbol{B}_n$ for $n = 8$ (lines 17–23). Then a 256×256 image subset is read from ENVI into the IDL variable `G`, converted to floating point (4 byte) format, and its size is printed out (lines 25–29). After the wavelet transformation in Equation 3.23 has been applied to the rows and columns (lines 32–34, the result of which is returned as an image to ENVI), the IDL function `SPRSIN()` is invoked to convert the wavelet coefficient image into row indexed sparse storage format, overwriting `G` and retaining only elements with an absolute value greater than a specified threshold, in this instance 1.0 (line 36). The sparse image is then written to disk with the `WRITE_SPR` procedure so that its file size can be queried and printed out. Finally the compressed image is restored to normal format with the function `FULSTR()`, and the wavelet transform is inverted, again overwriting `G` (lines 41–43). The result is returned to ENVI for comparison with the original (see Figure 3.6). Here is the printout for spectral band 1 of the Jülich image:

```
ENVI> ex3_2
Size original image:       262144 bytes
Size compressed image:        60424 bytes
```

Listing 3.2
Image compression with the Haar wavelet transform.

```
FUNCTION psi_m, x
   IF x LT 0.0 THEN RETURN, 0.0
   IF x LT 0.5 THEN RETURN, 1.0
   IF x LT 1.0 THEN RETURN, -1.0
   RETURN, 0.0
END
FUNCTION psi, m, k, n
   c = 2^n
   result = fltarr(c)
   x = findgen(c)/c
   FOR i=0,c-1 DO result[i]=psi_m(2^m*x[i]-k)
   RETURN, result
END

PRO ex3_2
; generate wavelet basis B_8
   n = 8L
   B = fltarr(2^n,2^n)+1.0
   i = 1
   FOR m=0,n-1 DO FOR k=0,2^m-1 DO BEGIN
      B[i,*] = psi(m,k,n)
      i++
   ENDFOR
; get a 256x256 grayscale image
   envi_select, title='Choose␣multispectral␣band', $
             fid=fid, dims=dims,pos=pos, /band_only
   G = envi_get_data(fid=fid,dims=dims,pos=pos)
   G = float(G[0:255,0:255])
   PRINT, 'Size␣original␣image:',256L*256L*4L,'␣bytes'
   envi_enter_data, G + 0.0
; transform the columns and rows
   FOR i=0,255 DO G[i,*] = invert(B)##G[i,*]
   FOR j=0,255 DO G[*,j] = invert(B)##transpose(G[*,j])
   envi_enter_data, G + 0.0
; convert to sparse format
   G = sprsin(G,thresh=1.0)
   write_spr, G, 'sparse.dat'
   OPENR, 1,'sparse.dat' & status = fstat(1) & close,1
   PRINT, 'Size␣compressed␣image:',status.size,'␣bytes'
; invert the transformation
   G = fulstr(G)
   FOR j=0,255 DO G[*,j] = B##transpose(G[*,j])
   FOR i=0,255 DO G[i,*]  = B##G[i,*]
   envi_enter_data, G
END
```

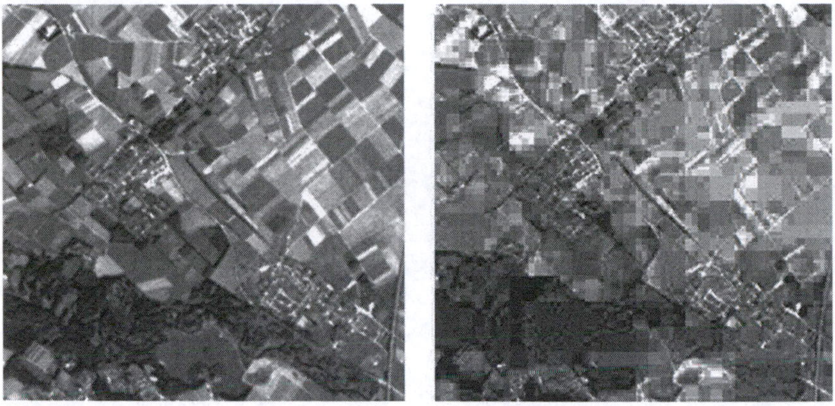

(a) (b)

FIGURE 3.6
(a) A 256 × 256 spatial subset of band 1 of the Jülich ASTER image of Figure 1.1. (b) The same subset after transformation to the Haar wavelet basis B_8, compression with `SPRSIN()`, and restoration with the inverse transformation (see Listing 3.2).

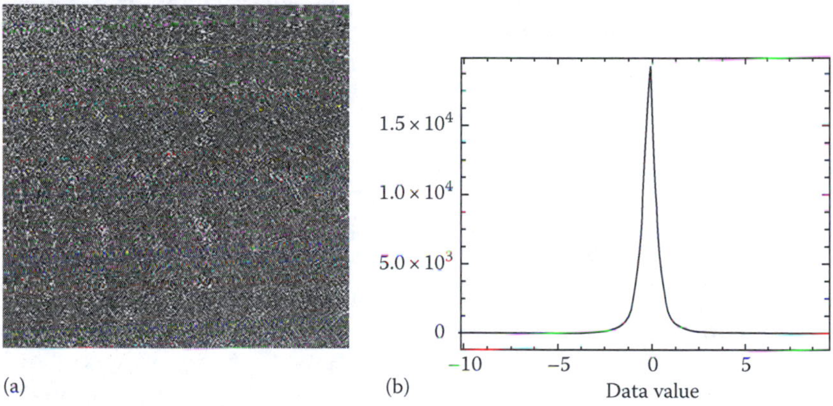

(a) (b)

FIGURE 3.7
(a) The Haar wavelet coefficients for the image of Figure 3.6. (b) Histogram of the wavelet coefficient image.

A compression of about a factor 4.3 is achieved at the cost of the (considerable) loss of image quality visible in Figure 3.6. (The program is very slow due to the many floating point operations involved in the transformations. This will be remedied in Chapter 4 when we treat filters and the fast wavelet transform.) The statistics of the wavelet coefficients themselves are illustrated in Figure 3.7, where it is apparent that they are approximately normally distributed and that, in this case, most have absolute magnitudes less than 1.

3.2.3 Multiresolution Analysis

So far we have represented only functions on the interval [0, 1] with the standard basis $\phi_{n,k}(x) = \phi(2^n x - k)$, $k = 1 \dots 2^n - 1$. We can extend this to functions defined on all real numbers in a straightforward way, still restricting ourselves, however, to functions with *compact support*. These are zero everywhere outside a closed, bounded interval. Thus

$$\{\phi(x-k) \mid k \in \mathbb{Z}\},$$

where $\mathbb{Z}$ is the set of *all* integers, is a basis for the space V_0 of all piecewise constant functions with compact support having possible breaks at integer values. Note that $\phi(x-k) = \phi_{0,k}$ is an *orthonormal* basis for V_0, that is,

$$\langle \phi(x-k), \phi(x-k') \rangle = \delta_{k,k'},$$

whereas $\phi(2^n x - k) = \phi_{n,k}$ with $n > 0$ is only an orthogonal basis for V_n, since the inner products for equal k are not unity. Quite generally then, an orthogonal basis for the set V_n of piecewise constant functions with possible breaks at $j \cdot 2^{-n}$ and compact support is

$$\{\phi(2^n x - k) \mid k \in \mathbb{Z}\}. \tag{3.24}$$

One can even allow $n < 0$. For example $n = -1$ means that the possible breaks are at even integer values.

We can think of the collection of nested subspaces of piecewise constant functions as being *generated* by the Haar scaling function ϕ. Such a collection is an example of a *multiresolution analysis* (MRA). There are many other possible scaling functions that define or generate an MRA. Although the subspaces will no longer consist of simple piecewise constant functions, nevertheless, based on our experience with the Haar wavelets, we can appreciate the following definition (Aboufadel and Schlicker, 1999).

Definition 3.2: An MRA is a collection of nested subspaces

$$\cdots \subseteq V_{-1} \subseteq V_0 \subseteq V_1 \subseteq V_2 \subseteq \cdots \subseteq L_2(\mathbb{R}),$$

with the following properties:

1. For any function $f \in L_2(\mathbb{R})$ there exists a series of functions, one in each V_n, which converges to f.
2. The only function common to all V_n is $f(x) = 0$.

3. The function $f(x) \in V_n$ if and only if $f(2^{-n}x) \in V_0$.
4. The scaling function ϕ is an orthonormal basis for the function space V_0, that is, $\langle \phi(x-k), \phi(x-k') \rangle = \delta_{kk'}$.

Clearly property 1 is met for the Haar MRA, since any function in $L_2(\mathbb{R})$ can be approximated to arbitrary accuracy with successively finer piecewise constant functions. Being common to all V_n means being piecewise constant on all intervals. The only function in $L_2(\mathbb{R})$ with this property and compact support is $f(x) = 0$, so property 2 is also satisfied for the Haar MRA. If $f(x) \in V_1$ then it is piecewise constant on intervals of length $1/2$. Therefore the function $f(2^{-1}x)$ is piecewise constant on intervals of length 1, that is $f(2^{-1}x) \in V_0$, etc., and so property 3 is satisfied as well. Finally, property 4 also holds for the Haar scaling function.

3.2.3.1 Dilation Equation and Refinement Coefficients

In the following, we will think of $\phi(x)$ as any scaling function which generates an MRA in the sense of Definition 3.2. Since $\{\phi(x-k) \mid k \in \mathbb{Z}\}$ is an orthonormal basis for V_0, it follows that $\{\phi(2x-k) \mid k \in \mathbb{Z}\}$ is an orthogonal basis for V_1. That is, let $f(x) \in V_1$. Then by property 3, $f(x/2) \in V_0$, hence

$$f(x/2) = \sum_k a_k \phi(x-k),$$

which implies that

$$f(x) = \sum_k a_k \phi(2x-k).$$

In particular, since $\phi(x) \in V_0 \subset V_1$, we have the *dilation equation*

$$\phi(x) = \sum_k c_k \phi(2x-k). \tag{3.25}$$

The constants c_k are called the *refinement coefficients*. For example, the dilation equation for the Haar scaling function is

$$\phi(x) = \phi(2x) + \phi(2x-1)$$

so that the refinement coefficients are $c_0 = c_1 = 1$, $c_k = 0$ otherwise. Note that $c_0^2 + c_1^2 = 2$. This is a general property of the refinement coefficients,

$$1 = \langle \phi(x), \phi(x) \rangle = \left\langle \sum_k c_k \phi(2x-k), \sum_{k'} c_{k'} \phi(2x-k') \right\rangle = \frac{1}{2} \sum_k c_k^2$$

and therefore,

$$\sum_{k=-\infty}^{\infty} c_k^2 = 2, \tag{3.26}$$

which is also called *Parseval's formula*. In a similar way one can show (Exercise 7)

$$\sum_{k=-\infty}^{\infty} c_k c_{k-2j} = 0 \quad \text{for all } j \neq 0. \tag{3.27}$$

3.2.3.2 Cascade Algorithm

Some of the scaling functions which generate an MRA cannot be expressed as simple, analytical functions. Nevertheless we can work with an MRA even when there is no simple representation for the scaling function which generates it. For instance, once we have the refinement coefficients for a scaling function, it can be approximated to any desired degree of accuracy using the dilation equation. The idea is to iterate the refinement equation with a so-called *cascade algorithm* until it converges to a sequence of points which approximates $\phi(x)$.

The following recursive scheme can be used to estimate a scaling function with up to five nonzero refinement coefficients $c_0, c_1 \dots c_4$:

$$\begin{aligned} f_0(x) &= \delta_{x,0} \\ f_i(x) &= c_0 f_{i-1}(2x) + c_1 f_{i-1}(2x-1) + c_2 f_{i-1}(2x-2) + c_3 f_{i-1}(2x-3) \\ &\quad + c_4 f_{i-1}(2x-4). \end{aligned}$$

In this scheme, x takes on values $j/2^n$, where j, n are any integers. The first definition is the termination condition for the recursion and approximates the scaling function to zeroth order as the delta function

$$\delta_{x,0} = \begin{cases} 1 & \text{if } x = 0 \\ 0 & \text{otherwise.} \end{cases}$$

The second relation defines the ith approximation to the scaling function in terms of the $(i-1)$th approximation using the dilation equation. We can calculate the set of values

$$\phi \approx f_n\left(\frac{j}{2^n}\right) \quad \text{for } j = 0 \dots 4 \cdot 2^n$$

for some $n > 1$ as a pointwise approximation of ϕ on the interval $[0, 4]$. Listing 3.3 uses a recursive IDL function `F(x,i)` to approximate a scaling

Listing 3.3
Cascade algorithm approximation to the scaling function.

```
 1 FUNCTION F, x, i
 2    COMMON refinement, c0,c1,c2,c3,c4
 3    IF (i EQ 0) THEN IF (x EQ 0) THEN $
 4       RETURN, 1.0 ELSE RETURN, 0.0 ELSE $
 5       RETURN, c0*f(2*x,i-1)+c1*f(2*x-1,i-1)+ $
 6          c2*f(2*x-2,i-1)+c3*f(2*x-3,i-1)+c4*f(2*x-4,i-1)
 7 END
 8
 9 PRO ex3_3
10
11    COMMON refinement, c0,c1,c2,c3,c4
12 ; refinement coefficients for Haar scaling function
13    c0=1 & c1=1 & c2=0 & c3=0 & c4=0
14 ; refinement coefficients for D4 scaling function
15 ;    c0=(1+sqrt(3))/4 & c1=(3+sqrt(3))/4
16 ;    c2=(3-sqrt(3))/4 & c3=(1-sqrt(3))/4 & c4=0
17
18 ; fourth order approximation
19    n=4
20    x = findgen(4*2^n)
21    ff=fltarr(4*2^n)
22    FOR i=0,4*2^n-1 DO ff[i]=F(x[i]/2^n,n)
23
24 ; output as EPS file
25    thisDevice =!D.Name
26    set_plot, 'PS'
27    Device, Filename='fig3_8.eps',xsize=3,ysize=2, $
28             /inches,/encapsulated
29    PLOT,x/2^n,ff,yrange=[-1,2]
30    device,/close_file
31    set_plot,thisDevice
32
33 END
```

function in this way. The result using the Haar refinement coefficients is shown in Figure 3.8 and is seen to be an approximation of Figure 3.2.

3.2.3.3 Mother Wavelet

For a general MRA we also require a generalization of Equation 3.20, which relates the Haar mother wavelet to the scaling function. Let some MRA have a scaling function ϕ defined by the dilation Equation 3.25. Since

$$\langle\phi(2x-k),\phi(2x-k)\rangle = \frac{1}{2}\cdot\langle\phi(x),\phi(x)\rangle = \frac{1}{2},$$

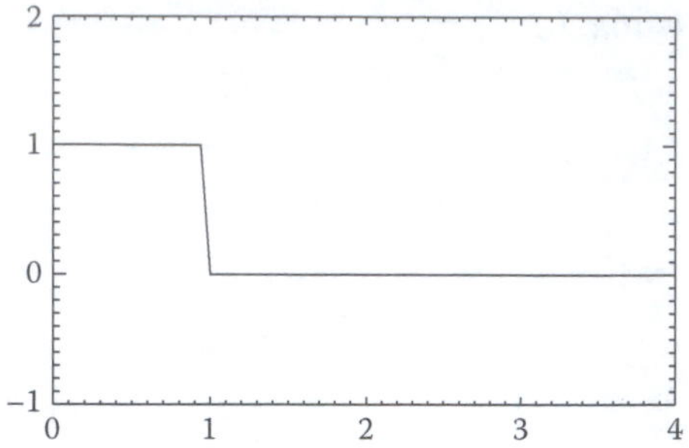

FIGURE 3.8
Approximation to the Haar scaling function with the program of Listing 3.3 after $n = 4$ iterations.

the functions $\sqrt{2}\phi(2x-k)$ are both normalized and orthogonal. We can write Equation 3.25 in the form

$$\phi(x) = \sum_k h_k \sqrt{2}\phi(2x-k), \tag{3.28}$$

where

$$h_k = \frac{c_k}{\sqrt{2}}.$$

It follows from Equation 3.26 that

$$\sum_k h_k^2 = 1. \tag{3.29}$$

Now we assume, in analogy to Equation 3.28, that the mother wavelet ψ can also be expressed in terms of the scaling function as*

$$\psi(x) = \sum_k g_k \sqrt{2}\phi(2x-k). \tag{3.30}$$

Since $\phi \in V_0$ and $\psi \in V_0^{\perp}$, we have

$$\langle \phi, \psi \rangle = \sum_k h_k g_k = 0. \tag{3.31}$$

Similarly, with some simple index manipulations,

$$\langle \psi(x-k), \psi(x-m) \rangle = \sum_i g_i g_{i-2(k-m)} = \delta_{k,m}. \tag{3.32}$$

* The coefficients g_k should not be confused with pixel intensities.

A set of coefficients that satisfies Equations 3.31 and 3.32 is given by

$$g_k = (-1)^k h_{1-k}. \tag{3.33}$$

So we obtain, finally, the general relationship between the mother wavelet and the scaling function:

$$\psi(x) = \sum_k (-1)^k h_{1-k}\sqrt{2}\phi(2x-k) = \sum_k (-1)^k c_{1-k}\phi(2x-k). \tag{3.34}$$

3.2.3.4 Daubechies D4 Scaling Function

A family of MRAs which is very useful in digital image analysis is generated by the Daubechies scaling functions and their associated wavelets (Daubechies, 1988). The Daubechies D4 scaling function, for example, can be derived by placing the following two additional requirements on an MRA (Aboufadel and Schlicker, 1999).

1. *Compact support:* The scaling function $\phi(x)$ is required to be zero outside the interval $0 < x < 3$. This means that the refinement coefficients c_k vanish for $k < 0$ and for $k > 3$. To see this, note that

   $$c_{-3} = 2\langle\phi(x), \phi(2x+3)\rangle = \int_0^3 \phi(x)\phi(2x+3)dx = 0$$

   and similarly for $k \le -4$ and for $k \ge 6$. Therefore, from the dilation equation,

   $$\phi(-1/2) = 0 = c_{-2}\phi(-1+2) + c_{-1}\phi(-1+1) + \cdots \quad \text{implying } c_{-2} = 0$$

   and similarly for $k = -1, 4, 5$. Thus from Equation 3.26, we can conclude that

   $$c_0^2 + c_1^2 + c_2^2 + c_3^2 = 2 \tag{3.35}$$

   and from Equation 3.27 with $j = 1$, that

   $$c_0c_2 + c_1c_3 = 0. \tag{3.36}$$

2. *Regularity:* All constant and linear polynomials can be written as a linear combination of the basis $\{\phi(x-k) \mid k \in \mathbb{Z}\}$ for V_0. This implies

that there is no residual in the orthogonal decomposition of $f(x) = 1$ and $f(x) = x$ onto the basis, that is,

$$\int_{-\infty}^{\infty} 1 \cdot \psi(x)dx = \int_{-\infty}^{\infty} x \cdot \psi(x)dt = 0. \tag{3.37}$$

With Equation 3.34 the mother wavelet is

$$\begin{aligned}\psi(x) &= -c_0\phi(2x-1) + c_1\phi(2x) - c_2\phi(2x+1) + c_3\phi(2x+2)\\ &= \sum_{k=0}^{3}(-1)^{k+1}c_k\phi(2x-1+k).\end{aligned} \tag{3.38}$$

The first requirement in Equation 3.37 gives immediately

$$-c_0 + c_1 - c_2 + c_3 = 0. \tag{3.39}$$

From the second requirement we have

$$\begin{aligned}0 = \int_{-\infty}^{\infty} x\psi(x)dx &= \sum_{k=0}^{3}(-1)^{k+1}c_k \int_{-\infty}^{\infty} x\phi(2x-1+k)dx\\ &= \sum_{k=0}^{3}(-1)^{k+1}c_k \int_{-\infty}^{\infty} \frac{u+1-k}{4}\phi(u)du\\ &= \frac{0}{4} \cdot \int_{-\infty}^{\infty} u\phi(u)du + \frac{-c_0+c_2-2c_3}{4}\int_{-\infty}^{\infty}\phi(u)du,\end{aligned}$$

using Equation 3.39. Thus

$$-c_0 + c_2 - 2c_3 = 0. \tag{3.40}$$

Equations 3.35, 3.36, 3.39, and 3.40 comprise a system of four equations in four unknowns. A solution is given by

$$c_0 = \frac{1+\sqrt{3}}{4},\quad c_1 = \frac{3+\sqrt{3}}{4},\quad c_2 = \frac{3-\sqrt{3}}{4},\quad c_3 = \frac{1-\sqrt{3}}{4},$$

which are known as the D4 refinement coefficients. Figure 3.9 shows the corresponding scaling function, determined with the cascade algorithm described earlier (Listing 3.3).

The D4 scaling function and the subspaces that it generates are thus anything but simple. The scaling function is continuous but not everywhere

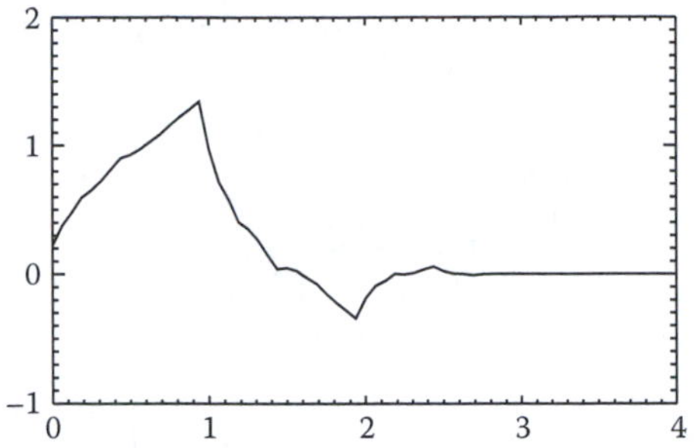

FIGURE 3.9
Approximation to the Daubechies D4 scaling function with the program of Listing 3.3 after $n = 4$ iterations.

differentiable and also self-similar (the tail is an exact but rescaled copy of the entire function). Nevertheless, the D4 wavelets provide a much more useful representation of digital images than the Haar wavelets.* We will return to them in Chapter 4 when we treat the fast wavelet transform and examine pyramid algorithms for image processing.

3.3 Principal Components

The principal components transformation generates linear combinations of multispectral pixel intensities which are mutually uncorrelated and which have maximum variance. Specifically, consider a multispectral image represented by the random vector $\boldsymbol{G}$ and assume that $\langle \boldsymbol{G} \rangle = \boldsymbol{0}$, so that the covariance matrix is given by $\boldsymbol{\Sigma} = \langle \boldsymbol{G}\boldsymbol{G}^\top \rangle$. Let us seek a linear combination $Y = \boldsymbol{w}^\top \boldsymbol{G}$ whose variance $\boldsymbol{w}^\top \boldsymbol{\Sigma} \boldsymbol{w}$ is maximum. This quantity can trivially be made as large as we like by choosing $\boldsymbol{w}$ sufficiently large, so that the maximization only makes sense if we restrict $\boldsymbol{w}$ in some way. A convenient constraint is $\boldsymbol{w}^\top \boldsymbol{w} = 1$. According to the discussion in Section 1.6, we can solve this problem by maximizing the unconstrained Lagrange function

$$L = \boldsymbol{w}^\top \boldsymbol{\Sigma} \boldsymbol{w} - \lambda(\boldsymbol{w}^\top \boldsymbol{w} - 1).$$

This leads directly (see Equation 1.68) to the eigenvalue problem

$$\boldsymbol{\Sigma} \boldsymbol{w} = \lambda \boldsymbol{w}. \tag{3.41}$$

Denote the orthogonal and normalized eigenvectors of $\boldsymbol{\Sigma}$ obtained by solving the above problem by $\boldsymbol{w}_1 \dots \boldsymbol{w}_N$, sorted according to decreasing eigenvalue $\lambda_1 \geq \cdots \geq \lambda_N$. These eigenvectors are the *principal axes* and

* The IDL function `WTN()` implements the discrete wavelet transform with various wavelets from the Daubechies family (see Chapter 4, Exercise 5).

the corresponding linear combinations $\boldsymbol{w}_i^\top \boldsymbol{G}$ are projections along those principal axes, called the *principal components* of $\boldsymbol{G}$. The individual principal components

$$Y_1 = \boldsymbol{w}_1^\top \boldsymbol{G},\ Y_2 = \boldsymbol{w}_2^\top \boldsymbol{G},\ \ldots,\ Y_N = \boldsymbol{w}_N^\top \boldsymbol{G}$$

can be expressed more compactly as a random vector $\boldsymbol{Y}$ by writing

$$\boldsymbol{Y} = \boldsymbol{W}^\top \boldsymbol{G}, \tag{3.42}$$

where $\boldsymbol{W}$ is the matrix whose columns comprise the eigenvectors, that is,

$$\boldsymbol{W} = (\boldsymbol{w}_1 \ldots \boldsymbol{w}_N).$$

Since the eigenvectors are orthogonal and normalized, $\boldsymbol{W}$ is an orthonormal matrix:

$$\boldsymbol{W}\boldsymbol{W}^\top = \boldsymbol{I}.$$

If the covariance matrix of the principal components vector $\boldsymbol{Y}$ is called $\boldsymbol{\Sigma}'$, then we have

$$\begin{aligned}\boldsymbol{\Sigma}' &= \langle \boldsymbol{Y}\boldsymbol{Y}^\top \rangle = \langle \boldsymbol{W}^\top \boldsymbol{G}\boldsymbol{G}^\top \boldsymbol{W} \rangle \\ &= \boldsymbol{W}^\top \boldsymbol{\Sigma} \boldsymbol{W} = \begin{pmatrix} \lambda_1 & 0 & \cdots & 0 \\ 0 & \lambda_2 & \cdots & 0 \\ \vdots & \vdots & \ddots & \vdots \\ 0 & 0 & \cdots & \lambda_N \end{pmatrix} = \boldsymbol{\Lambda}.\end{aligned} \tag{3.43}$$

The eigenvalues are thus seen to be the variances of the principal components, and all of the covariances are zero. The first principal component Y_1 has maximum variance $\text{var}(Y_1) = \lambda_1$, the second principal component Y_2 has maximum variance $\text{var}(Y_2) = \lambda_2$ subject to the condition that it is uncorrelated with Y_1, and so on. The fraction of the total variance in the original multispectral image which is accounted for by the first i principal components is

$$\frac{\lambda_1 + \cdots + \lambda_i}{\lambda_1 + \cdots + \lambda_i + \cdots + \lambda_N}.$$

If the original multispectral channels are highly correlated, as is often the case, then the first few principal components will usually account for a very high percentage of the total variance in the image. For example, a color composite of the first three principal components of a LANDSAT 7 ETM+ scene displays essentially all of the information contained in the six nonthermal spectral components in one single image (see Figure 3.10). The principal

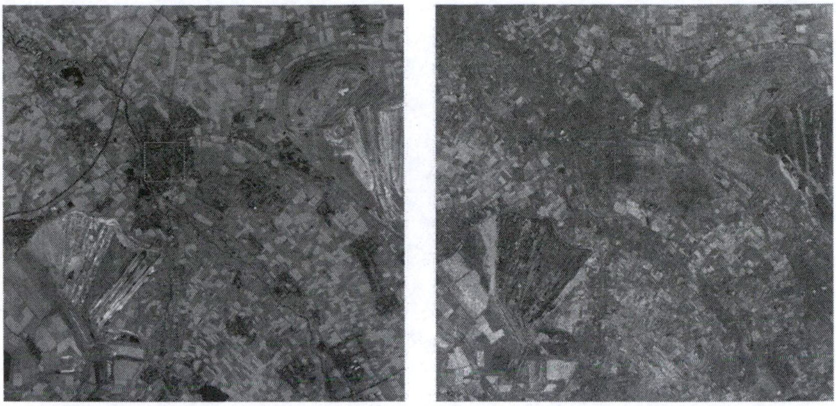

(a) (b)

FIGURE 3.10
RGB color composites (2% linear histogram stretch) of the first three (a) and last three (b) principal components of the six nonthermal bands of a LANDSAT 7 ETM+ image over Jülich acquired on August 29, 2001.

components transformation, also called *principal components analysis* (PCA), is therefore often used for dimensionality reduction of imagery prior, for instance, to land cover classification (Chapter 6).

3.3.1 Primal Solution

In practice, one estimates the covariance matrix $\boldsymbol{\Sigma}$ in terms of the data matrix $\mathcal{G}$, Equation 2.39, and then solves the eigenvalue problem, Equation 3.41, in the form

$$\mathcal{C}w = \frac{1}{m-1}\mathcal{G}^\top \mathcal{G}w = \lambda w. \tag{3.44}$$

This is the *primal* problem for PCA (see the discussion of primal and dual formulations for ridge regression in Chapter 1). Solution of the primal problem can be performed directly from the ENVI main menu:

```
Transform/Principal Components/Forward PC Rotation
```

The IDL program in Listing 1.4 illustrated the procedure explicitly.

3.3.2 Dual Solution

The normalized eigenvectors of $\mathcal{C}$ are w_i, $i = 1 \ldots N$. These, as explained, are the principal axes, and we now show how to express them in terms of the eigenvectors of the Gram matrix $\mathcal{G}\mathcal{G}^\top$ which was initially introduced in

Section 2.6.3 in connection with ridge regression. The Gram matrix, we recall, is symmetric, positive semidefinite.

Assume that $m > N$. Consider an eigenvector–eigenvalue pair (v_i, λ_i) for the Gram matrix $\mathcal{G}\mathcal{G}^\top$. Then we can write

$$\mathcal{C}(\mathcal{G}^\top v_i) = \frac{1}{m-1}(\mathcal{G}^\top \mathcal{G})(\mathcal{G}^\top v_i) = \frac{1}{m-1}\mathcal{G}^\top(\mathcal{G}\mathcal{G}^\top)v_i = \frac{1}{m-1}\lambda_i(\mathcal{G}^\top v_i), \tag{3.45}$$

so that $(\mathcal{G}^\top v_i, \lambda_i/(m-1))$ is an eigenvector–eigenvalue pair for $\mathcal{C}$. The norm of the eigenvector $\mathcal{G}^\top v_i$ is

$$\|\mathcal{G}^\top v_i\| = \sqrt{v_i^\top \mathcal{G}\mathcal{G}^\top v_i} = \sqrt{\lambda_i}. \tag{3.46}$$

For $\lambda_1 > \lambda_2 > \cdots \lambda_N > 0$, the normalized principal vectors w_i can thus be expressed equivalently in the form

$$w_i = \lambda_i^{-1/2}\mathcal{G}^\top v_i, \quad i = 1 \dots N.$$

In fact the Gram matrix $\mathcal{G}\mathcal{G}^\top$ has exactly N positive eigenvalues, the rest being zero ($\mathcal{G}\mathcal{G}^\top$ has rank N). Informally, every positive eigenvalue for $\mathcal{G}\mathcal{G}^\top$ generates, via Equation 3.45, an eigenvector–eigenvalue pair for $\mathcal{C}$, and $\mathcal{C}$ has only N eigenvectors. For example, the program

```
PRO dual_pca
; column centered design matrix for random 2D data
   m = 100
   G = 2*randomu(seed,2,m,/double)-1
; covariance matrix
   C = transpose(G)##G/(m-1)
; Gram matrix
   K = G##transpose(G)
   PRINT,  eigenql(C)
   PRINT, (eigenql(K))[0:3]/(m-1)
END
```

generates the output

```
ENVI> dual_pca
0.38706425     0.30646066
0.38706425     0.30646066   8.1496153e-017   6.5566325e-017
```

The first $N = 2$ eigenvalues of $\mathcal{G}\mathcal{G}^\top$ are equal to $(m-1)\times$ the eigenvalues of $\mathcal{C}$. The remaining $m - N$ eigenvalues of $\mathcal{G}\mathcal{G}^\top$ are zero.

In terms of m-dimensional *dual vectors* $\boldsymbol{\alpha}_i = \lambda^{-1/2}\boldsymbol{v}_i$, we have

$$\boldsymbol{w}_i = \mathcal{G}^\top \boldsymbol{\alpha}_i = \sum_{\nu=1}^{m} (\boldsymbol{\alpha}_i)_\nu \boldsymbol{g}(\nu). \tag{3.47}$$

So, just as for ridge regression, we get the dual form by expressing the parameter vector as a linear combination of the observations. The projection of any observation $\boldsymbol{g}$ along a principal axis is then

$$\boldsymbol{w}_i^\top \boldsymbol{g} = \sum_{\nu=1}^{m} (\boldsymbol{\alpha}_i)_\nu (\boldsymbol{g}(\nu)^\top \boldsymbol{g}), \quad i = 1 \ldots N. \tag{3.48}$$

Thus we can alternatively perform PCA by finding eigenvalues and eigenvectors of the Gram matrix. The observations $\boldsymbol{g}(\nu)$ appear only in the form of inner products, both in the determination of the Gram matrix as well as in the projection of any new observations, Equation 3.48. This forms the starting point for nonlinear, or *kernel* PCA. Nonlinear kernels will be introduced in Chapter 4 with application to nonlinear PCA and appear again in connection with support vector machine classification in Chapter 6 and unsupervised classification in Chapter 8. Chapter 9 gives an example of kernel PCA for change detection.

3.4 Minimum Noise Fraction

PCA maximizes variance. This does not always lead to images of the desired quality (e.g., having minimal noise). The *minimum noise fraction* (MNF) transformation (Green et al., 1988) can be used to maximize the *signal to noise ratio* (SNR) rather than maximizing variance, so, if this is the desired criterion, it is to be preferred over PCA. In the following we derive the MNF transformation directly by maximizing the ratio of signal variance to noise variance. Then we relate our procedure to the algorithm used in ENVI to perform the MNF transformation.

3.4.1 Additive Noise

A noisy multispectral image $\boldsymbol{G}$ may be represented in terms of an *additive noise model*, that is, as a sum of signal and noise contributions

$$\boldsymbol{G} = \boldsymbol{S} + \boldsymbol{N}. \tag{3.49}$$

The signal S is understood as the component carrying the information of interest. Noise, introduced most often by the sensors, corrupts the signal and masks that information. If both components are assumed to be normally distributed with respective covariance matrices $\mathbf{\Sigma}_S$ and $\mathbf{\Sigma}_N$, to have zero mean and, furthermore, to be uncorrelated, then the covariance matrix $\mathbf{\Sigma}$ for the image $\boldsymbol{G}$ is given by

$$\mathbf{\Sigma} = \langle \boldsymbol{G}\boldsymbol{G}^\top \rangle = \langle (\boldsymbol{S}+\boldsymbol{N})(\boldsymbol{S}+\boldsymbol{N})^\top \rangle = \langle \boldsymbol{S}\boldsymbol{S}^\top \rangle + \langle \boldsymbol{N}\boldsymbol{N}^\top \rangle,$$

the covariance $\langle \boldsymbol{N}\boldsymbol{S}^\top \rangle$ being zero by assumption. Thus the image covariance matrix is simply the sum of signal and noise contributions,

$$\mathbf{\Sigma} = \mathbf{\Sigma}_S + \mathbf{\Sigma}_N. \tag{3.50}$$

The signal to noise ratio in the ith band of a multispectral image is usually expressed by the ratio of the variance of the signal and noise components:

$$\text{SNR}_i = \frac{\text{var}(S_i)}{\text{var}(N_i)}.$$

Let us now seek a linear combination $Y = \boldsymbol{a}^\top \boldsymbol{G}$ of image bands for which this ratio is maximum. That is, we wish to maximize

$$\text{SNR} = \frac{\text{var}(\boldsymbol{a}^\top \boldsymbol{S})}{\text{var}(\boldsymbol{a}^\top \boldsymbol{N})} = \frac{\boldsymbol{a}^\top \mathbf{\Sigma}_S \boldsymbol{a}}{\boldsymbol{a}^\top \mathbf{\Sigma}_N \boldsymbol{a}}. \tag{3.51}$$

The ratio of quadratic forms on the right is referred to as a *Rayleigh quotient*. With Equation 3.50 we can write Equation 3.51 equivalently in the form

$$\text{SNR} = \frac{\boldsymbol{a}^\top \mathbf{\Sigma} \boldsymbol{a}}{\boldsymbol{a}^\top \mathbf{\Sigma}_N \boldsymbol{a}} - 1. \tag{3.52}$$

Setting its vector derivative with respect to $\boldsymbol{a}$ equal to zero, we get

$$\frac{\partial}{\partial \boldsymbol{a}}\text{SNR} = \frac{1}{\boldsymbol{a}^\top \mathbf{\Sigma}_N \boldsymbol{a}} 2\mathbf{\Sigma}\boldsymbol{a} - \frac{\boldsymbol{a}^\top \mathbf{\Sigma} \boldsymbol{a}}{(\boldsymbol{a}^\top \mathbf{\Sigma}_N \boldsymbol{a})^2} 2\mathbf{\Sigma}_N \boldsymbol{a} = \boldsymbol{0}$$

or, equivalently,

$$(\boldsymbol{a}^\top \mathbf{\Sigma}_N \boldsymbol{a})\mathbf{\Sigma}\boldsymbol{a} = (\boldsymbol{a}^\top \mathbf{\Sigma} \boldsymbol{a})\mathbf{\Sigma}_N \boldsymbol{a}.$$

This condition is met when $\boldsymbol{a}$ solves the *symmetric generalized eigenvalue problem*

$$\mathbf{\Sigma}_N \boldsymbol{a} = \lambda \mathbf{\Sigma} \boldsymbol{a}, \tag{3.53}$$

as can easily be seen by substitution. In Equation 3.53, both $\mathbf{\Sigma}_N$ and $\mathbf{\Sigma}$ are symmetric and positive definite. The equation can be reduced to the standard eigenvalue problem that we are familiar with by performing a *Cholesky decomposition* of $\mathbf{\Sigma}$. As explained in Appendix A, Cholesky decomposition factors $\mathbf{\Sigma}$ as $\mathbf{\Sigma} = \boldsymbol{L}\boldsymbol{L}^\top$, where $\boldsymbol{L}$ is a lower triangular matrix. Substituting this into Equation 3.53 gives

$$\mathbf{\Sigma}_N \boldsymbol{a} = \lambda \boldsymbol{L}\boldsymbol{L}^\top \boldsymbol{a}$$

or, multiplying both sides of the equation from the left by $\boldsymbol{L}^{-1}$ and inserting the identity $(\boldsymbol{L}^\top)^{-1}\boldsymbol{L}^\top$,

$$\boldsymbol{L}^{-1}\mathbf{\Sigma}_N(\boldsymbol{L}^\top)^{-1}\boldsymbol{L}^\top\boldsymbol{a} = \lambda\boldsymbol{L}^\top\boldsymbol{a}.$$

Now let $\boldsymbol{b} = \boldsymbol{L}^\top\boldsymbol{a}$. From the commutivity of the operations of inverse and transpose, it follows that

$$[\boldsymbol{L}^{-1}\mathbf{\Sigma}_N(\boldsymbol{L}^{-1})^\top]\boldsymbol{b} = \lambda\boldsymbol{b}, \tag{3.54}$$

a standard eigenvalue problem for the symmetric matrix $\boldsymbol{L}^{-1}\mathbf{\Sigma}_N(\boldsymbol{L}^{-1})^\top$. Let its (orthogonal and normalized) eigenvectors be $\boldsymbol{b}_i$, $i = 1\ldots N$. Then

$$\boldsymbol{b}_i^\top\boldsymbol{b}_j = \boldsymbol{a}_i^\top\boldsymbol{L}\boldsymbol{L}^\top\boldsymbol{a}_j = \boldsymbol{a}_i^\top\mathbf{\Sigma}\boldsymbol{a}_j = \delta_{ij}.$$

Therefore we see that the variances of the transformed components $Y_i = \boldsymbol{a}_i^\top\boldsymbol{G}$ are all unity:

$$\text{var}(Y_i) = \boldsymbol{a}_i^\top\mathbf{\Sigma}\boldsymbol{a}_i = 1, \quad i = 1\ldots N.$$

And that they are mutually uncorrelated:

$$\text{cov}(Y_i, Y_j) = \boldsymbol{a}_i^\top\mathbf{\Sigma}\boldsymbol{a}_j = 0, \quad i,j = 1\ldots N,\ i \neq j.$$

Listing 3.4 shows an IDL routine for solving the generalized eigenvalue problem with Cholesky decomposition. This routine is used in some of the ENVI/IDL extensions described in Appendix C.

With the definition $\boldsymbol{A} = (\boldsymbol{a}_1, \boldsymbol{a}_2 \ldots \boldsymbol{a}_N)$, the complete MNF transformation can be represented as

$$\boldsymbol{Y} = \boldsymbol{A}^\top\boldsymbol{G}, \tag{3.55}$$

similarly to the principal components transformation, Equation 3.42. The covariance matrix of $\boldsymbol{Y}$ is (compare with Equation 3.43)

$$\mathbf{\Sigma}' = \boldsymbol{A}^\top\mathbf{\Sigma}\boldsymbol{A} = \boldsymbol{I}, \tag{3.56}$$

where $\boldsymbol{I}$ is the $N \times N$ identity matrix.

Listing 3.4
Solving the generalized eigenvalue problem by Cholesky decomposition.

```
PRO gen_eigenproblem, C,B,A,lambda
; solve the generalized eigenproblem C##a = lambda*B##a
   choldc, B, P, /double
   FOR i=1L,(size(B))[1]-1 DO B[i,0:i]=[fltarr(i),P[i]]
   B[0,0]=P[0]
   Li  = invert(B,/double)
   D = Li ## C ## transpose(Li)
; ensure symmetry after roundoff errors
   D = (D+transpose(D))/2
   lambda = eigenql(D,/double,eigenvectors=A)
   A = transpose(A##Li)
END
```

It follows from Equation 3.52 that the SNR for eigenvalue λ_i is just

$$\mathrm{SNR}_i = \frac{\boldsymbol{a}_i^\top \boldsymbol{\Sigma} \boldsymbol{a}_i}{\boldsymbol{a}_i^\top(\lambda_i \boldsymbol{\Sigma} \boldsymbol{a}_i)} - 1 = \frac{1}{\lambda_i} - 1. \tag{3.57}$$

Thus the projection $Y_i = \boldsymbol{a}_i^\top \boldsymbol{G}$ corresponding to the *smallest* eigenvalue λ_i will have the largest signal to noise ratio. Note that with Equation 3.53 we can write

$$\boldsymbol{\Sigma}_N \boldsymbol{A} = \boldsymbol{\Sigma} \boldsymbol{A} \boldsymbol{\Lambda}, \tag{3.58}$$

where $\boldsymbol{\Lambda} = \mathrm{Diag}(\lambda_1 \ldots \lambda_N)$.

3.4.2 Minimum Noise Fraction Transformation in ENVI

The MNF transformation is available in the ENVI environment:

```
Transform/MNF Rotation/Forward MNF
```

It is carried out somewhat differently in two steps which are, as we shall now show, equivalent to the above derivation.

In the first step the noise contribution to the observation $\boldsymbol{G}$ is "whitened," that is, a transformation is performed after which the noise component $\boldsymbol{N}$ has covariance matrix $\boldsymbol{\Sigma}_N = \boldsymbol{I}$, the identity matrix. This can be accomplished by first doing a transformation which diagonalizes $\boldsymbol{\Sigma}_N$. Suppose that the transformation matrix for this operation is $\boldsymbol{C}$ and that $\boldsymbol{Z}$ is the resulting random vector. Then

$$\boldsymbol{Z} = \boldsymbol{C}^\top \boldsymbol{G}, \quad \boldsymbol{C}^\top \boldsymbol{\Sigma}_N \boldsymbol{C} = \boldsymbol{\Lambda}_N,\ \boldsymbol{C}^\top \boldsymbol{C} = \boldsymbol{I}, \tag{3.59}$$

where $\mathbf{\Lambda}_N$ is a diagonal matrix, the diagonal elements of which are the variances of the transformed noise component $\mathbf{C}^\top\mathbf{N}$. Next, apply the transformation $\mathbf{\Lambda}_N^{-1/2}$ (the diagonal matrix whose diagonal elements are the inverse of the square roots of the diagonal elements of $\mathbf{\Lambda}_N$) to give a new random vector $\mathbf{X}$,

$$\mathbf{X} = \mathbf{\Lambda}_N^{-1/2}\mathbf{Z} = \mathbf{\Lambda}_N^{-1/2}\mathbf{C}^\top\mathbf{G}.$$

Then the covariance matrix of the noise component $\mathbf{\Lambda}_N^{-1/2}\mathbf{C}^\top\mathbf{N}$ is given by

$$\mathbf{\Lambda}_N^{-1/2}\mathbf{\Lambda}_N\mathbf{\Lambda}_N^{-1/2} = \mathbf{I},$$

as desired: the noise contribution has been "whitened." At this stage of affairs, the covariance matrix of the transformed random vector $\mathbf{X}$ is

$$\mathbf{\Sigma}_X = \mathbf{\Lambda}_N^{-1/2}\mathbf{C}^\top\mathbf{\Sigma}\mathbf{C}\mathbf{\Lambda}_N^{-1/2}. \tag{3.60}$$

In the second step an ordinary principal components transformation is performed on $\mathbf{X}$, leading finally to the random vector $\mathbf{Y}$ representing the MNF components:

$$\mathbf{Y} = \mathbf{B}^\top\mathbf{X}, \quad \mathbf{B}^\top\mathbf{\Sigma}_X\mathbf{B} = \mathbf{\Lambda}_X,\ \mathbf{B}^\top\mathbf{B} = \mathbf{I}. \tag{3.61}$$

The overall transformation is thus

$$\mathbf{Y} = \mathbf{B}^\top\mathbf{\Lambda}_N^{-1/2}\mathbf{C}^\top\mathbf{G} = \mathbf{A}^\top\mathbf{G}, \tag{3.62}$$

where $\mathbf{A} = \mathbf{C}\mathbf{\Lambda}_N^{-1/2}\mathbf{B}$. To see that this transformation is indeed equivalent to solving the generalized eigenvalue problem, Equation 3.58, consider

$$\begin{aligned}
\mathbf{\Sigma}_N\mathbf{A} &= \mathbf{\Sigma}_N\mathbf{C}\mathbf{\Lambda}_N^{-1/2}\mathbf{B}\\
&= \mathbf{C}\mathbf{\Lambda}_N\mathbf{\Lambda}_N^{-1/2}\mathbf{B} \quad \text{from Equation 3.59}\\
&= \mathbf{C}\mathbf{\Lambda}_N^{1/2}\mathbf{B}\\
&= \mathbf{C}\mathbf{\Lambda}_N^{1/2}(\mathbf{\Sigma}_X\mathbf{B}\mathbf{\Lambda}_X^{-1}) \quad \text{from Equation 3.61}\\
&= \mathbf{C}\mathbf{\Lambda}_N^{1/2}\mathbf{\Lambda}_N^{-1/2}\mathbf{C}^\top\mathbf{\Sigma}\mathbf{C}\mathbf{\Lambda}_N^{-1/2}\mathbf{B}\mathbf{\Lambda}_X^{-1} \quad \text{from Equation 3.60}\\
&= \mathbf{\Sigma}\mathbf{A}\mathbf{\Lambda}_X^{-1}.
\end{aligned} \tag{3.63}$$

This is the same as Equation 3.58 with $\mathbf{\Lambda}$ replaced by $\mathbf{\Lambda}_X^{-1}$, that is,

$$\lambda_{Xi} = \frac{1}{\lambda_i} = \mathrm{SNR}_i + 1, \qquad i = 1\ldots N,$$

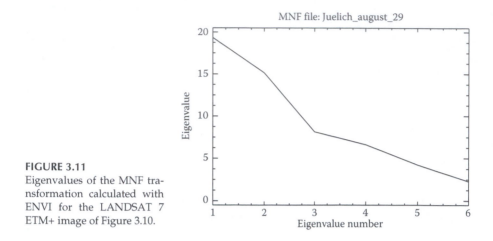

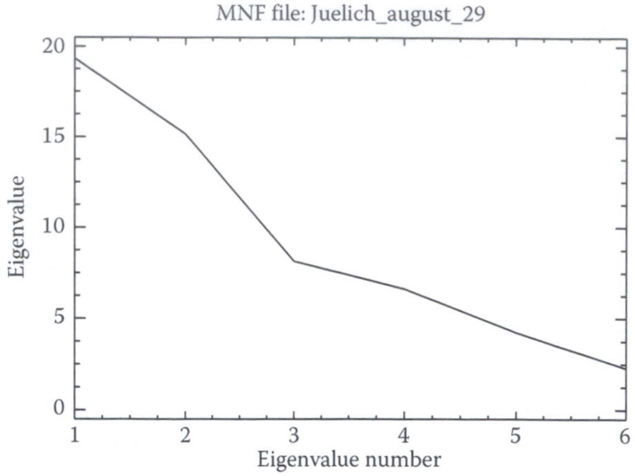

FIGURE 3.11
Eigenvalues of the MNF transformation calculated with ENVI for the LANDSAT 7 ETM+ image of Figure 3.10.

using Equation 3.57. The eigenvalues returned by ENVI's MNF transformation are those of the second (i.e., the principal components) transformation, namely the λ_{Xi} above.* They are equal to the SNR plus one, so that values equal to one correspond to "pure noise." Figure 3.11 shows an example.

3.5 Spatial Correlation

Before the MNF transformation can be performed, it is of course necessary to estimate both the image and noise covariance matrices $\boldsymbol{\Sigma}$ and $\boldsymbol{\Sigma}_N$. The former poses no problem, but how does one estimate the noise covariance matrix? The spatial characteristics of the image can be used to estimate $\boldsymbol{\Sigma}_N$, taking advantage of the fact that the intensity of neighboring pixels is usually approximately constant. This property is quantified as the *autocorrelation* of an image. We shall first find a spectral transformation that maximizes the autocorrelation and then see how to relate it to the image noise statistics.

3.5.1 Maximum Autocorrelation Factor

Let $\boldsymbol{x} \sim (i, j)$ represent the coordinates of a pixel within image $\boldsymbol{G}$, and assume that $\langle \boldsymbol{G} \rangle = 0$. The *spatial covariance* $\boldsymbol{\Gamma}(\boldsymbol{x}, \boldsymbol{\Delta})$ is defined as the covariance of the original image, represented by $\boldsymbol{G}(\boldsymbol{x})$, with itself, but shifted by the amount

* Note that the MNF components returned by ENVI, unlike those of Section 3.4.1, do not have unit variance. Their variances are the eigenvalues λ_{Xi}.

$\mathbf{\Delta} = (\Delta_1, \Delta_2)^\top$,

$$\mathbf{\Gamma}(x, \mathbf{\Delta}) = \langle G(x)G(x + \mathbf{\Delta})^\top \rangle. \tag{3.64}$$

We make the so-called *second order stationarity assumption*, namely that $\mathbf{\Gamma}(x, \mathbf{\Delta}) = \mathbf{\Gamma}(\mathbf{\Delta})$ is independent of x. Then $\Gamma(\mathbf{0}) = \langle GG^\top \rangle = \mathbf{\Sigma}$, and furthermore

$$\begin{aligned}\mathbf{\Gamma}(-\mathbf{\Delta}) &= \langle G(x)G(x - \mathbf{\Delta})^\top \rangle \\ &= \langle G(x + \mathbf{\Delta})G(x)^\top \rangle \\ &= \langle (G(x)G(x + \mathbf{\Delta})^\top)^\top \rangle \\ &= \mathbf{\Gamma}(\mathbf{\Delta})^\top.\end{aligned} \tag{3.65}$$

Now let us look at the spatial covariance of projections $Y = \boldsymbol{a}^\top G$ of the original and shifted images. This is given by

$$\begin{aligned}\text{cov}(\boldsymbol{a}^\top G(x), \boldsymbol{a}^\top G(x + \mathbf{\Delta})) &= \boldsymbol{a}^\top \langle G(x)G(x + \mathbf{\Delta})^\top \rangle \boldsymbol{a} \\ &= \boldsymbol{a}^\top \mathbf{\Gamma}(\mathbf{\Delta})\boldsymbol{a} \\ &= \boldsymbol{a}^\top \mathbf{\Gamma}(-\mathbf{\Delta})\boldsymbol{a} \\ &= \frac{1}{2}\boldsymbol{a}^\top (\mathbf{\Gamma}(\mathbf{\Delta}) + \mathbf{\Gamma}(-\mathbf{\Delta}))\boldsymbol{a},\end{aligned} \tag{3.66}$$

where we have used Equation 3.65. Introducing the quantity $\mathbf{\Sigma}_\Delta$, defined as the covariance matrix of the difference image $G(x) - G(x + \mathbf{\Delta})$,

$$\begin{aligned}\mathbf{\Sigma}_\Delta &= \langle (G(x) - G(x + \mathbf{\Delta}))(G(x) - G(x + \mathbf{\Delta})^\top \rangle \\ &= \langle G(x)G(x)^\top \rangle + \langle G(x + \mathbf{\Delta})G(x + \mathbf{\Delta})^\top \rangle \\ &\quad - \langle G(x)G(x + \mathbf{\Delta})^\top \rangle - \langle G(x + \mathbf{\Delta})G(x)^\top \rangle \\ &= 2\mathbf{\Sigma} - \mathbf{\Gamma}(\mathbf{\Delta}) - \mathbf{\Gamma}(-\mathbf{\Delta}),\end{aligned} \tag{3.67}$$

we see that $\mathbf{\Gamma}(\mathbf{\Delta}) + \mathbf{\Gamma}(-\mathbf{\Delta}) = 2\mathbf{\Sigma} - \mathbf{\Sigma}_\Delta$, and so we can write Equation 3.66 in the form

$$\text{cov}(\boldsymbol{a}^\top G(x), \boldsymbol{a}^\top G(x + \mathbf{\Delta})) = \boldsymbol{a}^\top \mathbf{\Sigma}\boldsymbol{a} - \frac{1}{2}\boldsymbol{a}^\top \mathbf{\Sigma}_\Delta \boldsymbol{a}. \tag{3.68}$$

The *spatial autocorrelation* of the projections is therefore given by

$$\begin{aligned}\mathrm{corr}(\boldsymbol{a}^\top G(x), \boldsymbol{a}^\top G(x+\boldsymbol{\Delta})) &= \frac{\boldsymbol{a}^\top \boldsymbol{\Sigma}\boldsymbol{a} - \frac{1}{2}\boldsymbol{a}^\top \boldsymbol{\Sigma}_\Delta \boldsymbol{a}}{\sqrt{\mathrm{var}(\boldsymbol{a}^\top G(x))\mathrm{var}(\boldsymbol{a}^\top G(x+\boldsymbol{\Delta}))}} \\ &= \frac{\boldsymbol{a}^\top \boldsymbol{\Sigma}\boldsymbol{a} - \frac{1}{2}\boldsymbol{a}^\top \boldsymbol{\Sigma}_\Delta \boldsymbol{a}}{\sqrt{(\boldsymbol{a}^\top \boldsymbol{\Sigma}\boldsymbol{a})(\boldsymbol{a}^\top \boldsymbol{\Sigma}\boldsymbol{a})}} \\ &= 1 - \frac{1}{2}\frac{\boldsymbol{a}^\top \boldsymbol{\Sigma}_\Delta \boldsymbol{a}}{\boldsymbol{a}^\top \boldsymbol{\Sigma}\boldsymbol{a}}. \end{aligned} \tag{3.69}$$

The *maximum autocorrelation factor* (MAF) transformation determines that vector $\boldsymbol{a}$, which maximizes Equation 3.69. We obtain it by minimizing the Rayleigh quotient

$$R(\boldsymbol{a}) = \frac{\boldsymbol{a}^\top \boldsymbol{\Sigma}_\Delta \boldsymbol{a}}{\boldsymbol{a}^\top \boldsymbol{\Sigma}\boldsymbol{a}}.$$

Setting the vector derivative equal to zero gives

$$\frac{\partial R}{\partial \boldsymbol{a}} = \frac{1}{\boldsymbol{a}^\top \boldsymbol{\Sigma}\boldsymbol{a}}\frac{1}{2}\boldsymbol{\Sigma}_\Delta \boldsymbol{a} - \frac{\boldsymbol{a}^\top \boldsymbol{\Sigma}_\Delta \boldsymbol{a}}{(\boldsymbol{a}^\top \boldsymbol{\Sigma}\boldsymbol{a})^2}\frac{1}{2}\boldsymbol{\Sigma}\boldsymbol{a} = \boldsymbol{0}$$

or

$$(\boldsymbol{a}^\top \boldsymbol{\Sigma}\boldsymbol{a})\boldsymbol{\Sigma}_\Delta \boldsymbol{a} = (\boldsymbol{a}^\top \boldsymbol{\Sigma}_\Delta \boldsymbol{a})\boldsymbol{\Sigma}\boldsymbol{a}.$$

This condition is met when $\boldsymbol{a}$ solves the generalized eigenvalue problem

$$\boldsymbol{\Sigma}_\Delta \boldsymbol{a} = \lambda \boldsymbol{\Sigma}\boldsymbol{a}, \tag{3.70}$$

which is seen to have the same form as Equation 3.53, with $\boldsymbol{\Sigma}_\Delta$ replacing the noise covariance matrix $\boldsymbol{\Sigma}_N$. Again, both $\boldsymbol{\Sigma}_\Delta$ and $\boldsymbol{\Sigma}$ are symmetric, and the latter is also positive definite. We obtain as before, via Cholesky decomposition, the standard eigenvalue problem

$$[\boldsymbol{L}^{-1}\boldsymbol{\Sigma}_\Delta(\boldsymbol{L}^{-1})^\top]\boldsymbol{b} = \lambda \boldsymbol{b}, \tag{3.71}$$

for the symmetric matrix $\boldsymbol{L}^{-1}\boldsymbol{\Sigma}_\Delta(\boldsymbol{L}^{-1})^\top$ with $\boldsymbol{b} = \boldsymbol{L}^\top \boldsymbol{a}$.

Let the eigenvalues of Equation 3.71 be ordered from smallest to largest, $\lambda_1 \le \cdots \le \lambda_N$, and the corresponding (orthogonal) eigenvectors be $\boldsymbol{b}_i$. We have

$$\boldsymbol{b}_i^\top \boldsymbol{b}_j = \boldsymbol{a}_i^\top \boldsymbol{L}\boldsymbol{L}^\top \boldsymbol{a}_j = \boldsymbol{a}_i^\top \boldsymbol{\Sigma}\boldsymbol{a}_j = \delta_{ij} \tag{3.72}$$

so that, like the components of the MNF transformation, the MAF components $Y_i = \boldsymbol{a}_i^\top \boldsymbol{G},\ \ i = 1 \ldots N$, are orthogonal (uncorrelated) with unit variance. Moreover, with Equation 3.69,

$$\mathrm{corr}(\boldsymbol{a}_i^\top \boldsymbol{G}(\boldsymbol{x}), \boldsymbol{a}_i^\top \boldsymbol{G}(\boldsymbol{x}+\boldsymbol{\Delta})) = 1 - \frac{1}{2}\lambda_i, \quad i = 1 \dots N, \tag{3.73}$$

and the first MAF component has maximum autocorrelation.

3.5.2 Noise Estimation

The similarity of Equations 3.70 and 3.53 is a result of the fact that $\boldsymbol{\Sigma}_\Delta$ is, under fairly general circumstances, proportional to $\boldsymbol{\Sigma}_N$. We can demonstrate this as follows (Green et al., 1988). Let

$$\boldsymbol{G}(\boldsymbol{x}) = \boldsymbol{S}(\boldsymbol{x}) + \boldsymbol{N}(\boldsymbol{x})$$

and assume

$$\begin{aligned} \langle \boldsymbol{S}(\boldsymbol{x})\boldsymbol{N}(\boldsymbol{x})^\top \rangle &= \boldsymbol{0} \\ \langle \boldsymbol{S}(\boldsymbol{x})\boldsymbol{S}(\boldsymbol{x} \pm \boldsymbol{\Delta})^\top \rangle &= b_\Delta \boldsymbol{\Sigma}_S \\ \langle \boldsymbol{N}(\boldsymbol{x})\boldsymbol{N}(\boldsymbol{x} \pm \boldsymbol{\Delta})^\top \rangle &= c_\Delta \boldsymbol{\Sigma}_N, \end{aligned} \tag{3.74}$$

where b_Δ and c_Δ are constants. Under these assumptions, $\boldsymbol{\Gamma}(\Delta) = \boldsymbol{\Gamma}(-\Delta)$ from Equation 3.65 and, from Equation 3.67, we can conclude that

$$\boldsymbol{\Sigma}_\Delta = 2(\boldsymbol{\Sigma} - \boldsymbol{\Gamma}(\Delta)). \tag{3.75}$$

But with Equation 3.64

$$\boldsymbol{\Gamma}(\Delta) = \Big\langle \big(\boldsymbol{S}(\boldsymbol{x}) + \boldsymbol{N}(\boldsymbol{x})\big)\big(\boldsymbol{S}(\boldsymbol{x}+\boldsymbol{\Delta}) + \boldsymbol{N}(\boldsymbol{x}+\boldsymbol{\Delta})\big)^\top \Big\rangle$$

or, with Equation 3.74,

$$\boldsymbol{\Gamma}(\Delta) = b_\Delta \boldsymbol{\Sigma}_S + c_\Delta \boldsymbol{\Sigma}_N. \tag{3.76}$$

Finally, combining Equations 3.50, 3.75, and 3.76 gives

$$\frac{1}{2}\boldsymbol{\Sigma}_\Delta = (1 - b_\Delta)\boldsymbol{\Sigma} + (b_\Delta - c_\Delta)\boldsymbol{\Sigma}_N. \tag{3.77}$$

For a signal with high spatial coherence and for random (salt and pepper) noise, we expect that in Equation 3.74

$$b_\Delta \approx 1 \gg c_\Delta$$

and therefore, from Equation 3.77, that

$$\boldsymbol{\Sigma}_N \approx \frac{1}{2}\boldsymbol{\Sigma}_\Delta. \tag{3.78}$$

Thus we can obtain an estimate for the noise covariance matrix by estimating

$$\boldsymbol{\Sigma}_\Delta = \langle (G(x) - G(x+\boldsymbol{\Delta}))(G(x) - G(x+\boldsymbol{\Delta})^\top \rangle$$

and dividing the result by 2. This is illustrated in Listing 3.5. There, the matrix $\boldsymbol{\Sigma}_\Delta$ is determined by averaging the correlations from two difference

Listing 3.5

Estimation of the noise covariance matrix from the difference of one-pixel shifts.

```
PRO ex3_4

envi_select, title='Choose␣multispectral␣image', $
             fid=fid, dims=dims,pos=pos
IF (fid EQ -1) THEN BEGIN
   PRINT, 'Canceled'
   RETURN
ENDIF

IF n_elements(pos) LT 2 THEN BEGIN
   PRINT, 'Aborted'
   Message, 'Spectral␣subset␣size␣must␣be␣at␣least␣2'
ENDIF

envi_file_query, fid, fname=fname
num_cols = dims[2]-dims[1]+1
num_rows = dims[4]-dims[3]+1
num_pixels = (num_cols*num_rows)
num_bands = n_elements(pos)

; create difference images
dd = fltarr(num_bands,num_pixels)
FOR i=0,num_bands-1 DO BEGIN
   temp=float(envi_get_data(fid=fid,dims=dims, $
              pos=pos[i]))
   dd[i,*]=temp-(shift(temp,1,0)+shift(temp,0,1))/2.0
ENDFOR

sigma_D = correlate(dd,/covariance,/double)

PRINT,'Noise␣covariance␣matrix,␣file␣', $
     file_basename(fname)
PRINT,sigma_D/2

END
```

images, one for a horizontal shift of one pixel, the other for a vertical shift of one pixel.* Here is the calculation for the first 3 bands of the LANDSAT 7 ETM+ image of Figure 3.10:

```
Noise covariance matrix, file juelich_august_29
        7.3732025        7.5563052        11.989391
        7.5563052        10.240908        15.273025
        11.989391        15.273025        28.044577
```

3.6 Exercises

1. Show for $g(x) = \sin(2\pi x)$ in Equation 3.1, that the corresponding frequency coefficients in Equation 3.4 are given by

$$\hat{g}(-1) = -\frac{1}{2\mathrm{i}}, \quad \hat{g}(1) = \frac{1}{2\mathrm{i}},$$

   and $\hat{g}(k) = 0$ otherwise.

2. Demonstrate Equation 3.6 with the help of Equation 3.7.
3. Calculate the discrete Fourier transform of the sequence $2, 4, 6, 8$ from Equation 3.5. You have to solve four simultaneous equations, the first of which is

$$2 = \hat{g}(0) + \hat{g}(1) + \hat{g}(2) + \hat{g}(3).$$

   Verify your result in IDL with the command

```
PRINT, FFT([2,4,6,8])
```

4. Prove the Fourier translation property, Equation 3.11.
5. Derive the discrete form of *Parseval's theorem*,

$$\sum_{k=0}^{c-1} |\hat{g}(k)|^2 = \frac{1}{c}\sum_{j=0}^{c-1} |g(j)|^2,$$

   using the orthogonality property, Equation 3.7.

6. Show that

$$B_2 = \{\phi_{0,0}(x), \psi_{0,0}(x), \psi_{1,0}(x), \psi_{1,1}(x)\},$$

   where

$$\psi_{1,0}(x) = \psi(2x), \quad \psi_{1,1}(x) = \psi(2x-1),$$

   is an orthogonal basis for the subspace $V_1^{\top}$.

---

* This will tend to overestimate the noise in an image in which the signal itself varies considerably. The correlations should be restricted to regions having as little detailed structure as possible. The MNF procedure in ENVI allows spatial subsetting for precisely this reason.

7. Prove Equation 3.27.

8. It can be shown that, for any MRA, $\int_{-\infty}^{\infty} \phi(x)dx \neq 0$. Show that this implies that the refinement coefficients satisfy

$$\sum_k c_k = 2.$$

9. The cubic B-spline wavelet has the refinement coefficients $c_0 = c_4 = 1/8$, $c_1 = c_3 = 1/2$, $c_2 = 3/4$. Use the cascade algorithm of Listing 3.3 to display the scaling function.

10. (a) (Strang and Nguyen, 1997) Given the dilation Equation 3.25 with n nonzero refinement coefficients $c_0 \ldots c_{n-1}$, argue on the basis of the cascade algorithm that the scaling function $\phi(x)$ must be zero outside the interval $[0, n-1]$.

    (b) Prove that $\phi(x)$ is supported on (extends over) the entire interval $[0, n-1]$.

11. The routine in Listing 3.6 simulates two classes of observations B_1 and B_2 in a two-dimensional feature space and calculates the

Listing 3.6

Simulation of two classes of observations.

```
 1 PRO ex3_5
 2 ; generate two classes
 3     n1 = randomu(seed,1000,/normal)
 4     n2 = n1 + randomu(seed,1000,/normal)
 5     B1 = [[n1],[n2]]
 6     B2 = [[n1+4],[n2]]
 7     image = [B1,B2]
 8     center_x = mean(image[*,0])
 9 ; principal components analysis
10     C = correlate(transpose(image),/covariance,/double)
11     void = eigenql(C, eigenvectors=U, /double)
12 ; slopes of the principal axes
13     a1 = U[1,0]/U[0,0]
14     a2 = U[1,1]/U[0,1]
15 ; scatterplot and principal axes
16     thisDevice =!D.Name
17     set_plot, 'PS'
18     Device, Filename='c:\temp\fig3_12.eps',xsize=3, $
19             ysize=3,/inches,/encapsulated
20     PLOT, image[*,0],image[*,1],psym=4,/isotropic
21     oplot, [center_x-10,center_x+10], [a1*(-10),a1*(10)]
22     oplot, [center_x-10,center_x+10], [a2*(-10),a2*(10)]
23     device,/close_file
24     set_plot,thisDevice
25 END
```

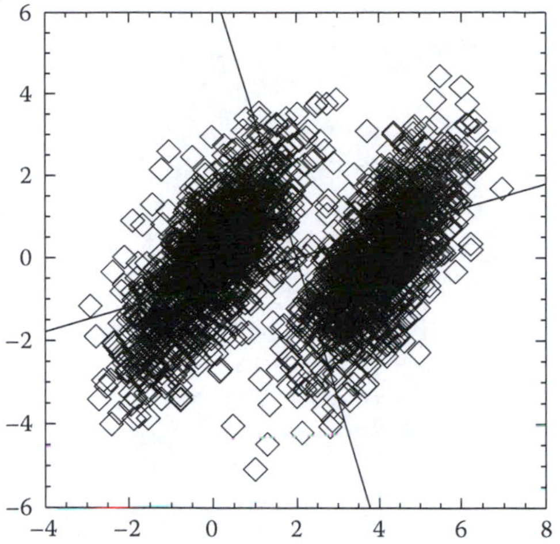

FIGURE 3.12
Two classes of observations in a two-dimensional feature space. The solid lines are the principal axes (see Listing 3.6).

principal axes of the combined data (see Figure 3.12). While the classes are nicely separable in two dimensions, their one-dimensional projections along the x- or y-axes or along either of the principal axes are obviously not. A dimensionality reduction with PCA would thus result in a considerable loss of information about the class structure. *Fisher's linear discriminant* projects the observations $\boldsymbol{g}$ onto a direction w, $v = \boldsymbol{w}^\top \boldsymbol{g}$, such that the ratio $J(w)$ of the squared difference of the class means of the projections v to their overall variance is maximized (Duda and Hart, 1973). Specifically, define

$$\boldsymbol{m}_i = \frac{1}{n_i}\sum_{g\in B_i} \boldsymbol{g}, \quad \boldsymbol{C}_i = \frac{1}{n_i}\sum_{g\in B_i}(\boldsymbol{g}-\boldsymbol{m}_i)(\boldsymbol{g}-\boldsymbol{m}_i)^\top, \quad i = 1,2.$$

(a) Show that the objective function can be written in the form

$$J(w) = \frac{\boldsymbol{w}^\top \boldsymbol{C}_B \boldsymbol{w}}{\boldsymbol{w}^\top(\boldsymbol{C}_1+\boldsymbol{C}_2)\boldsymbol{w}}, \tag{3.79}$$

where $\boldsymbol{C}_B = (\boldsymbol{m}_1 - \boldsymbol{m}_2)(\boldsymbol{m}_1 - \boldsymbol{m}_2)^\top$.

(b) Show that the desired projection direction is given by

$$\boldsymbol{w} = (\boldsymbol{C}_1 + \boldsymbol{C}_2)^{-1}(\boldsymbol{m}_1 - \boldsymbol{m}_2). \tag{3.80}$$

(c) Modify the program in Listing 3.6 to calculate and plot the projection direction $\boldsymbol{w}$.

12. Using the code in Listings 3.4 and 3.5 as a starting point, write an IDL procedure to perform the MAF transformation on a multispectral image.
13. Show from Equation 3.48 that the variance of the principal components is given in terms of the eigenvalues, λ_i, of the Gram matrix by

$$\mathrm{var}(\boldsymbol{w}_i^\top \boldsymbol{g}) = \frac{\lambda_i}{m-1}.$$

14. Formulate the primal and dual problems for the MNF transformation. (*Hint:* Similarly to the case for PCA, Equation 3.47, a dual problem can be obtained by writing $\boldsymbol{a} \propto \mathcal{G}^\top \boldsymbol{\alpha}$.) Write the dual formulation in the form of a symmetric generalized eigenvalue problem. Can it be solved with Cholesky decomposition?

4

Filters, Kernels, and Fields

This chapter is somewhat of a catchall, intended mainly to consolidate and extend material presented in the preceding chapters and to help lay the foundation for the rest of the book. In Sections 4.1 and 4.2, building on the discrete Fourier transform introduced in Chapter 3, the concept of discrete convolution is introduced, and filtering, both in the spatial and in the frequency domain, is discussed. Frequent reference to filtering will be made in Chapter 5 when we treat enhancement, and geometric and radiometric corrections of multispectral imagery. Then, in Section 4.3, it is shown that the discrete wavelet transform of Chapter 3 is equivalent to a recursive application of low- and high-pass filters (a filter bank), and a pyramid algorithm for multiscale image representation is described and programmed in IDL. Pyramid representations are applied in Chapter 5 for panchromatic sharpening and in Chapter 8 for contextual clustering. Section 4.4 then introduces so-called *kernelization*, in which the dual representations of linear problems described in Chapters 2 and 3 can be modified to treat nonlinear data. Kernel methods are illustrated with a nonlinear version of the principal components transformation, and an ENVI/IDL extension, the first of many, is provided to implement it. Kernelization will be met again in Chapter 6 when we consider support vector machines for supervised classification, and in Chapter 8 in the kernel K-means clustering algorithm. The chapter closes (Section 4.5) with a brief introduction to Gibbs–Markov random fields. These will be invoked in Chapter 8 for including spatial context in unsupervised image classification.

4.1 Convolution Theorem

The *convolution* of two continuous functions, $g(x)$ and $h(x)$, denoted by $h * g$, is defined by the integral:

$$(h * g)(x) = \int_{-\infty}^{\infty} h(t)g(x-t)dt. \tag{4.1}$$

This definition is symmetric, that is, $h * g = g * h$, but often one function, $g(x)$ for example, is considered to be a "signal" and the other, $h(x)$, an instrument "response" or *kernel*, which is more local than $g(x)$ and which "smears" the signal according to the above prescription.

In the analysis of digital images, of course, we are dealing with discrete signals. In order to define the discrete analog of Equation 4.1, we will again make reference to a signal consisting of a row of pixels, $g(j)$, $j = 0 \dots c-1$. The discrete convolution kernel is any array of values $h(\ell)$, $\ell = 0 \dots m-1$, where $m < c$. The array h is referred to as a *finite impulse response* (FIR) filter kernel having duration m. The discrete convolution, $f = h * g$, is then defined as

$$f(j) = \begin{cases} \sum_{\ell=0}^{m-1} h(\ell)g(j-\ell) & \text{for } m-1 \le j \le c-1 \\ 0 & \text{otherwise.} \end{cases} \tag{4.2}$$

Discrete convolution can be performed in IDL with the function `CONVOL()` with the keyword `CENTER` explicitly set to zero. The restriction on j in Equation 4.2 is necessary because of edge effects: $g(j)$ is not defined for $j < 0$. This can be circumvented in `CONVOL()` by setting the keyword `EDGE_WRAP` or `EDGE_TRUNCATE`. In the former case, the subscripts are wrapped as if the signal were periodic; in the latter case, the endpoints are repeated for $j < 0$ as often as necessary, thereby allowing the calculation of $f(j)$ for all values of $j = 0 \dots c-1$.

Let us extend the kernel $h(\ell)$ to have the same length, $m = c$, as the signal $g(j)$ by "padding" it with zeroes, that is,

$$h(\ell) = 0, \quad \ell = m \dots c-1.$$

Then we can write Equation 4.2, assuming edge effects have been accommodated, simply as

$$f(j) = \sum_{\ell=0}^{c-1} h(\ell)g(j-\ell), \quad j = 0 \dots c-1. \tag{4.3}$$

The following theorem provides us with a useful alternative to perform this calculation.

THEOREM 4.1 (Convolution Theorem)

In the frequency domain, convolution is replaced by multiplication, that is, $h * g \Leftrightarrow c \cdot \hat{h} \cdot \hat{g}$.

Proof

Taking the Fourier transform of Equation 4.3, we have

$$\hat{f}(k) = \frac{1}{c}\sum_{j=0}^{c-1} f(j)e^{-\mathbf{i}2\pi kj/c} = \frac{1}{c}\sum_{\ell=0}^{c-1} h(\ell)\sum_{j=0}^{c-1} g(j-\ell)e^{-\mathbf{i}2\pi kj/c}.$$

But from the translation property (Equation 3.11),

$$\frac{1}{c}\sum_{j=0}^{c-1} g(j-\ell)e^{-\mathbf{i}2\pi kj/c} = \hat{g}(k)e^{-\mathbf{i}2\pi k\ell/c},$$

therefore,

$$\hat{f}(k) = \sum_{\ell=0}^{c-1} h(\ell)e^{-\mathbf{i}2\pi k\ell/c} \cdot \hat{g}(k) = c \cdot \hat{h}(k) \cdot \hat{g}(k).$$

□

A full statement of the theorem includes the fact that $h \cdot g \Leftrightarrow c \cdot \hat{h} * \hat{g}$, but that need not concern us here. Theorem 4.1 says that we can carry out the convolution operation (Equation 4.3) by

1. Doing a Fourier transform on the signal and on the (padded) filter
2. Multiplying the two transforms together (and multiplying with c)
3. Performing an inverse Fourier transform on the result

The fast Fourier transform, as the name implies, is very fast, and ordinary array multiplication is much faster than convolution. So, depending on the size of the arrays involved, convolving them in the frequency domain may be a better alternative. A pitfall when performing convolution in this fashion has to do with the so-called *wraparound error*. The discrete Fourier transform assumes that both arrays are periodic. This means that the signal might overlap (at the edges) with a preceding or a following period of the kernel, thus falsifying the result. This problem can be avoided by padding *both* arrays to $c+m-1$ (see, e.g., Gonzalez and Woods, 2002, Chapter 4). The two alternative convolution procedures are illustrated in Listing 4.1 and Figure 4.1.

Convolution of a two-dimensional array is a straightforward extension of Equation 4.3. For a two-dimensional kernel, $h(k, \ell)$, which has been appropriately padded, the convolution with a $c \times r$ pixel array, $g(i, j)$, is given by

$$f(i,j) = \sum_{k=0}^{c-1}\sum_{\ell=0}^{r-1} h(k,\ell)g(i-k, j-\ell). \tag{4.4}$$

Listing 4.1
Illustrating convolution in spatial and frequency domains.

```
PRO EX4_1

; get an image band from ENVI
envi_select, title='Choose␣multispectral␣band', $
             fid=fid, dims=dims,pos=pos, /band_only
IF (fid EQ -1) THEN BEGIN
   PRINT, 'cancelled'
   RETURN
ENDIF
num_cols = (c = dims[2]-dims[1]+1)
num_rows = dims[4]-dims[3]+1

; pick out the center row of pixels
image = envi_get_data(fid=fid,dims=dims,pos=pos[0])
g = float(image[*,num_rows/2])

; define a FIR kernel of length m = 5
h = [1,2,3,2,1]

thisDevice = !D.Name
set_plot, 'PS'
Device, FileName = 'fig4_1.eps',xsize=5,ysize=3, $
       /inches,/encapsulated

; convolve in the spatial domain
PLOT, convol(g,h,center=0)

; pad the arrays to c + m - 1
g = [g,[0,0,0,0]]
hp = g*0
hp[0:4] = h

; convolve in the frequency domain
oplot, fft(c*fft(g)*fft(hp),1)-200

Device, /close_file
set_plot,thisDevice

END
```

The convolution theorem now reads

$$h * g \Leftrightarrow c \cdot r \cdot \hat{h} \cdot \hat{g}, \tag{4.5}$$

so that convolution can be carried out in the frequency domain using Fourier and inverse Fourier transforms in two dimensions (Equations 3.8 and 3.9).

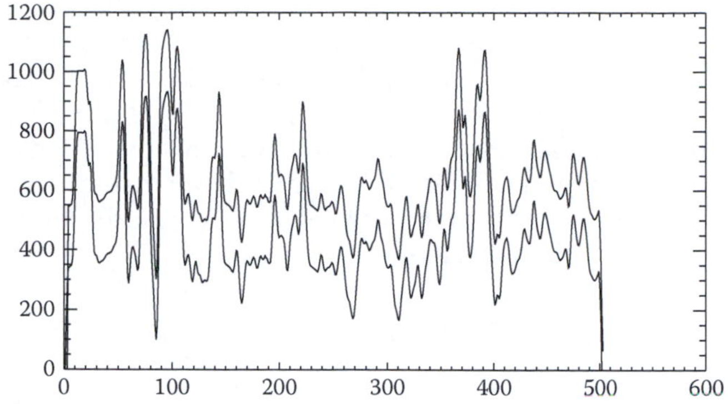

FIGURE 4.1
Illustrating the equivalence of convolution in spatial (upper curve) and frequency (lower curve) domains.

4.2 Linear Filters

Linear filtering of images in the spatial domain generally involves moving a template across the image array, forming some specified linear combination of the pixel intensities within the template, and associating the result with the coordinates of the pixel at the template's center. Specifically, for a rectangular template, h, of dimension $(2m+1) \times (2n+1)$,

$$f(i,j) = \sum_{k=-m}^{m} \sum_{\ell=-n}^{n} h(k,\ell) g(i+k, j+\ell), \tag{4.6}$$

where g represents the original image array and f the filtered result. The similarity of Equation 4.6 to convolution (Equation 4.4) is readily apparent, and spatial filtering can be carried out in the frequency domain if desired. Whether or not the convolution theorem should be used to evaluate Equation 4.6 depends again on the size of the arrays involved. Richards and Jia (2006) give a detailed discussion and calculate a cost factor,

$$F = \frac{m^2}{2\log_2 c + 1},$$

for an $m \times m$ template on a $c \times c$ image. If $F > 1$, it is economical to convolve in the frequency domain.

Linear smoothing templates are usually normalized so that $\sum_{k,\ell} h(k,\ell) = 1$. For example, the 3×3 "weighted average" filter,

$$h = \frac{1}{16}\begin{pmatrix} 1 & 2 & 1 \\ 2 & 4 & 2 \\ 1 & 2 & 1 \end{pmatrix}, \tag{4.7}$$

might be used to suppress uninteresting small details or random noise in an image prior to intensity thresholding in order to identify larger objects. However, the convolution theorem suggests an alternative approach of designing filters in the frequency domain right from the beginning. This is often more intuitive, since suppressing fine details in an image, $g(i,j)$, is equivalent to attenuating high spatial frequencies in its Fourier representation, $\hat{g}(k,\ell)$ (low-pass filtering). Conversely, enhancing details, for instance, for performing edge detection, can be done by attenuating low frequencies in $\hat{g}(k,\ell)$ (high-pass filtering). Both effects are achieved by transforming $g(i,j)$ to $\hat{g}(k,\ell)$, choosing an appropriate form for $\hat{h}(k,\ell)$ in Equation 4.5 *without reference to a spatial filter h*, multiplying the two together, and then doing the inverse transformation. Listing 4.2 illustrates the procedure in the case of a Gaussian filter. Figure 4.2 shows the Gaussian low-pass filter,

$$\hat{h}(k,\ell) = \exp(d^2/\sigma^2), \quad d^2 = \left((k-c/2)^2 + (\ell - r/2)^2\right),$$

generated by the program. The high-pass filter is its complement, $1 - \hat{h}(k,\ell)$. Figures 4.3 and 4.4 display the result of applying them to an image band. We

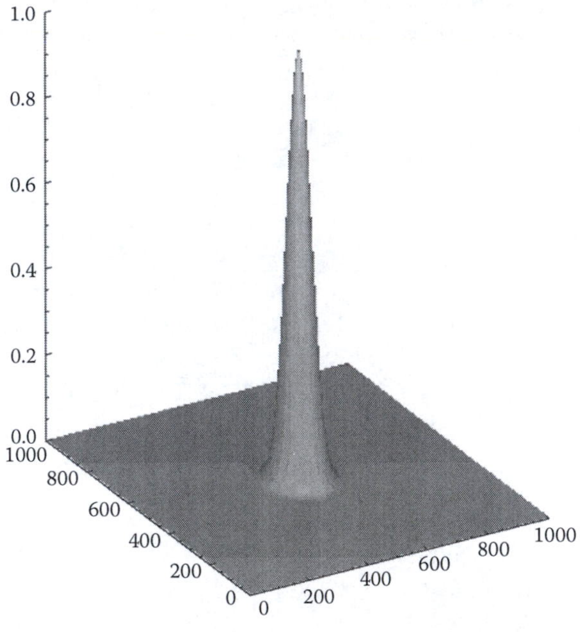

FIGURE 4.2
Gaussian filter in the frequency domain with $\sigma = 50$. Zero frequency is at the center.

Listing 4.2
Illustrating filtering in the frequency domain.

```
PRO EX4_2

envi_select, title='Choose␣multispectral␣band', $
             fid=fid, dims=dims,pos=pos, /band_only
IF (fid EQ -1) THEN BEGIN
   PRINT, 'cancelled' & RETURN
ENDIF

num_cols = dims[2]-dims[1]+1
num_rows = dims[4]-dims[3]+1

; get the image band from ENVI
g = envi_get_data(fid=fid,dims=dims,pos=pos[0])

; transform it
g_hat = fft(g)

; create a Gauss filter in the frequency domain
sigma = 50
d = dist(num_cols,num_rows)
h_hat =  exp(-d^2/sigma^2)

; output surface plot of filter as EPS file
thisDevice =!D.Name
set_plot, 'PS'
Device, Filename='fig4_2.eps',xsize=3,ysize=3, $
         /inches,/encapsulated
shade_surf, shift(h_hat,num_cols/2,num_rows/2)
device,/close_file
set_plot,thisDevice

; multiply, do inverse FFT and return result to ENVI
envi_enter_data, fft(g_hat*h_hat,1)        ; low-pass
envi_enter_data, fft(g_hat*(1-h_hat),1)    ; high-pass

END
```

will return to the subject of low- and high-pass filters in Chapter 5, where we discuss image enhancement.

4.3 Wavelets and Filter Banks

Using the Haar scaling function of Section 3.2.1, we were able to carry out the wavelet transformation in an equivalent vector space. In general, as was

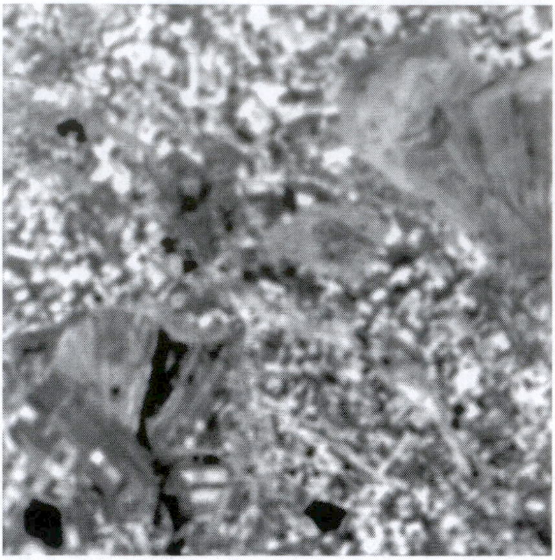

FIGURE 4.3
The 3N band of the Jülich ASTER image after filtering with the low-pass Gaussian filter of Figure 4.2.

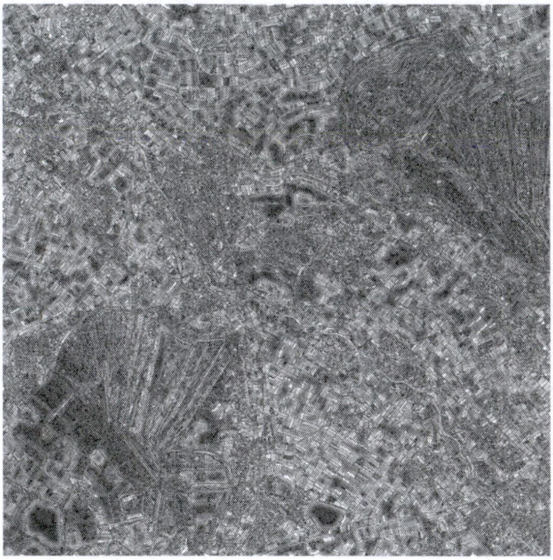

FIGURE 4.4
The 3N band of the Jülich ASTER image after filtering with a high-pass Gaussian filter (complement of Figure 4.2).

pointed out there, one cannot represent scaling functions and wavelets in this way. In fact, usually all that we have to work with are the refinement coefficients of Equation 3.25 or Equation 3.28. So how does one perform the wavelet transformation in this case?

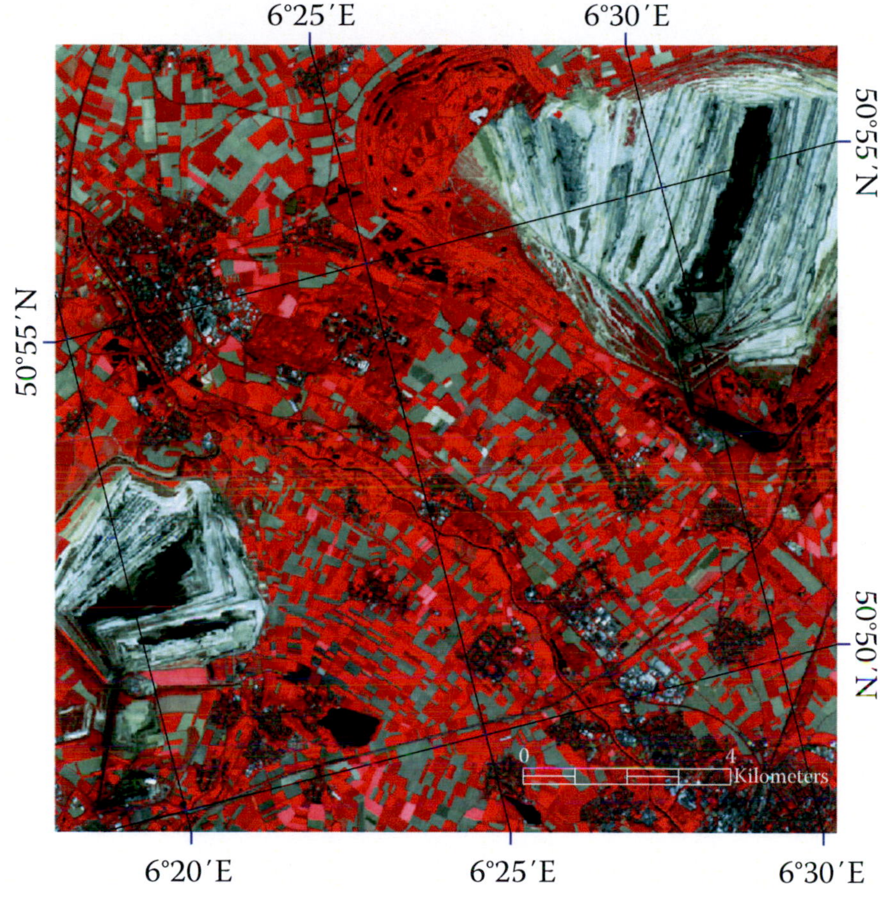

FIGURE 1.1
ASTER color composite image (1000 × 1000 pixels) of VNIR bands 1 (blue), 2 (green), and 3N (red) over the town of Jülich in Germany, acquired on May 1, 2007. The bright areas are open cast coal mines.

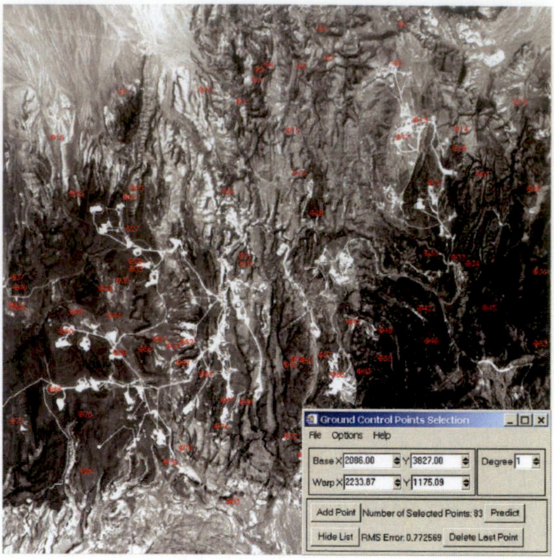

FIGURE 5.29
85 tie-points obtained by matching the contours of Figure 5.28 with those obtained from a similar image acquired in June 2001. The RMS error is 0.77 pixel for first-order polynomial warping (see text).

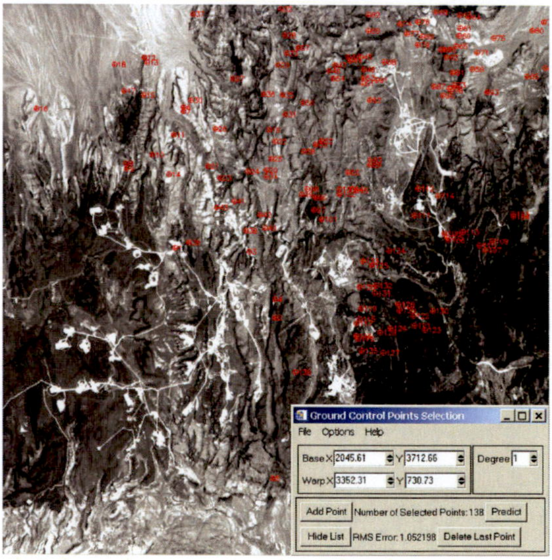

FIGURE 5.30
138 tie-points obtained with ENVI's feature-based matching procedure applied to the July 2003 and June 2001 ASTER scenes. The RMS error is 1.05 pixel for first-order polynomial warping.

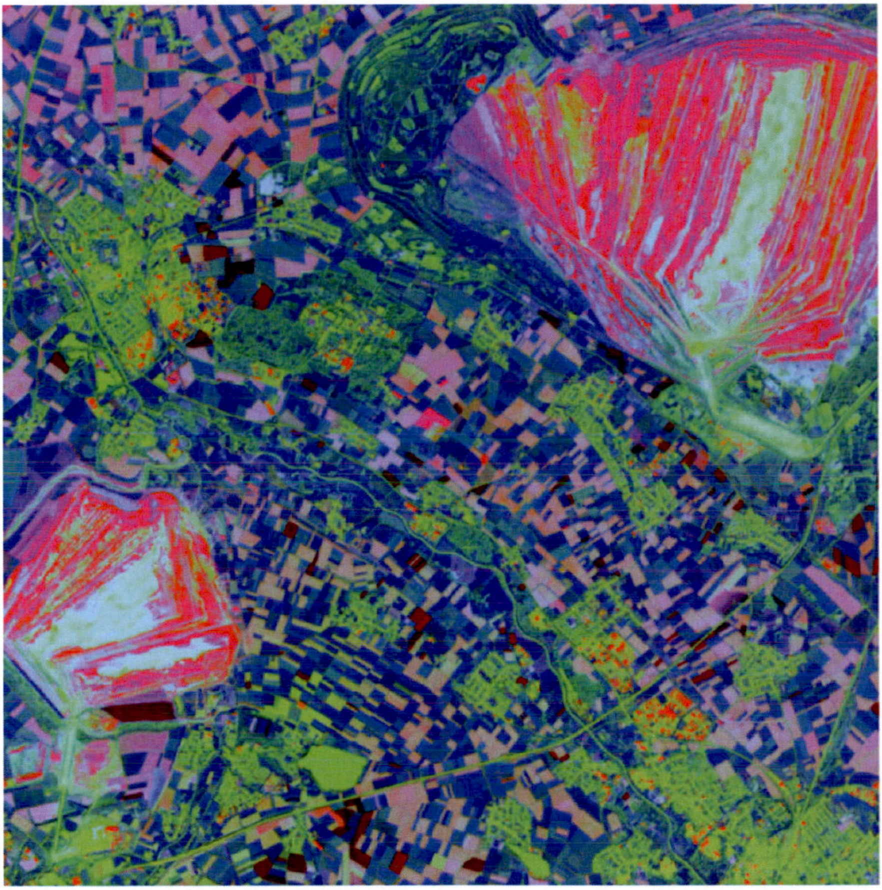

FIGURE 6.1
An RGB color composite (1000 × 1000 pixels, linear 2% saturation stretch) of the first three principal components, 1 (red), 2 (green), and 3 (blue), of the nine nonthermal bands of the ASTER scene acquired over Jülich, Germany, on May 1, 2007.

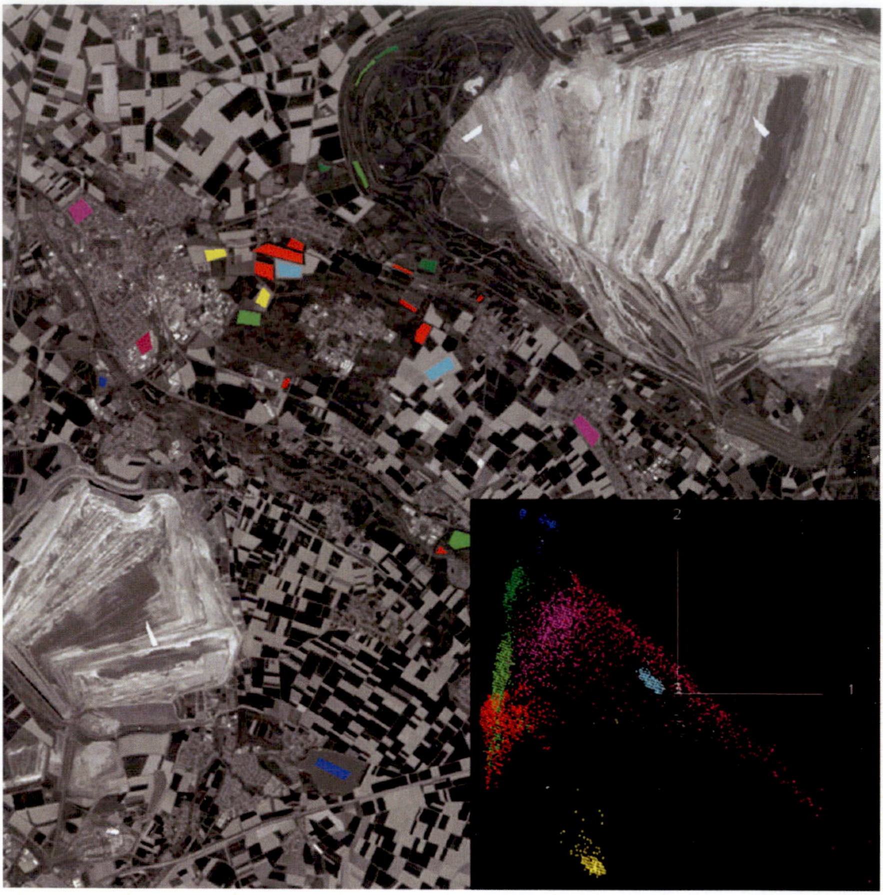

FIGURE 6.3
ROIs for supervised classification. The inset shows the training observations projected onto the plane of the first two principal axes.

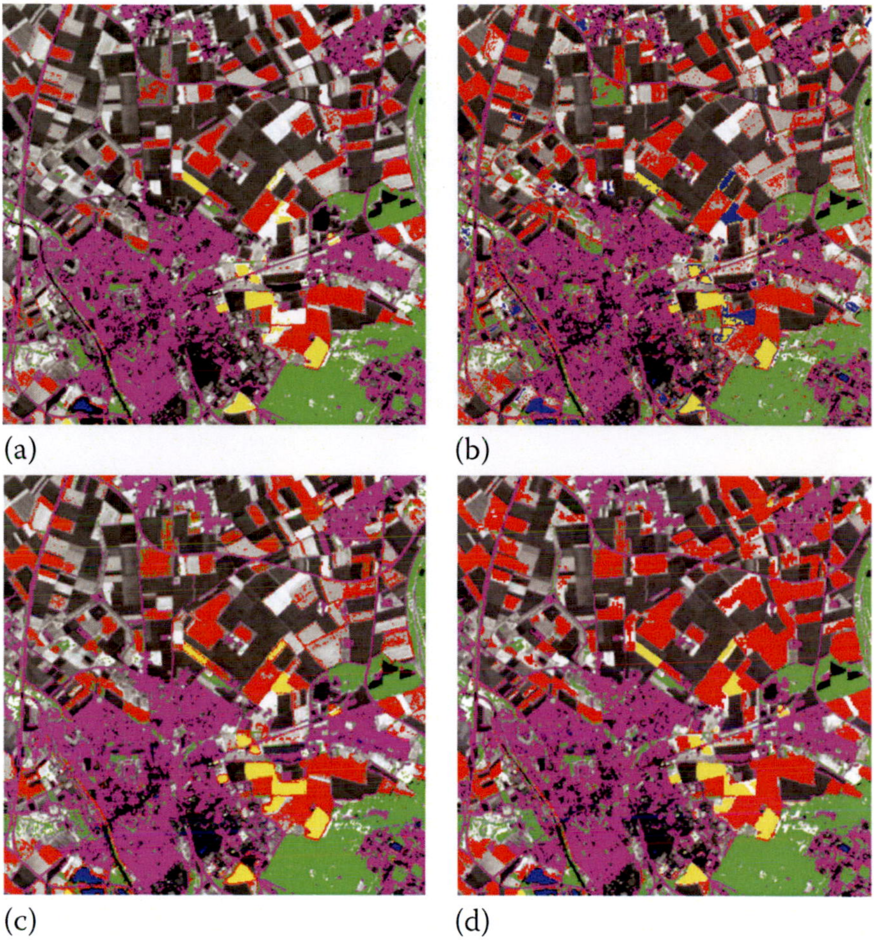

FIGURE 6.4
Supervised classification of a portion of the Jülich ASTER scene: (a) maximum likelihood; (b) Gaussian kernel; (c) neural network; (d) SVM. Five land cover categories (water: blue, settlement: magenta, herbifierous forest: green, rapeseed: yellow, cereal grain: red) are superimposed on VNIR band 3.

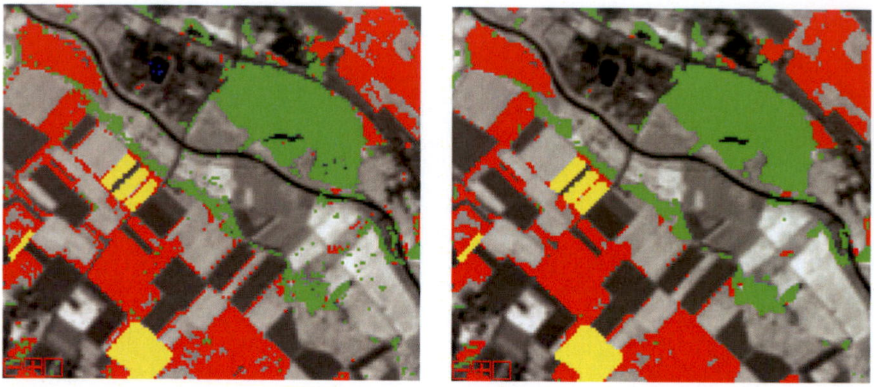

(a) (b)

FIGURE 7.1
An example of postclassification processing. (a) Original classification of a portion of the Jülich ASTER scene with a neural network. The classes shown are coniferous forest (green), rapeseed (yellow), and cereal grain (red). (b) After three iterations of PLR.

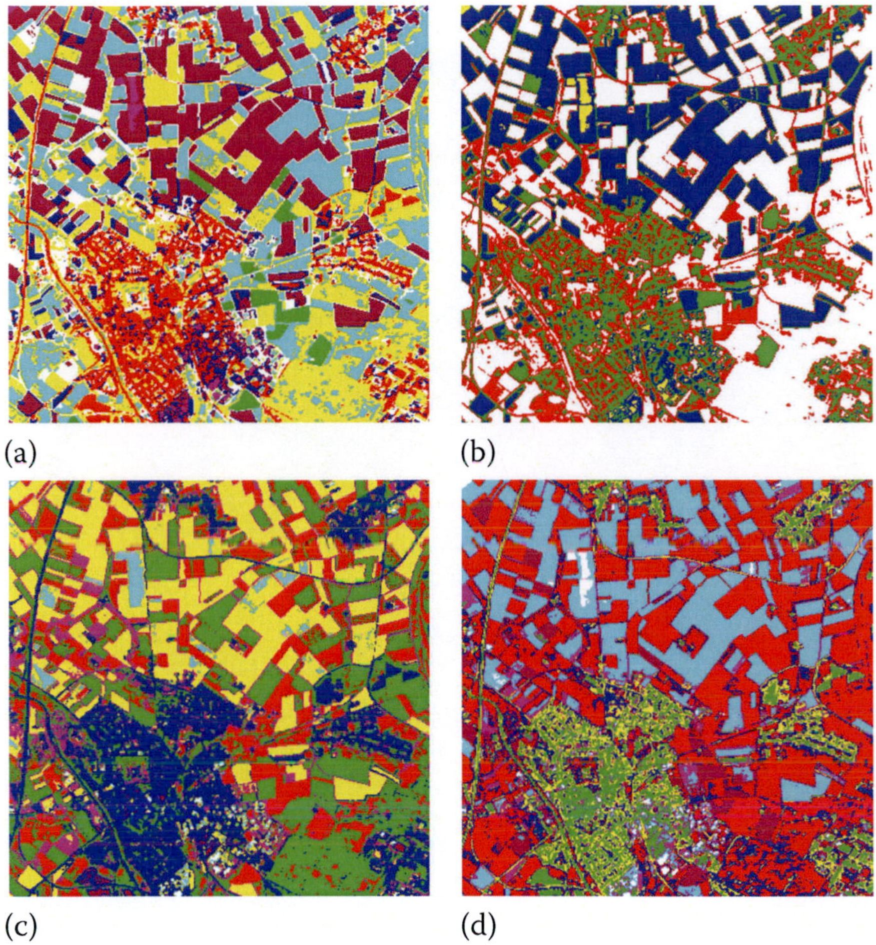

FIGURE 8.1
Unsupervised classification of a spatial subset of the principal components of the fused May 1, 2007 ASTER image over Jülich (Figure 6.1) using eight classes. (a) Kernel K-means clustering on the first four components; (b) extended K-means clustering on the first component with $\tilde{K} = 8$ (see Equation 8.22); (c) agglomerative hierarchical clustering on the first four components; and (d) fuzzy K-means clustering on the first four components.

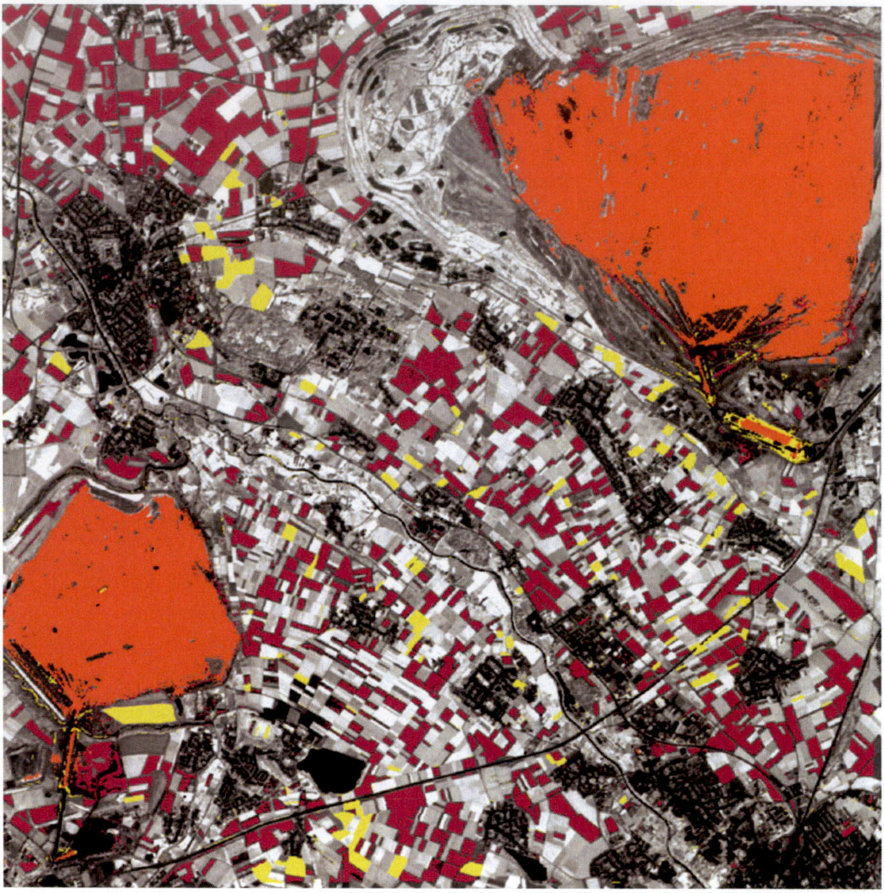

FIGURE 8.2
Gaussian mixture clustering of the first four principal components of the Jülich ASTER scene, eight clusters. Three clusters associated with new sugarbeet plantation (maroon), rapeseed (yellow), and open cast coal mining (coral) are shown superimposed on VNIR spectral band 3N (gray).

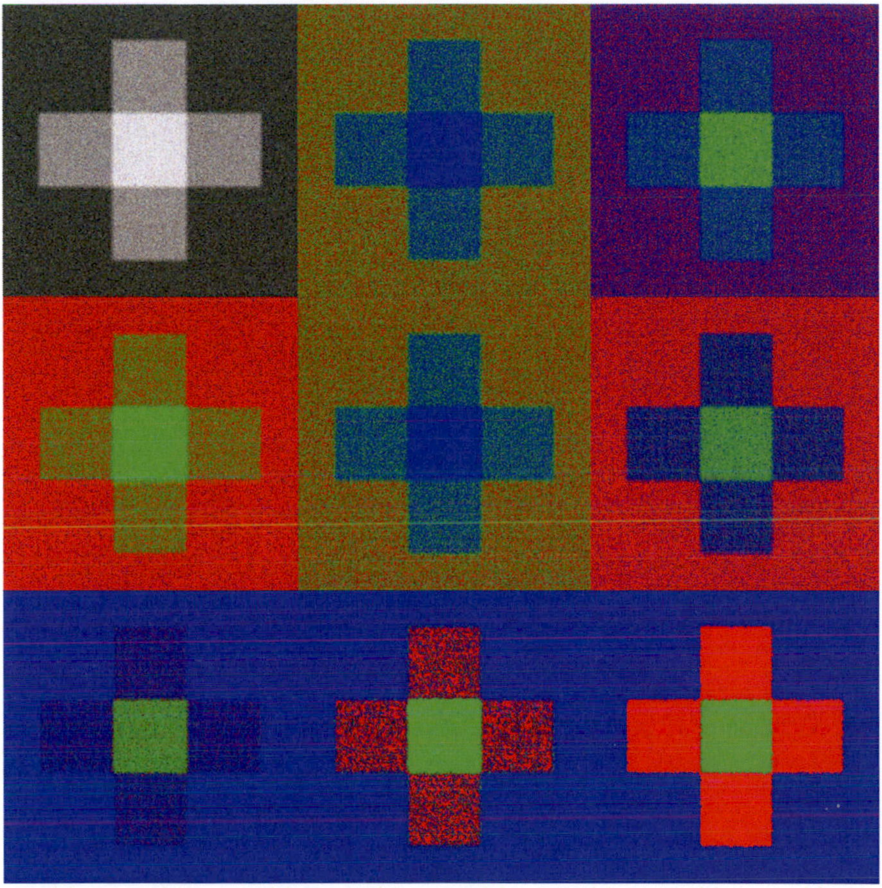

FIGURE 8.3
Unsupervised classification of a toy image. Row-wise, left to right, top to bottom: The toy image, K-means, kernel K-means, EKM (on first principal component), FKM, agglomerative hierarchical, Gaussian mixture with depth 0 and $\beta = 1.0$, Gaussian mixture with depth 2 and $\beta = 0.0$, and Gaussian mixture with depth 2 and $\beta = 1.0$.

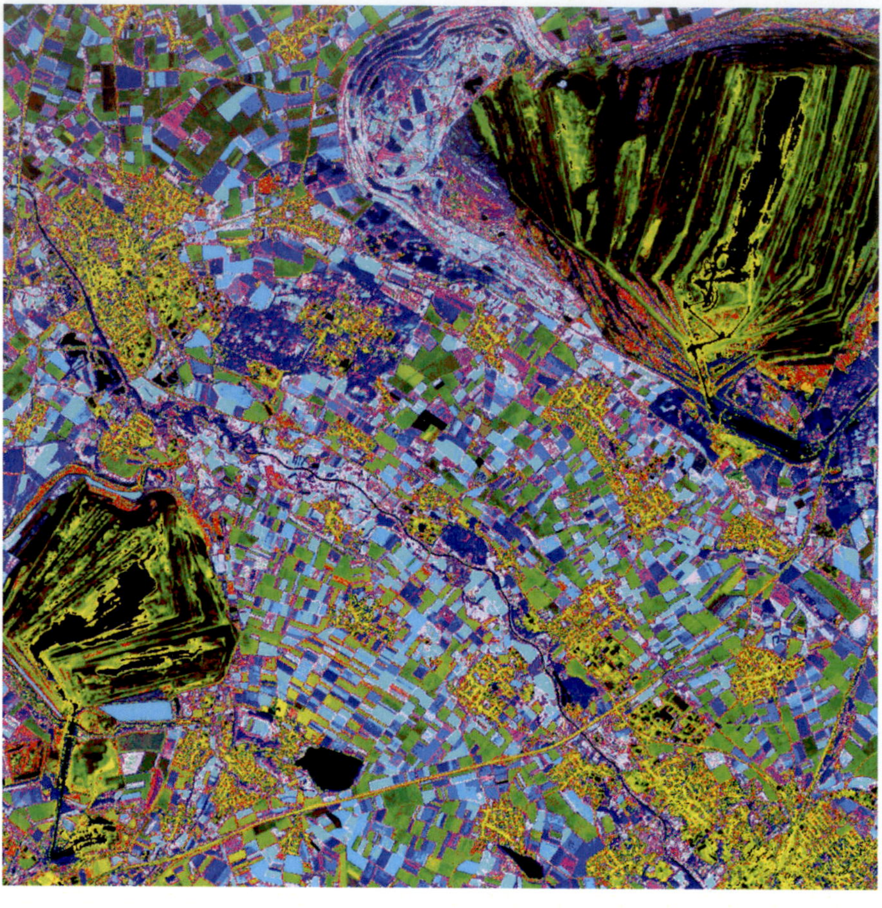

FIGURE 8.6
Kohonen SOM of the nine VNIR and sharpened SWIR spectral bands of the Jülich ASTER scene. The network is a cube having dimensions $6 \times 6 \times 6$.

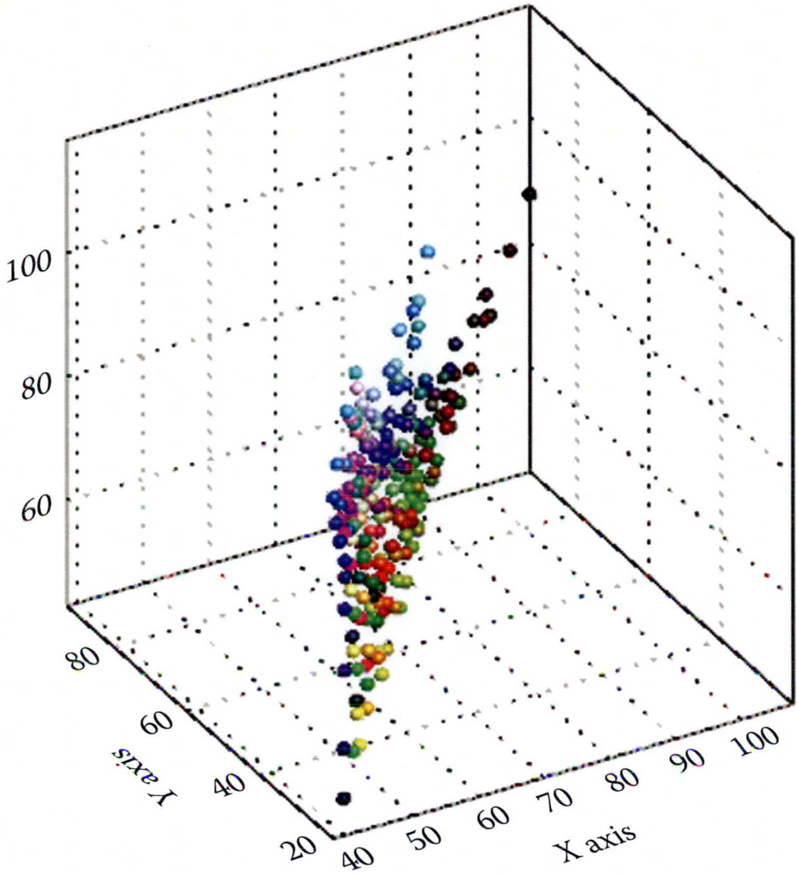

FIGURE 8.7
The $6^3 = 216$ SOM neurons in the feature space of the three VNIR spectral bands of the Jülich scene (see Figure 8.6).

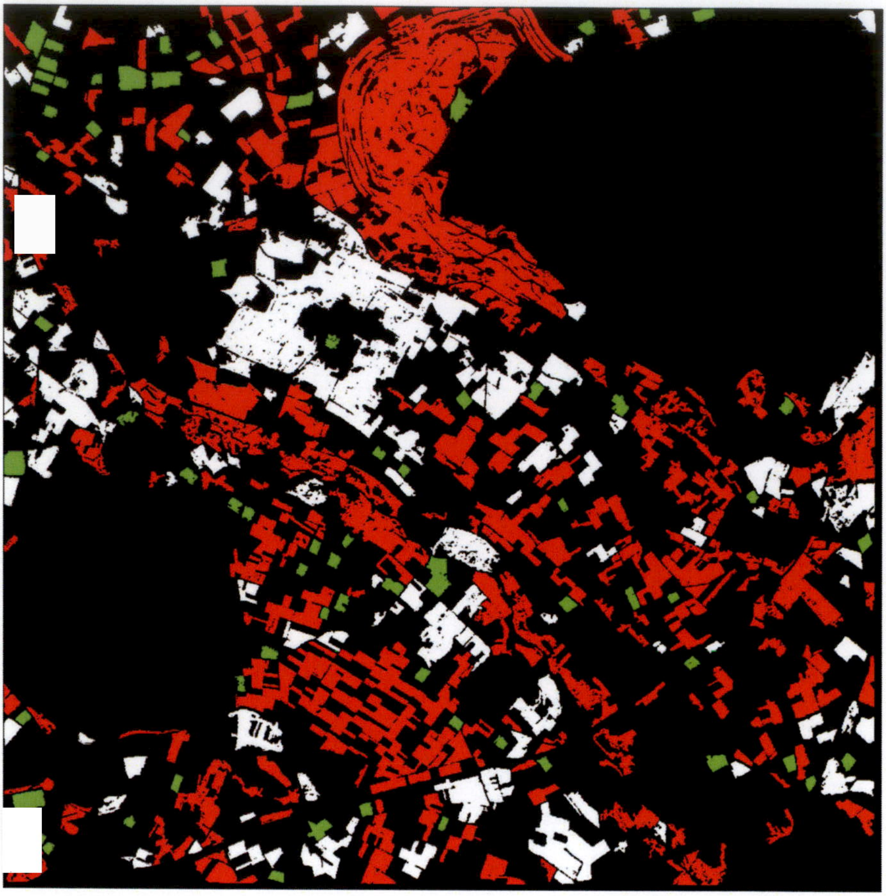

FIGURE 8.9
Classification of the segments in Figure 8.8 on the basis of their invariant moments.

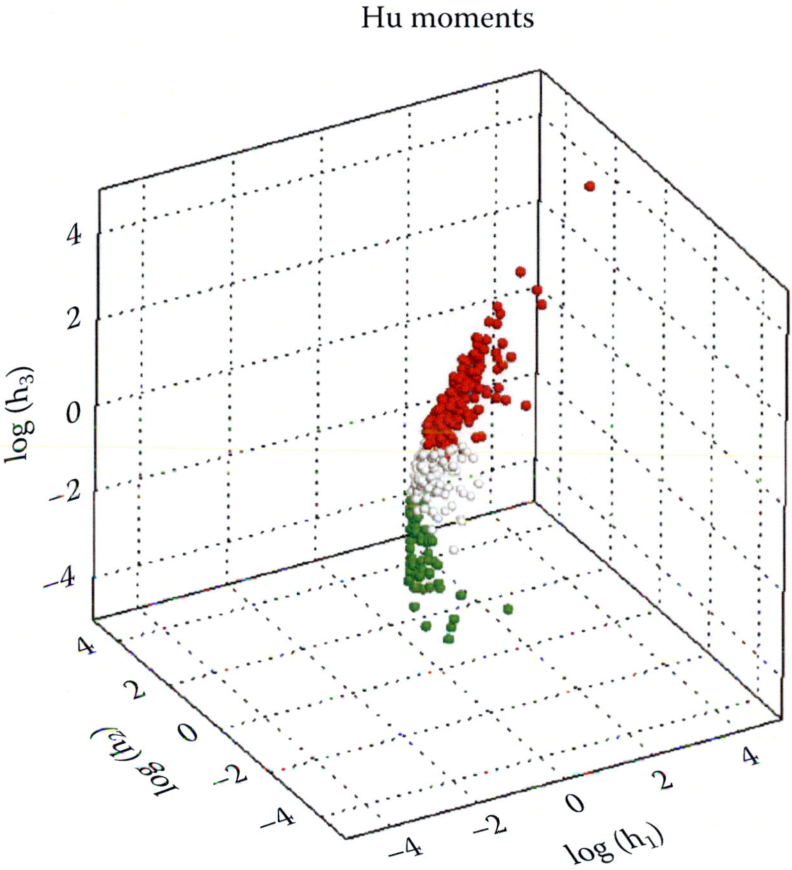

FIGURE 8.10
Standardized logarithms of the first three Hu moments of the classified segments of Figure 8.9.

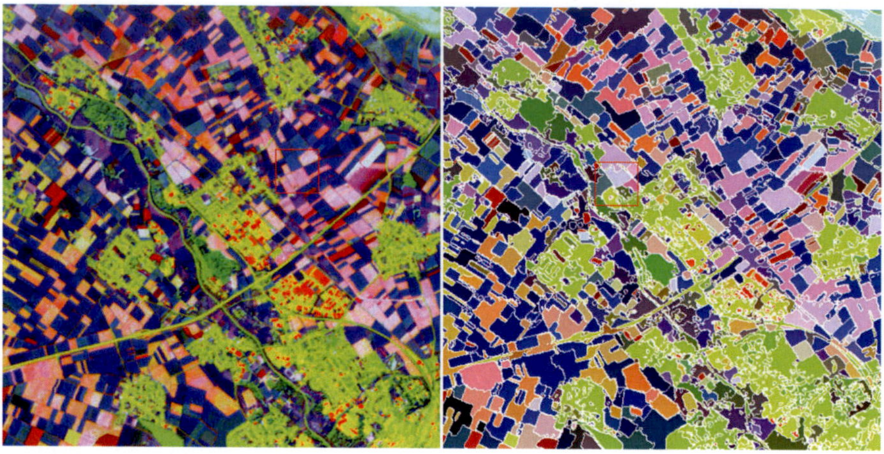

FIGURE 8.11
Left, a spatial subset of the Jülich principal component image of Figure 6.1. Right, the mean shift filtered image and the segment boundaries.

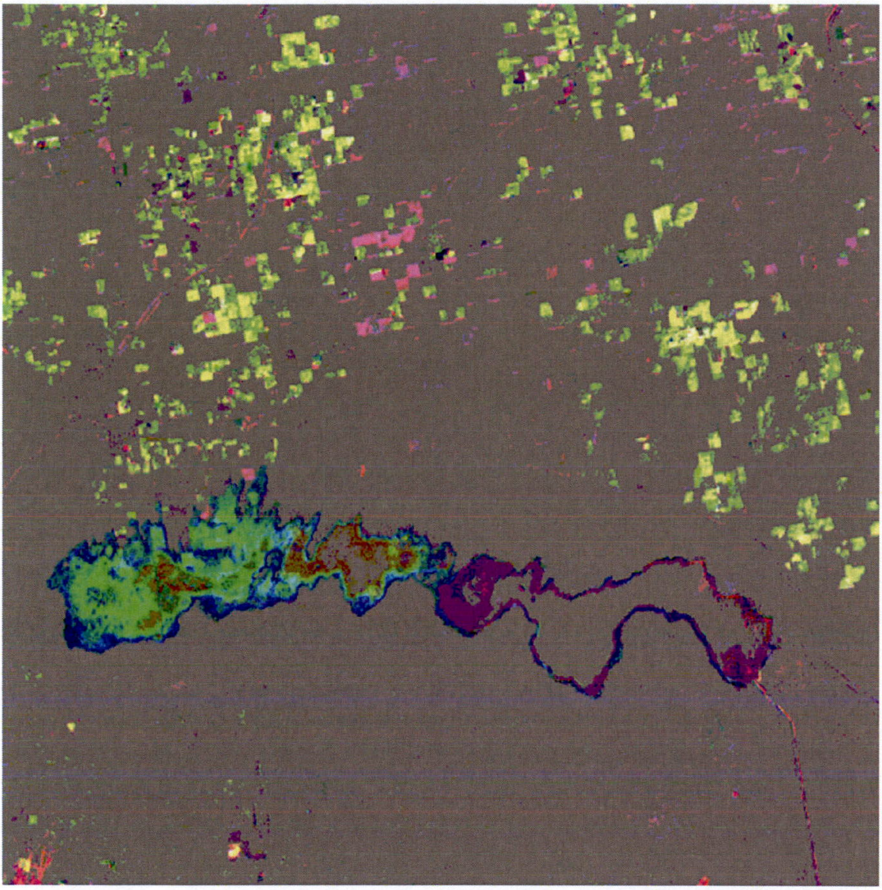

FIGURE 9.15
Color composite of MAD/MNF components 1 (red), 2 (green), and 3 (blue) for the bitemporal image of Figure 9.1, generated with `MAD_VIEW` using decision thresholds obtained from the Gaussian mixture model fit, linear stretch over ±16 standard deviations of the no-change observations.

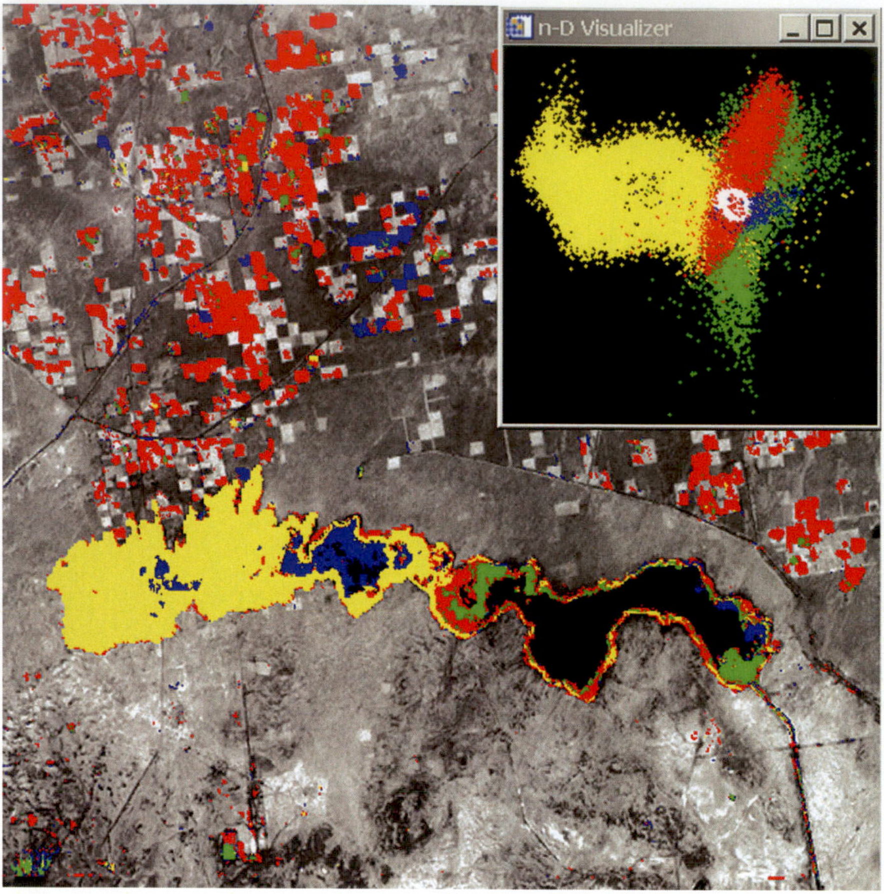

FIGURE 9.16
Unsupervised classification of changes for the reservoir scenes of Figure 9.1. Five clusters were used, including no change. The four change clusters are shown superimposed onto spectral band 5 of the first image. The inset displays the clusters projected onto the MAD/MNF 1–2 plane using ENVI's *n*-dimensional visualizer. The partially obscured white cluster near the center is no change.

4.3.1 One-Dimensional Arrays

To answer this question, consider again a row of pixel intensities in a satellite image, which we now write in the form of a row vector, f*:

$$f = (f_0, f_1 \ldots f_{c-1}),$$

where we assume that $c = 2^n$ for some integer n.

In the multiresolution analysis (MRA) (see Definition 3.2), generated by a scaling function, ϕ, such as the Daubechies D_4 scaling function of Figure 3.9, the signal f defines a function, $f_n(x)$, in the subspace, V_n, on the interval $[0, 1]$ according to

$$f_n(x) = \sum_{j=0}^{c-1} f_j \phi_{n,j}(x) = \sum_{j=0}^{c-1} f_j \phi(2^n x - j). \tag{4.8}$$

Assume that the V_n basis functions, $\phi_{n,j}(x)$, are appropriately normalized:

$$\langle \phi_{n,j}(x), \phi_{n,j'}(x) \rangle = \delta_{j,j'}.$$

Now let us project $f_n(x)$ onto the subspace V_{n-1}, which has a factor of 2 coarser resolution. The projection is given by

$$f_{n-1}(x) = \sum_{k=0}^{c/2-1} \langle f_n(x), \phi(2^{n-1}x - k) \rangle \phi(2^{n-1}x - k) = \sum_{k=0}^{c/2-1} (Hf)_k \phi(2^{n-1}x - k),$$

where we have introduced the quantity

$$(Hf)_k = \langle f_n(x), \phi(2^{n-1}x - k) \rangle, \quad k = 0 \ldots c/2 - 1. \tag{4.9}$$

This is the kth component of a vector Hf representing the row of pixels in V_{n-1}. This notation implies that H is an operator. Its effect is to average the

* Rather than our usual g, in order to avoid confusion with the wavelet coefficients, g_k.

pixel vector f and to reduce its length by a factor of 2. It is thus a kind of a low-pass filter. More specifically, we have from Equation 4.8,

$$(Hf)_k = \sum_{j=0}^{c-1} f_j \langle \phi(2^n x - j), \phi(2^{n-1}x - k) \rangle. \tag{4.10}$$

The dilation equation (Equation 3.28) with normalized basis functions can be written in the form

$$\phi(2^{n-1}x - k) = \sum_{k'} h_{k'} \phi(2^n x - 2k - k'). \tag{4.11}$$

Substituting this into Equation 4.10, we have

$$\begin{aligned}(Hf)_k &= \sum_{j=0}^{c-1} f_j \sum_{k'} h_{k'} \langle \phi(2^n x - j), \phi(2^n x - 2k - k') \rangle \\ &= \sum_{j=0}^{c-1} f_j \sum_{k'} h_{k'} \delta_{j,k'+2k}.\end{aligned}$$

Therefore, the filtered pixel vector has components

$$(Hf)_k = \sum_{j=0}^{c-1} h_{j-2k} f_j, \quad k = 0 \ldots \frac{c}{2} - 1 \ = 2^{n-1} - 1. \tag{4.12}$$

Let us examine this vector in the case of the four nonvanishing refinement coefficients,

$$h_0, h_1, h_2, h_3,$$

of the Daubechies D4 wavelet. The elements of the filtered signal are

$$\begin{aligned}(Hf)_0 &= h_0 f_0 + h_1 f_1 + h_2 f_2 + h_3 f_3 \\ (Hf)_1 &= h_0 f_2 + h_1 f_3 + h_2 f_4 + h_3 f_5 \\ (Hf)_2 &= h_0 f_4 + h_1 f_5 + h_2 f_6 + h_3 f_7 \\ &\vdots\end{aligned}$$

We recognize the above as the convolution $H * f$ of the low-pass filter kernel, $H = (h_3, h_2, h_1, h_0)$ (note the order!), with the vector f (see Equation 4.3), except that *only every second result is retained*. This is referred to as *downsampling* or *decimation*, and is illustrated in Figure 4.5.

FIGURE 4.5
Schematic representation of Equation 4.12. The symbol ↓ indicates downsampling by a factor of 2.

The residual, that is, the difference between the original vector f and its projection Hf onto V_{n-1}, resides in the orthogonal subspace, $V_{n-1}^{\perp}$. It is the projection of $f_n(x)$ onto the wavelet basis:

$$\psi_{n-1,j}(x), \quad j = 0 \ldots 2^{n-1} - 1.$$

A similar argument (Exercise 4) then leads us to the high-pass filter, G, which projects $f_n(x)$ onto $V_{n-1}^{\perp}$ according to

$$(Gf)_k = \sum_{j=0}^{c-1} g_{j-2k} f_j, \quad k = 0 \ldots \frac{c}{2} - 1 = 2^{n-1} - 1. \tag{4.13}$$

The g_k are related to the h_k by Equation 3.33, that is,

$$g_k = (-1)^k h_{1-k},$$

so that the nonzero high-pass filter coefficients are actually

$$g_{-2} = h_3, \; g_{-1} = -h_2, \; g_0 = h_1, \; g_1 = -h_0. \tag{4.14}$$

The *concatenated vector*,

$$(Hf, Gf) = (f^1, d^1), \tag{4.15}$$

is thus the projection of $f_n(x)$ onto $V_{n-1} \oplus V_{n-1}^{\perp}$. It has the same length as the original vector f and is an alternative representation of that vector. Its generation is illustrated in Figure 4.6 as a *filter bank*.

The projections can be repeated on $f^1 = Hf$ to obtain the projection

$$(Hf^1, Gf^1, Gf) = (f^2, d^2, d^1) \tag{4.16}$$

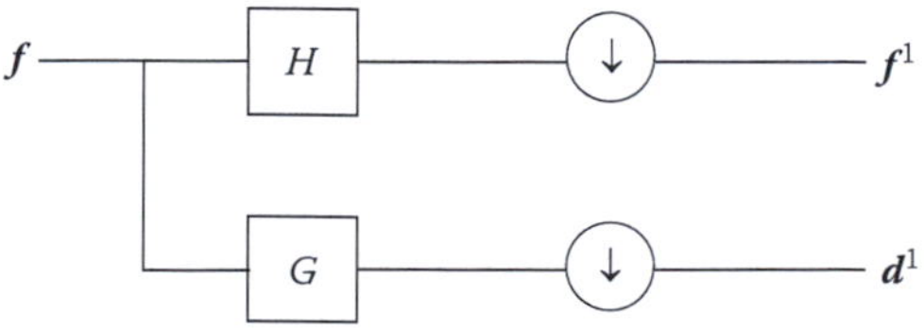

FIGURE 4.6
Schematic representation of the filter bank, H, G.

onto $V_{n-2} \oplus V_{n-2}^{\perp} \oplus V_{n-1}^{\perp}$, and so on, until the complete wavelet transformation has been obtained. Since the filtering process is applied recursively to arrays which are, at each application, reduced by a factor of 2, the procedure is very fast. It constitutes *Mallat's algorithm* (Mallat, 1989), and is also referred to as the *fast wavelet transform*, the *discrete wavelet transform*, or the *pyramid algorithm*.

The original vector can be reconstructed at any stage by applying inverse operators, H^* and G^*. For recovery from $(\boldsymbol{f}^1, \boldsymbol{d}^1)$, these are defined by

$$(H^* \boldsymbol{f}^1)_k = \sum_{j=0}^{c/2-1} h_{k-2j} f_j^1, \quad k = 0 \ldots c-1 = 2^n - 1, \tag{4.17}$$

$$(G^* \boldsymbol{d}^1)_k = \sum_{j=0}^{c/2-1} g_{k-2j} d_j^1, \quad k = 0 \ldots c-1 = 2^n - 1, \tag{4.18}$$

with analogous definitions for the other stages. To understand what is happening, consider the elements of the filtered vector in Equation 4.17. These are

$$\begin{aligned}
(H^* \boldsymbol{f}^1)_0 &= h_0 f_0^1 \\
(H^* \boldsymbol{f}^1)_1 &= h_1 f_0^1 \\
(H^* \boldsymbol{f}^1)_2 &= h_2 f_0^1 + h_0 f_1^1 \\
(H^* \boldsymbol{f}^1)_3 &= h_3 f_0^1 + h_1 f_1^1 \\
(H^* \boldsymbol{f}^1)_4 &= h_2 f_1^1 + h_0 f_2^1 \\
(H^* \boldsymbol{f}^1)_5 &= h_3 f_1^1 + h_1 f_2^1 \\
&\vdots
\end{aligned}$$

This is just the convolution of the filter, $H^* = (h_0, h_1, h_2, h_3)$, with the vector,

$$f_0^1, 0, f_1^1, 0, f_2^1, 0 \ldots f_{c/2-1}^1, 0,$$

which is called an *upsampled* array. The filter of Equation 4.17 is represented schematically in Figure 4.7.

FIGURE 4.7
Schematic representation of the filter H^*. The symbol ↑ indicates upsampling by a factor of 2.

Equation 4.18 is interpreted in a similar way. Finally, we add the two results to get the original pixel vector:

$$H^* \boldsymbol{f}^1 + G^* \boldsymbol{d}^1 = \boldsymbol{f}. \tag{4.19}$$

To see this, write the equation out for a particular value of k:

$$(H^* \boldsymbol{f}^1)_k + (G^* \boldsymbol{d}^1)_k = \sum_{j=0}^{c/2-1} h_{k-2j} \left[\sum_{j'=0}^{c-1} h_{j'-2j} f_{j'} + g_{k-2j} \sum_{j'=0}^{c-1} g_{j'-2j} f_{j'} \right].$$

Combining terms and interchanging the summations, we get

$$(H^* \boldsymbol{f}^1)_k + (G^* \boldsymbol{d}^1)_k = \sum_{j'=0}^{c-1} f_{j'} \sum_{j=0}^{c/2-1} [h_{k-2j} h_{j'-2j} + g_{k-2j} g_{j'-2j}].$$

Now, using $g_k = (-1)^k h_{1-k}$,

$$(H^* \boldsymbol{c}^1)_k + (G^* \boldsymbol{d}^1)_k = \sum_{j'=0}^{c-1} f_{j'} \sum_{j=0}^{c/2-1} [h_{k-2j} h_{j'-2j} + (-1)^{k+j'} h_{1-k+2j} h_{1-j'+2j}].$$

With the help of Equations 3.26 and 3.27, it is easy to show that the second summation above is just $\delta_{j'k}$. For example, suppose k is even. Then,

$$\sum_{j=0}^{c/2-1} [h_{k-2j} h_{j'-2j} + (-1)^{k+j'} h_{1-k+2j} h_{1-j'+2j}]$$

$$= h_0 h_{j'-k} + h_2 h_{j'-k+2} + (-1)^{j'} [h_1 h_{1-j'+k} + h_3 h_{3-j'+k}].$$

If $j' = k$, the right-hand side reduces to

$$h_0^2 + h_1^2 + h_2^2 + h_3^2 = 1,$$

from Equation 3.26 and the fact that $h_k = c_k/\sqrt{2}$. For any other value of j', the expression is zero. Therefore, we can write

$$(H^* \boldsymbol{f}^1)_k + (G^* \boldsymbol{d}^1)_k = \sum_{j'=0}^{c-1} f_{j'} \delta_{j'k} = f_k, \quad k = 0 \ldots c-1, \tag{4.20}$$

as claimed. The reconstruction of the original vector from $\boldsymbol{s}^1$ and $\boldsymbol{d}^1$ is shown in Figure 4.8 as a *synthesis bank*.

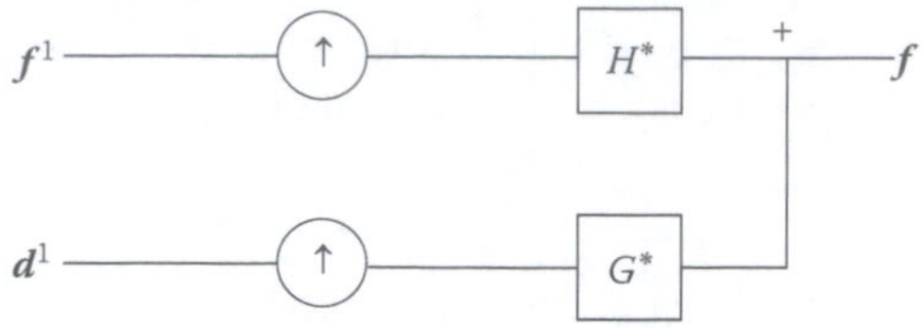

FIGURE 4.8
Schematic representation of the synthesis bank H^*, G^*.

4.3.2 Two-Dimensional Arrays

The extension of the procedure to two-dimensional arrays is straightforward. Figure 4.9 shows an application of the filters H and G to the rows and columns of an image array, $f_n(i,j)$, at scale n. The image is filtered and downsampled into four quadrants:

- f_{n-1}: the result of applying the low-pass filter, H, to both the rows and the columns plus downsampling
- C^H_{n-1}: the result of applying the low-pass filter, H, to the columns and then the high-pass filter, G, to the rows plus downsampling
- C^V_{n-1}: the result of applying the high-pass filter, G, to the columns and then the low-pass filter, H, to the rows plus downsampling
- C^D_{n-1}: the result of applying the high-pass filter, G, to both the columns and the rows plus downsampling

The original image can be losslessly recovered by inverting the filter, as in Figure 4.8.

The filter bank of Figure 4.9 (and the corresponding synthesis filter bank) is implemented in IDL as the object class `DWT`, which is described in Appendix C.2.2. Listing 4.3 illustrates its application to an image band using the Daubechies D4 MRA. The images generated are shown in Figures 4.10 and 4.11. Thus `DWT` produces *pyramid representations* of image bands, which occupy the same amount of storage as the original image array. An excerpt from the program is shown in Listing 4.4 in which the forward transformation (the method `COMPRESS`) is implemented. The object class provides similar functionality to IDL built-in function `WV_DWT()` (discrete wavelet

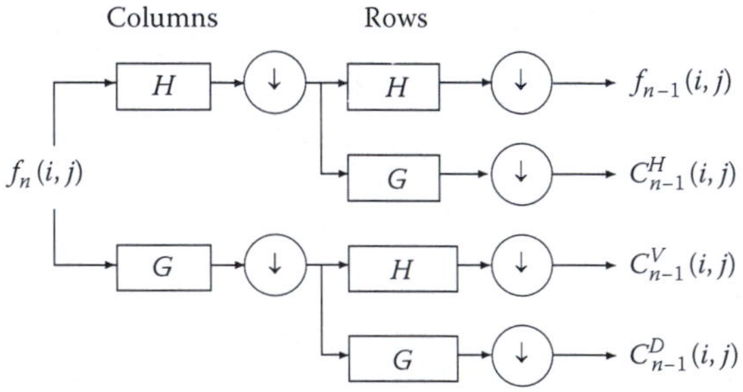

FIGURE 4.9
Wavelet filter bank. H is a low-pass filter and G a high-pass filter derived from the refinement coefficients of the wavelet transformation. The symbol ↓ indicates downsampling by a factor of 2.

FIGURE 4.10
Application of the filter bank of Figure 4.9 to the 3N band of the Jülich ASTER image (see Listing 4.3). The original 1000×1000 image, padded out to 1024×1024 ($1024 = 2^{10}$), is a two-dimensional function, $f(x, y)$, in $V_{10} \otimes V_{10}$. The result of low-pass filtering rows and columns is in the upper left-hand quadrant, the projection of f onto $V_9 \otimes V_9$. The other three quadrants represent the projections onto the orthogonal subspaces $V_9^{\perp} \otimes V_9$ (upper right), $V_9 \otimes V_9^{\perp}$ (lower left), and $V_9^{\perp} \otimes V_9^{\perp}$ (lower right).

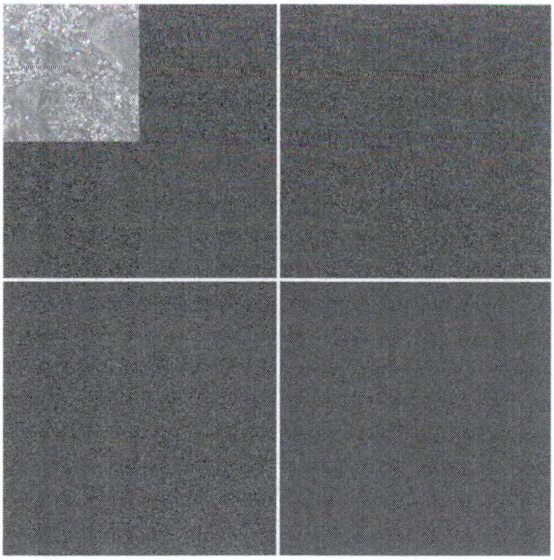

FIGURE 4.11
Recursive application of the filter bank of Figure 4.9 to the upper left quadrant in Figure 4.10 (see Listing 4.3).

Listing 4.3
Application of two iterations of the discrete wavelet transform to an image band with the object class DWT.

```
PRO EX4_3

envi_select, title='Choose␣multispectral␣band', $
             fid=fid, dims=dims,pos=pos, /band_only
IF (fid EQ -1) THEN BEGIN
   PRINT, 'cancelled'
   RETURN
ENDIF

; get the image band from ENVI
g = envi_get_data(fid=fid,dims=dims,pos=pos[0])

; create an instance of the DWT class with image band
aDWT = Obj_New('DWT', g)

; apply the filter bank once
aDWT->compress

; return the result to ENVI
envi_enter_data, aDWT->get_image()

; apply again
aDWT->compress

; return the result to ENVI
envi_enter_data, aDWT->get_image()

; remove the class instance from the heap
Obj_Destroy, aDWT

END
```

transform) with the N_LEVELS keyword, but includes methods for normalization and substitution of wavelet coefficients. DWT will be used in Chapter 5 for image sharpening and in Chapter 8 for unsupervised classification.

4.4 Kernel Methods

In Section 2.6.3, it was shown that regularized linear regression (ridge regression) possesses a dual formulation in which observation vectors $g(\nu)$,

Listing 4.4

The COMPRESS method for the object class DWT.

```
PRO DWT::Compress
IF self.compressions LT self.max_compressions $
THEN BEGIN
; single application of filter bank
   H = *self.H &  G = *self.G
   m = self.num_rows/2^(self.compressions)
   n = self.num_cols/2^(self.compressions)
   f0 = (*self.image)[0:n-1,0:m-1]

; temporary arrays for wavelet coefficients
   f1 = fltarr(n,m/2)
   g1 = f1*0
   ff1 = fltarr(n/2,m/2)
   fg1 = ff1*0 & gf1 = ff1*0 & gg1 = ff1*0

; filter columns and downsample
   ds = indgen(m/2)*2+1
   FOR i=0,n-1 DO BEGIN
      temp = convol(transpose(f0[i,*]),H,/edge_wrap)
      f1[i,*] = temp[ds]
      temp = convol(transpose(f0[i,*]),G,/edge_wrap)
      g1[i,*] = temp[ds]
   ENDFOR

; filter rows and downsample
   ds = indgen(n/2)*2+1
   FOR i=0,m/2-1 DO BEGIN
      temp = convol(f1[*,i],H,/edge_wrap)
      ff1[*,i] = temp[ds]
      temp = convol(f1[*,i],G,/edge_wrap)
      fg1[*,i] = temp[ds]
      temp = convol(g1[*,i],H,/edge_wrap)
      gf1[*,i] = temp[ds]
      temp = convol(g1[*,i],G,/edge_wrap)
      gg1[*,i] = temp[ds]
   ENDFOR
   f0[0:n/2-1,0:m/2-1] = ff1[*,*]
   f0[n/2:n-1,0:m/2-1] = fg1[*,*]
   f0[0:n/2-1,m/2:m-1] = gf1[*,*]
   f0[n/2:n-1,m/2:m-1] = gg1[*,*]
   (*self.image)[0:n-1,0:m-1] = f0
   self.compressions = self.compressions+1
ENDIF
END
```

$\nu = 1 \dots m$, only enter in the form of inner products, $g(\nu)^\top g(\nu')$. A similar dual representation was found in Section 3.3.2 for the principal components transformation. In both cases, the symmetric, positive semi-definite Gram matrix,

$$\mathcal{G}\mathcal{G}^\top \quad (\mathcal{G} = \text{data matrix}),$$

played a central role. It turns out that many linear methods in pattern recognition have dual representations, ones that can be exploited to extend well-established linear theories to treat nonlinear data. One speaks in this context of *kernel methods* or *kernelization*. An excellent reference for kernel methods is Shawe-Taylor and Cristianini (2004). In the following, we outline the basic ideas and then illustrate them with a nonlinear, or kernelized, version of principal components analysis.

4.4.1 Valid Kernels

Suppose that $\boldsymbol{\phi}(g)$ is some nonlinear function that maps the original N-dimensional Euclidean input space of the observation vectors g (image pixels) to some nonlinear (usually higher-dimensional) inner product space $\mathcal{H}$,

$$\boldsymbol{\phi} : \mathbb{R}^N \mapsto \mathcal{H}. \tag{4.21}$$

The mapping takes a data matrix, $\mathcal{G}$, into a new data matrix, Φ, given by

$$\Phi = \begin{pmatrix} \phi(g(1))^\top \\ \phi(g(2))^\top \\ \vdots \\ \phi(g(m))^\top \end{pmatrix}. \tag{4.22}$$

This matrix has a dimension $m \times P$, where $P \geq N$ is the (possibly infinite) dimension of the nonlinear feature space, $\mathcal{H}$. We have the following definition:

Definition 4.1: A *valid kernel* is a function κ that, for all $g, g' \in \mathbb{R}^N$, satisfies

$$\kappa(g, g') = \phi(g)^\top \phi(g'), \tag{4.23}$$

where ϕ is given by Equation 4.21. For a set of observations, $g(\nu)$, $\nu = 1 \dots m$, the $m \times m$ matrix, $\mathcal{K}$, with elements $\kappa(g(\nu), g(\nu'))$ is called the *kernel matrix*.

Note that with Equation 4.22, we can write the *kernel matrix* equivalently in the form

$$\mathcal{K} = \boldsymbol{\Phi}\boldsymbol{\Phi}^\top. \tag{4.24}$$

In Section 2.6.3, we saw that the Gram matrix is symmetric and positive semi-definite. This is also the case for kernel matrices since, for any m-component vector $\boldsymbol{x}$,

$$\begin{aligned}
\boldsymbol{x}^\top \mathcal{K}\boldsymbol{x} &= \sum_{\nu,\nu'} x_\nu x_{\nu'} \mathcal{K}_{\nu\nu'} \\
&= \sum_{\nu,\nu'} x_\nu x_{\nu'} \boldsymbol{\phi}_\nu^\top \boldsymbol{\phi}_{\nu'} \\
&= \left(\sum_\nu x_\nu \boldsymbol{\phi}_\nu\right)^\top \left(\sum_{\nu'} x_{\nu'} \boldsymbol{\phi}_{\nu'}\right) \\
&= \left\| \sum_\nu x_\nu \boldsymbol{\phi}_\nu \right\|^2 \geq 0.
\end{aligned}$$

Positive semi-definiteness of the kernel matrix is in fact a necessary and sufficient condition for any symmetric function $\kappa(\boldsymbol{g},\boldsymbol{g}')$ to be a valid kernel in the sense of Definition 4.1. This is stated in Theorem 4.2 (Shawe-Taylor and Cristianini, 2004).

THEOREM 4.2

Let $k(\boldsymbol{g},\boldsymbol{g}')$ be a symmetric function on the space of observations, that is, $k(\boldsymbol{g},\boldsymbol{g}') = k(\boldsymbol{g}',\boldsymbol{g})$. Let $\{\boldsymbol{g}(\nu) \mid \nu = 1 \ldots m\}$ be any finite subset of the input space $\mathbb{R}^N$ and define the matrix $\mathcal{K}$ with elements

$$(\mathcal{K})_{\nu\nu'} = k(\boldsymbol{g}(\nu), \boldsymbol{g}(\nu')), \quad \nu, \nu' = 1 \ldots m.$$

Then $k(\boldsymbol{g},\boldsymbol{g}')$ is a valid kernel if and only if $\mathcal{K}$ is positive semi-definite.

The motivation for using valid kernels is that it allows us to apply known linear methods to nonlinear data simply by replacing the inner products in the dual formulation by an appropriate nonlinear, valid kernel. One implication of Theorem 4.2 is that one can obtain valid kernels without even specifying the nonlinear mapping $\boldsymbol{\phi}$ at all. In fact, it is possible to build up whole families of valid kernels in which the associated mappings are defined implicitly and are otherwise unknown. An important example, one which we will use shortly to illustrate kernel methods, is the *Gaussian kernel* given by

$$\kappa_{\text{rbf}}(\boldsymbol{g},\boldsymbol{g}') = \exp(-\gamma \|\boldsymbol{g} - \boldsymbol{g}'\|^2). \tag{4.25}$$

This is an instance of a *homogeneous kernel*, also called a *radial basis kernel*, one which depends only on the Euclidean distance between the observations. It

Listing 4.5
Calculating a kernel matrix.

```
FUNCTION gausskernel_matrix,G1,G2,GMA=gma,NSCALE=nscale
   IF n_params() EQ 1 THEN G2 = G1
   IF n_elements(nscale) EQ 0 THEN nscale = 1.0
   m = n_elements(G1[0,*])
   n = n_elements(G2[0,*])
   K = transpose(total(G1^2,1))##(intarr(n)+1)
   K = K + transpose(intarr(m)+1)##total(G2^2,1)
   K = K - 2*G1##transpose(G2)
   IF n_elements(gma) EQ 0 THEN BEGIN
     scale = total(sqrt(abs(K)))/(m^2-m)
     gma = 1/(2*(nscale*scale)^2)
   ENDIF
   RETURN, exp(-gma*K)
END
```

is equivalent to the inner product of two infinite-dimensional feature vector mappings, $\boldsymbol{\phi}(g)$ (Exercise 7). Listing 4.5 shows an IDL function to calculate a Gaussian kernel matrix (see Appendix C.2.1).

If we wish to make use of kernels that, like the Gaussian kernel, imply nonlinear mappings, $\boldsymbol{\phi}(g)$, to which we have no direct access, then clearly we must work exclusively with inner products

$$\kappa(g,g') = \phi(g)^\top \phi(g').$$

Apart from dual formulations that only involve these inner products, it turns out that many other properties of the mapped observations, $\boldsymbol{\phi}(g)$, can also be expressed purely in terms of the elements $\kappa(g,g')$ of the kernel matrix. Thus, the norm or the length of the mapping of g in the nonlinear feature space, $\mathcal{H}$, is

$$\|\phi(g)\| = \sqrt{\phi(g)^\top \phi(g)} = \sqrt{\kappa(g,g)}. \tag{4.26}$$

The squared distance between any two points in $\mathcal{H}$ is given by

$$\begin{aligned}\|\phi(g) - \phi(g')\|^2 &= (\phi(g) - \phi(g'))^\top (\phi(g) - \phi(g')) \\ &= \phi(g)^\top \phi(g) + \phi(g')^\top \phi(g') - 2\phi(g)^\top \phi(g') \\ &= \kappa(g,g) + \kappa(g',g') - 2\kappa(g,g').\end{aligned} \tag{4.27}$$

The $1 \times P$ row vector, $\bar{\phi}^\top$, of column means of the mapped data matrix, $\mathbf{\Phi}$, (Equation 4.22) can be written in a matrix form as

$$\bar{\phi}^\top = \mathbf{1}_m^\top \Phi / m, \tag{4.28}$$

where $\mathbf{1}_m$ is a column vector of m ones. The norm of the mean vector, $\bar{\boldsymbol{\phi}}$, is thus

$$\|\bar{\phi}\| = \sqrt{\bar{\phi}^\top \bar{\phi}} = \frac{1}{m}\sqrt{\mathbf{1}_m^\top \mathbf{\Phi}\mathbf{\Phi}^\top \mathbf{1}_m} = \frac{1}{m}\sqrt{\mathbf{1}_m^\top \mathcal{K} \mathbf{1}_m}. \tag{4.29}$$

We are even able to determine the elements of the kernel matrix, $\tilde{\mathcal{K}}$, which corresponds to a column-centered data matrix, $\tilde{\mathbf{\Phi}}$ (see Section 2.3.1). The rows of $\tilde{\mathbf{\Phi}}$ are

$$\tilde{\phi}(g(\nu)) = (\phi(g(\nu)) - \bar{\phi})^\top = \phi(g(\nu))^\top - \bar{\phi}^\top, \quad \nu = 1 \ldots m. \tag{4.30}$$

With Equation 4.28, the $m \times P$ matrix of m repeated column means of $\mathbf{\Phi}$ is given by

$$\begin{pmatrix} \bar{\phi}^\top \\ \bar{\phi}^\top \\ \vdots \\ \bar{\phi}^\top \end{pmatrix} = \mathbf{1}_{mm} \mathbf{\Phi} / m,$$

where $\mathbf{1}_{mm}$ is an $m \times m$ matrix of ones, so that Equation 4.30 can be written in the matrix form:

$$\tilde{\mathbf{\Phi}} = \mathbf{\Phi} - \mathbf{1}_{mm} \mathbf{\Phi} / m.$$

Therefore,

$$\begin{aligned} \tilde{\mathcal{K}} = \tilde{\mathbf{\Phi}}\tilde{\mathbf{\Phi}}^\top &= (\mathbf{\Phi} - \mathbf{1}_{mm}\mathbf{\Phi}/m)(\mathbf{\Phi} - \mathbf{1}_{mm}\mathbf{\Phi}/m)^\top \\ &= \mathbf{\Phi}\mathbf{\Phi}^\top - \mathbf{\Phi}\mathbf{\Phi}^\top \mathbf{1}_{mm}/m - \mathbf{1}_{mm}\mathbf{\Phi}\mathbf{\Phi}^\top/m - 1_{mm}\mathbf{\Phi}\mathbf{\Phi}^\top \mathbf{1}_{mm}/m^2 \\ &= \mathcal{K} - \mathcal{K}\mathbf{1}_{mm}/m - \mathbf{1}_{mm}\mathcal{K}/m + \mathbf{1}_{mm}\mathcal{K}\mathbf{1}_{mm}/m^2. \end{aligned} \tag{4.31}$$

Examining this equation componentwise, we see that, to center the kernel matrix, we subtract from each element $(\mathcal{K})_{ij}$ the mean of the ith row and the mean of the jth column, and add to it the mean of the entire matrix. IDL function CENTER(), shown in Listing 4.6, takes a kernel matrix, $\mathcal{K}$, as an input and returns the column-centered kernel matrix, $\tilde{\mathcal{K}}$.

4.4.2 Kernel PCA

In Section 3.3.2, it was shown that, in the dual formulation of the principal components transformation, the projection $P_i[g]$ of an observation, g, along

Listing 4.6
Centering a kernel matrix.

```
FUNCTION center, K
   m = (size(K))[1]
   ones=dblarr(m,m)+1
   RETURN, K-(ones##K+K##ones-total(K)/m)/m
END
```

a principal axis, w_i, can be expressed as

$$P_i[g] = w_i^\top g = \sum_{\nu=1}^{m} (\alpha_i)_\nu \, g(\nu)^\top g, \quad i = 1 \dots N. \tag{4.32}$$

In this equation, the dual vectors $\boldsymbol{\alpha}_i$ are determined by the eigenvectors v_i and the eigenvalues λ_i of the Gram matrix, $\mathcal{G}\mathcal{G}^\top$, according to

$$\boldsymbol{\alpha}_i = \lambda^{-1/2} v_i, \quad i = 1 \dots m. \tag{4.33}$$

Kernelization of principal components analysis simply involves replacing the Gram matrix, $\mathcal{G}\mathcal{G}^\top$, by the (centered) kernel matrix, $\tilde{\mathcal{K}}$ (Equation 4.31), and the inner products, $g(\nu)^\top g$, in Equation 4.32, by the corresponding kernel function. The projection along the ith principal axes in the nonlinear feature space, $\mathcal{H}$ (the ith nonlinear principal component), is then

$$P_i[\phi(g)] = \sum_{\nu=1}^{m} (\alpha_i)_\nu \kappa(g(\nu), g), \quad i = 1 \dots m, \tag{4.34}$$

where the dual vectors $\boldsymbol{\alpha}_i$ are still given by Equation 4.33, but v_i and λ_i, $i = 1 \dots m$, are the eigenvectors and the eigenvalues of the kernel matrix, $\tilde{\mathcal{K}}$. The variance of the projections is given by (see Chapter 3, Exercise 13)

$$\mathrm{var}(P_i[\phi(g)]) = \frac{\lambda_i}{m-1}, \quad i = 1 \dots m.$$

Schölkopf et al. (1998) were the first to introduce kernel PCA, and Shawe-Taylor and Cristianini (2004) analyzed the method in detail. In the context of remote sensing image analysis, the number of observations is in general very large (order 10^6–10^8), so that diagonalization of the kernel matrix is only feasible if the image is sampled. The sampled pixel vectors, $g(\nu)$, $\nu = 1 \dots m$, are then referred to as *training data*, and the calculation of the centered kernel matrix and its diagonalization constitute the *training phase*. A realistic upper limit on m would appear to be about 2000. (The diagonalization of a 2000×2000 symmetric matrix on, for example, a PC workstation

Listing 4.7
Excerpt from the program GPUKPCA_RUN.PRO.

```
122 seed = 12345L
123 indices = randomu(seed,m,/long) MOD num_pixels
124 G = GG[*,indices]
125
126 start_time = systime(2)
127
128 ; centered radial basis kernel matrix
129 IF gpuDetectDeviceMode() EQ 1 THEN BEGIN
130    G_gpu = gpuputarr(G)
131    K_gpu = $
132     gpugausskernel_matrix(G_gpu,gma=gma,nscale=nscale)
133    K = gpugetarr(K_gpu)
134    gpufree,[G_gpu,K_gpu]
135 END $
136   ELSE K = gausskernel_matrix(G,gma=gma,nscale=nscale)
137 K = center(K)
```

with IDL will generally not require page swapping and takes the order of minutes.) After diagonalization, the *generalization phase* involves the projection of each image pixel vector, according to Equation 4.34. This means, for every pixel g, recalculation of the kernels $\kappa(g(\nu), g)$, for $\nu = 1 \ldots m$, is required.[*] For an image with n pixels, there are, therefore, $m \times n$ kernels involved, generally too large an array to be held in memory, so it is advisable to read in and project the image pixels row-by-row.

Appendix C.2.1 describes an ENVI extension, GPUKPCA_RUN, for performing kernel PCA on multispectral imagery using the Gaussian kernel (Equation 4.25). The parameter γ defaults to $1/2\sigma^2$, where σ is equal to the mean distance between the observations in input space, that is, the average over all test observations of $\|g(\nu) - g(\nu')\|$, $\nu \neq \nu'$. In the excerpt shown in Listing 4.7, sampling takes place in lines 123 and 124, while the centered kernel matrix is calculated in lines 129–137 (see also Listings 4.5 and 4.6). There are two equivalent code blocks in this extension, one of which is programmed in standard ENVI/IDL. The other makes use of Tech-X Corporation's GPULib,[†] taking advantage of the parallel computing power available in graphics hardware to speed up the projection step by an order of magnitude.

Figure 4.12 shows the logarithms of the eigenvalues for kernel PCA, performed on the LANDSAT 7 ETM+ image of Figure 3.10 with a sample of

[*] This is not the case for a linear PCA, where we do not require the training data to project new observations. For this reason, kernel PCA is said to be *memory based*.

[†] GPULib is a library of high-level language bindings (including IDL) to the so-called Compute Unified Device Architecture (CUDA) that allows access to graphics processors (GPUs) (see Appendix C for details).

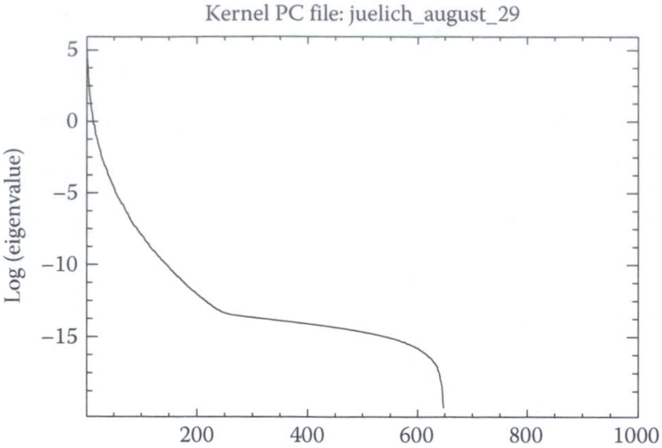

FIGURE 4.12
Logarithm of the eigenvalues for kernel PCA on the image of Figure 3.10. The size of the training data set is $m = 1000$.

1000 pixels. Only 84.3% of the variance is explained by the first three eigenvalues. In contrast, linear PCA concentrates 98.9% of the variance into the first three principal components, so there is evidently little reason to apply kernel PCA in this case. Chapter 9 will illustrate the use of kernel PCA for change detection with (simulated) nonlinear data.

4.5 Gibbs–Markov Random Fields

Random fields are frequently invoked to describe prior expectations in a Bayesian approach to image analysis (Winkler, 1995; Li, 2001). The following brief introduction will serve to make their use more plausible in the unsupervised land cover classification methods which are covered in Chapter 8. The development adheres closely to Li (2001), but in a notation specific to that used later in the treatment of image classification.

Image classification is a problem of *labeling*: Given an observation, that is, a pixel intensity vector, we ask, "Which class label should be assigned to it?" If the observations are assumed to have no spatial context, then the labeling will consist of partitioning the pixels into K disjoint subsets according to some decision criterion, K being the number of land cover categories present. If spatial context within the image is also to be taken into account, the labeling will take place on what is referred to as a *regular lattice*. A regular lattice representing an image with c columns and r rows is a discrete set of sites:

$$\mathcal{I} = \{(i,j) \mid 0 \le i \le c-1, 0 \le j \le r-1\}.$$

By re-indexing, we can write this in a more convenient, linear form:

$$\mathcal{I} = \{i \mid 1 \leq i \leq m\},$$

where $m = rc$ is the number of pixels. The interrelationship between the sites is governed by a *neighborhood system*:

$$\mathcal{N} = \{\mathcal{N}_i \mid i \in \mathcal{I}\},$$

where $\mathcal{N}_i$ is the set of pixels neighboring site i. Two frequently used neighborhood systems are shown in Figure 4.13.

The pair $(\mathcal{I}, \mathcal{N})$ may be thought of as constituting an *undirected graph* in which the pixels are nodes and the neighborhood system determines the edges between the nodes. Thus, any two neighboring pixels are represented by two nodes connected by an edge. A *clique* is a single node or a subset of nodes which are all directly connected to one another in the graph by edges, that is, they are mutual neighbors. Figure 4.14 shows the possible cliques for the neighborhoods of Figure 4.13. (Note that one also distinguishes their orientation.) We will denote by the symbol $\mathcal{C}$ the set of all cliques in $\mathcal{I}$.

Next, let us introduce a set of *class labels*:

$$\mathcal{K} = \{k \mid 1 \leq k \leq K\},$$

where K is the number of possible classes present in the image. Assigning a class label to a site on the basis of measurement is a random experiment, so we associate with the ith site a discrete random variable, L_i, representing its

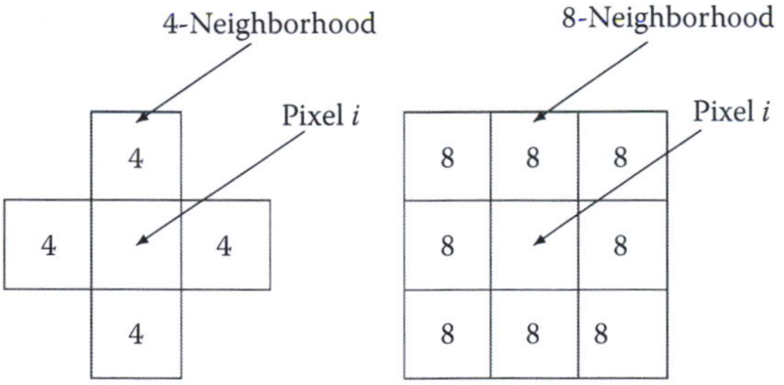

FIGURE 4.13
Pixel neighborhoods, $\mathcal{N}_i$.

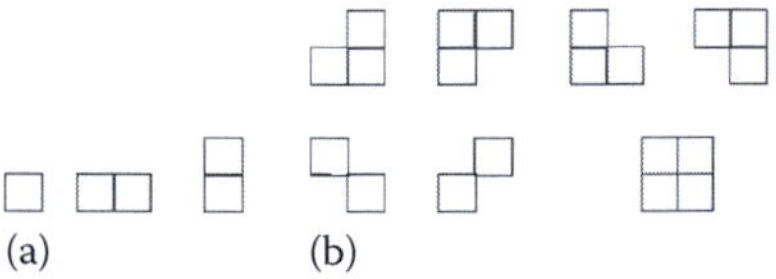

FIGURE 4.14
(a) Cliques for a 4-neighborhood and (b) cliques for an 8-neighborhood.

label. The set L of all such random variables,

$$L = \{L_1 \dots L_m\},$$

is called a *random field* on $\mathcal{I}$. The possible realizations of L_i are labels $\ell_i \in \mathcal{K}$. A specific realization for the entire lattice, for example, a classified image or a thematic map, is a *configuration*, ℓ, given by

$$\ell = \{\ell_1 \dots \ell_m\}, \quad \ell_i \in \mathcal{K},$$

and the space of all configurations is the Cartesian product set:

$$\mathcal{L} = \overbrace{\mathcal{K} \otimes \mathcal{K} \cdots \otimes \mathcal{K}}^{m \text{ times}}.$$

Thus, there are K^m possible configurations. For each site i, the probability that the site has the label ℓ_i is

$$\Pr(L_i = \ell_i) = \Pr(\ell_i),$$

and the joint probability for the configuration ℓ is

$$\Pr(L_1 = \ell_1, L_2 = \ell_2 \dots L_n = \ell_m) = \Pr(L = \ell) = \Pr(\ell).$$

Definition 4.2: The random field L is said to be a *Markov random field (MRF)* on $\mathcal{I}$ with respect to the neighborhood system $\mathcal{N}$ if and only if

$$\Pr(\ell) > 0 \quad \text{for all } \ell \in \mathcal{L}, \tag{4.35}$$

which is referred to as the positivity condition, and

$$\Pr(\ell_i \mid \ell_1 \dots \ell_{i-1}, \ell_{i+1} \dots \ell_m) = \Pr(\ell_i \mid \ell_j, j \in \mathcal{N}_i), \tag{4.36}$$

called the *Markovianity* condition.

The Markovianity condition in the above definition simply says that a label assignment can be influenced only by neighboring pixels.

Definition 4.3: The random field L constitutes a *Gibbs random field (GRF)* on $\mathcal{I}$ with respect to the neighborhood system $\mathcal{N}$ if and only if it obeys a

Gibbs distribution. A *Gibbs distribution* has the density function,

$$p(\ell) = \frac{1}{Z}\exp(-\beta U(\ell)), \tag{4.37}$$

where β is a constant and Z is the normalization factor,

$$Z = \sum_{\ell \in \mathcal{L}} \exp(-\beta U(\ell)), \tag{4.38}$$

and where the *energy function* $U(\ell)$ is represented as a sum of contributing terms for each clique:

$$U(\ell) = \sum_{c \in \mathcal{C}} V_c(\ell). \tag{4.39}$$

$V_c(\ell)$ is called a *clique potential* for clique c in the configuration ℓ. If clique potentials are independent of the location of the clique within the lattice, then the GRF is said to be *homogeneous*. If V_c is independent of the orientation of c then the GRF is *isotropic*.

According to this definition, configurations in a GRF with low clique potentials are more probable than those with high clique potentials. The parameter β is an "inverse temperature." For small β (high temperature), all configurations become equally probable, irrespective of their associated energy.

An MRF is characterized by its local property (Markovianity) and a GRF by its global property (Gibbs distribution). It turns out that the two are equivalent.

THEOREM 4.3 (Hammersley–Clifford Theorem)

A random field, L, is an MRF on $\mathcal{I}$ with respect to $\mathcal{N}$ if and only if L is a GRF on $\mathcal{I}$ with respect to $\mathcal{N}$.

Proof

The proof of the "if" part of the theorem that a random field is an MRF if it is a GRF, is straightforward.* Write the conditional probability on the left-hand side of Equation 4.36 as

$$\Pr(\ell_i \mid \ell_1 \ldots \ell_{i-1}, \ell_{i+1} \ldots \ell_m) = \Pr(\ell_i \mid \ell_{\mathcal{I}-\{i\}}).$$

* The "only if" part is more difficult (see Li, 2001).

Here, $\mathcal{I}-\{i\}$ is the set of all sites except the ith one. According to the definition of conditional probability (Equation 2.53),

$$\Pr(\ell_i \mid \ell_{\mathcal{I}-\{i\}}) = \frac{\Pr(\ell_i, \ell_{\mathcal{I}-\{i\}})}{\Pr(\ell_{\mathcal{I}-\{i\}})}.$$

Equivalently,

$$\Pr(\ell_i \mid \ell_{\mathcal{I}-\{i\}}) = \frac{\Pr(\ell)}{\sum_{\ell_i\in\mathcal{K}} \Pr(\ell')},$$

where $\ell' = \{\ell_1 \dots \ell_{i-1}, \ell'_i, \ell_{i+1} \dots \ell_m\}$ is any configuration that agrees with ℓ at all sites except, possibly, i. With Equations 4.37 and 4.39,

$$\Pr(\ell_i \mid \ell_{\mathcal{I}-\{i\}}) = \frac{\exp(-\sum_{c\in\mathcal{C}} V_c(\ell))}{\sum_{\ell_i\in\mathcal{K}} \exp(-\sum_{c\in\mathcal{C}} V_c(\ell'))}.$$

Now divide the set of cliques $\mathcal{C}$ into two sets, namely, $\mathcal{A}$, consisting of those cliques that contain site i, and $\mathcal{B}$, consisting of the rest. Then,

$$\Pr(\ell_i \mid \ell_{\mathcal{I}-\{i\}}) = \frac{[\exp(-\sum_{c\in\mathcal{A}} V_c(\ell))][\exp(-\sum_{c\in\mathcal{B}} V_c(\ell))]}{\sum_{\ell_i\in\mathcal{K}}[\exp(-\sum_{c\in\mathcal{A}} V_c(\ell'))][\exp(-\sum_{c\in\mathcal{B}} V_c(\ell'))]}.$$

But for all cliques $c \in \mathcal{B}$, $V_c(\ell) = V_c(\ell')$, and the second factors in the numerator and in the denominator cancel. Thus,

$$\Pr(\ell_i \mid \ell_{\mathcal{I}-\{i\}}) = \frac{\exp(-\sum_{c\in\mathcal{A}} V_c(\ell))}{\sum_{\ell_i\in\mathcal{K}} \exp(-\sum_{c\in\mathcal{A}} V_c(\ell'))}.$$

This shows that the probability for ℓ_i is conditional only on the potentials of the cliques containing site i; in other words, it shows that the random field is an MRF. □

The above theorem allows one to express the joint probability for a configuration, ℓ, of image labels in terms of the local clique potentials, $V_c(\ell)$. We shall encounter an example in Chapter 7 in connection with unsupervised image classification.

4.6 Exercises

1. Give an explicit expression for the convolution of the array $(g(0)\dots g(5))$ with the array $(h(0), h(1), h(2))$ as it would be calculated with IDL function `CONVOL()` with the `EDGE_WRAP` keyword set.
2. Modify the program in Listing 4.1 to convolve a two-dimensional array with the filter of Equation 4.7, both in spatial and frequency domains.
3. Modify the program in Listing 4.2 to implement the *Butterworth filter*:

$$h(k,\ell) = \frac{1}{1+(d/d_0)^{2n}}, \tag{4.40}$$

   where d_0 is a width parameter and $n = 1, 2\dots$.
4. Demonstrate Equation 4.13.
5. (Press et al., 2002) Consider a row of $c = 8$ pixels represented by the row vector

$$f = (f_0, f_1 \dots f_7).$$

   Assuming that f is periodic (repeats itself), the applications of low- and high-pass filter equations (Equations 4.9 and 4.13) can be accomplished by multiplication of the column vector $f^\top$ with the matrix

$$\boldsymbol{W} = \begin{pmatrix} h_0 & h_1 & h_2 & h_3 & 0 & 0 & 0 & 0 \\ h_3 & -h_2 & h_1 & -h_0 & 0 & 0 & 0 & 0 \\ 0 & 0 & h_0 & h_1 & h_2 & h_3 & 0 & 0 \\ 0 & 0 & h_3 & -h_2 & h_1 & -h_0 & 0 & 0 \\ 0 & 0 & 0 & 0 & h_0 & h_1 & h_2 & h_3 \\ 0 & 0 & 0 & 0 & h_3 & -h_2 & h_1 & -h_0 \\ h_2 & h_3 & 0 & 0 & 0 & 0 & h_0 & h_1 \\ h_1 & -h_0 & 0 & 0 & 0 & 0 & h_3 & -h_2 \end{pmatrix}.$$

   (a) Prove that $\boldsymbol{W}$ is an orthonormal matrix (its inverse is equal to its transpose).

   (b) In the transformed vector $\boldsymbol{W}f^\top$, the components of $f^1 = Hf$ and $\boldsymbol{d}^1 = Gf$ are interleaved. They can be sorted to give the vector

(f^1, d^1). When the matrix

$$W^1 = \begin{pmatrix} h_0 & h_1 & h_2 & h_3 \\ h_3 & -h_2 & h_1 & -h_0 \\ h_2 & h_3 & h_0 & h_1 \\ h_1 & -h_0 & h_3 & -h_2 \end{pmatrix}$$

is then applied to the smoothed vector $(f^1)^\top$ and the result again sorted, we obtain the complete discrete wavelet transformation of f, namely, (f^2, d^2, d^1). Precisely this transformation can be invoked in IDL with the function `WTN(F,4,inverse=inverse)` for the Daubechies D4 wavelet and for any vector `F` of length $c = 2^n$. Moreover, by applying the inverse transformation to a unit vector, the D4 wavelet itself can be generated. Write an IDL routine to plot the D4 wavelet by performing the inverse transformation on the vector $\underbrace{(0,0,0,1,0\ldots0)}_{1024}$.

6. Most filtering operations with the Fourier transform have their wavelet counterparts. Modify the program in Listing 4.3 to perform high-pass filtering with the discrete wavelet transformation, proceeding as follows:
   - Perform a single compression with the object method `COMPRESS`
   - Zero the upper left quadrant in the transformed image (object method `INJECT` with a null array; see Appendix C.2 and the source listing to understand how this method works)
   - Invert the transformation with the object method `EXPAND`
   - Display the result in ENVI
7. Show that the Gaussian kernel (Equation 4.25) is equivalent to the inner product of infinite-dimensional mappings, $\boldsymbol{\phi}(g)$.
8. Modify the program in Listing 4.5 to calculate the kernel matrix for the *polynomial kernel function*:

$$\kappa_{\text{poly}}(g_i, g_j) = (\gamma g_i^\top g_j + r)^d, \tag{4.41}$$

   where
   the parameter r is called the *bias*
   d is the *degree* of the kernel function

9. Shawe-Taylor and Cristianini (2004) prove (in their Proposition 5.2) that the centering operation minimizes the average eigenvalue of the kernel matrix. Write an IDL program to do the following:

(a) Generate m random N-dimensional observation vectors and, with the routine in Listing 4.5, calculate the $m \times m$ Gaussian kernel matrix.

(b) Determine its eigenvalues in order to confirm that the kernel matrix is positive semi-definite.

(c) Center it by calling the routine in Listing 4.6.

(d) Redetermine the eigenvalues and verify that their average value has decreased.

(e) Do the same using the polynomial kernel matrix from the preceding exercise.

10. A kernelized version of the dual formulation for the MNF transformation (see Chapter 3, Exercise 14) leads to a generalized eigenvalue problem having the form

$$\boldsymbol{A}\boldsymbol{x} = \lambda \boldsymbol{B}\boldsymbol{x}, \tag{4.42}$$

where $\boldsymbol{A}$ and $\boldsymbol{B}$ are symmetric but not full rank. Let the symmetric $m \times m$ matrix $\boldsymbol{B}$ have rank $r < m$, eigenvalues $\lambda_1 \dots \lambda_m$, and eigenvectors $\boldsymbol{u}_1 \dots \boldsymbol{u}_m$. Show that Equation 4.42 can be reformulated as an ordinary symmetric eigenvalue problem by writing $\boldsymbol{B}$ as a product of *matrix square roots*, $\boldsymbol{B} = \boldsymbol{B}^{1/2}\boldsymbol{B}^{1/2}$. The square root is defined as

$$\boldsymbol{B}^{1/2} = \boldsymbol{P}\boldsymbol{\Lambda}^{1/2}\boldsymbol{P}^\top,$$

where

$\boldsymbol{P} = (\boldsymbol{u}_1 \dots \boldsymbol{u}_r)$

$\boldsymbol{\Lambda} = \text{Diag}(\lambda_1 \dots \lambda_r)$

11. Does a 24-neighborhood (the 5×5 array centered at a location, i) have more clique types than are shown in Figure 4.14?

5

Image Enhancement and Correction

In preparation for the treatment of supervised/unsupervised classification and change detection, which will be the subjects of Chapters 6 through 9 of this book, the present chapter focuses on preprocessing methods. These fall into two general categories: *image enhancement* (Sections 5.1 through 5.3) and *geometric correction* (Sections 5.5 and 5.6).

5.1 Lookup Tables and Histogram Functions

Gray-level enhancements of an image are easily accomplished by means of *lookup tables*. For byte-encoded data for example, the pixel intensities $g(i,j)$ are used to index into the array

$$LUT[k], \quad k = 0 \ldots 255,$$

the entries of which also lie between 0 and 255. These entries can be chosen to implement simple histogram processing, such as linear stretching, saturation, and equalization. The pixel values $g(i,j)$ are replaced by

$$f(i,j) = LUT[g(i,j)], \quad 0 \le i \le r-1,\ 0 \le j \le c-1. \tag{5.1}$$

In deriving the appropriate transformations, it is convenient to think of the normalized histogram of pixel intensities, g, as a probability density, $p_g(g)$, of a continuous random variable, G, and the lookup table itself as a continuous transformation function, that is,

$$f(g) = LUT(g),$$

where both g and f are restricted to the interval $[0,1]$. We will illustrate the technique for the case of *histogram equalization*. First of all, we claim that the function

$$f(g) = LUT(g) = \int_0^g p_g(t)dt \tag{5.2}$$

corresponds to histogram equalization in the sense that a random variable, $F = LUT(G)$, has uniform probability density. To see this, note that the function LUT in Equation 5.2 satisfies the monotonicity condition for the function u of Theorem 2.1. Therefore, with Equation 2.13, the PDF for F is

$$p_f(f) = p_g(g)\left|\frac{dg}{df}\right|.$$

Differentiating Equation 5.2,

$$\frac{df}{dg} = p_g(g), \quad \frac{dg}{df} = \frac{1}{p_g(g)}$$

and, since all probabilities are positive on the interval [0,1],

$$p_f(f) = p_g(g)\left|\frac{1}{p_g(g)}\right| = 1,$$

so F indeed has a uniform density. Histogram equalization can be approximated for byte-encoded data by replacing $p_g(g)$ by the normalized histogram:

$$p_g(g_k) = \frac{n_k}{n}, \quad k = 0\ldots 255, \quad n = \sum_{j=0}^{255} n_j,$$

where n_k is the number of pixels with the gray value g_k. This leads to the lookup table

$$LUT(g_k) = 255 \cdot \sum_{j=0}^{k} p_g(g_k) = 255 \cdot \sum_{j=0}^{k} \frac{n_j}{n}, \quad k = 0\ldots 255.$$

The result is rounded down to the nearest integer. Because of the quantization, the resulting histogram will in general not be perfectly uniform; however, the desired effect of spreading the intensities to span the full range of grayscale values will be achieved.

The closely related procedure of *histogram matching*, which is the transformation of an image histogram to match the histogram of another image or some specified function, can be similarly derived using the probability density approximation (see Gonzalez and Woods, 2002, for a detailed discussion). Both histogram equalization and histogram matching are standard functions within the ENVI environment.

5.2 Filtering and Feature Extraction

In Section 4.2, we introduced filtering in the spatial domain, giving a simple example of a low-pass filter (see Equation 4.7). We shall now examine high-pass filtering for edge detection, a technique that will be used later in this chapter to implement low-level feature matching for image co-registration. Edge detection is also often used in conjunction with feature extraction for scene analysis and for image segmentation, one of the subjects treated in Chapter 8. Localized image features and/or segments can be conveniently characterized by their so-called *geometric moments*. A short description of geometric moments together with an IDL program to calculate them will also be presented in this section.

5.2.1 Edge Detection

We begin by introducing the *gradient operator*:

$$\nabla = \frac{\partial}{\partial \boldsymbol{x}} = \boldsymbol{i}\frac{\partial}{\partial x_1} + \boldsymbol{j}\frac{\partial}{\partial x_2}, \tag{5.3}$$

where $\boldsymbol{i}$ and $\boldsymbol{j}$ are unit vectors in the image plane in the horizontal and vertical directions, respectively. In a two-dimensional image represented by the continuous scalar function $g(x_1, x_2) = g(\boldsymbol{x})$, $\nabla g(\boldsymbol{x})$ is a vector in the direction of the maximum rate of change of grayscale intensity.

Of course the intensity values are actually discrete, so the partial derivatives must be approximated. For example, we can use the *Sobel operators*:

$$\begin{aligned}\left.\frac{\partial g(\boldsymbol{x})}{\partial x_1}\right|_{\boldsymbol{x}=(i,j)} &\approx [g(i-1,j-1) + 2g(i-1,j) + g(i-1,j+1)] \\ &\quad - [g(i+1,j-1) + 2g(i+1,j) + g(i+1,j+1)] =: \nabla_1(i,j)\end{aligned}$$

$$\begin{aligned}\left.\frac{\partial g(\boldsymbol{x})}{\partial x_2}\right|_{\boldsymbol{x}=(i,j)} &\approx [g(i-1,j-1) + 2g(i,j-1) + g(i+1,j-1)] \\ &\quad - [g(i-1,j+1) + 2g(i,j+1) + g(i+1,j+1)] =: \nabla_2(i,j),\end{aligned}$$

which are equivalent to the two-dimensional spatial filters,

$$h_1 = \begin{pmatrix} 1 & 0 & -1 \\ 2 & 0 & -2 \\ 1 & 0 & -1 \end{pmatrix} \quad \text{and} \quad h_2 = \begin{pmatrix} 1 & 2 & 1 \\ 0 & 0 & 0 \\ -1 & -2 & -1 \end{pmatrix}, \tag{5.4}$$

Listing 5.1
Calculating the power spectrum of the filter h_1 in Equation 5.4.

```
PRO EX5_1
; create filter
   g = fltarr(512,512)
   g[0:2,0:2] = [[1,0,-1],[2,0,-2],[1,0,-1]]
; shift Fourier transform to center
   a = lindgen(512,512)
   i = a MOD 512
   j = a/512
   g = (-1)^(i+j)*g
; output Fourier power spectrum as EPS file
   thisDevice =!D.Name
   set_plot, 'PS'
   Device, Filename='fig5_1.eps',xsize=3,ysize=3, $
            /inches,/encapsulated
   tvscl, (abs(FFT(g)))^2
   device,/close_file
   set_plot,thisDevice
END
```

respectively (see Equation 4.6). The magnitude of the gradient at pixel (i,j) is

$$\|\nabla(i,j)\| = \sqrt{\nabla_1(i,j)^2 + \nabla_2(i,j)^2}.$$

Edge detection can be achieved by calculating the filtered image,

$$f(i,j) = \|\nabla(i,j)\|,$$

and setting an appropriate threshold. The Sobel filters (Equation 5.4) involve *differences*, a characteristic of high-pass filters. They have the property of returning near zero values when traversing regions of constant intensity, and positive or negative values in regions of changing intensity. Listing 5.1 calculates the Fourier power spectrum of h_1, and the result is shown in Figure 5.1. The Fourier spectrum of h_2 is the same, but rotated by 90°. The Sobel filter, invoked from the ENVI main menu and applied to a spatial subset of the 3N spectral band from the Jülich ASTER image, generates the image shown in Figure 5.2. Note that the edge widths vary considerably, depending upon the contrast and abruptness (strength) of the edge.

The magnitude of the gradient reaches a maximum at an edge in a grayscale image. The second derivative, on the other hand, is zero at that maximum and has opposite signs immediately on either side. This offers the possibility to determine edge positions to the accuracy of one pixel by using second derivative filters. Thin edges, as we will see later, are useful in the automatic determination of invariant features for image registration.

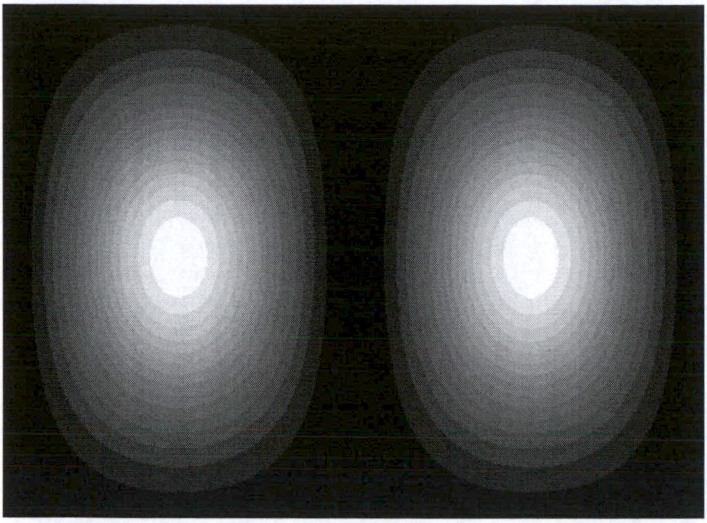

FIGURE 5.1
Power spectrum image of the Sobel filter h_1 of Equation 5.4. Low frequencies are at the center of the image.

The second derivatives of the image intensities can be calculated with the *Laplacian* operator:

$$\nabla^2 = \nabla^\top \nabla = \frac{\partial^2}{\partial x_1^2} + \frac{\partial^2}{\partial x_2^2}.$$

The Laplacian $\nabla^2 g(\boldsymbol{x})$ is an isotropic scalar function, which is zero whenever the gradient magnitude is maximum. Like the gradient operator, the Laplacian operator can also be approximated by a spatial filter, for example,

$$h = \begin{pmatrix} 0 & 1 & 0 \\ 1 & -4 & 1 \\ 0 & 1 & 0 \end{pmatrix}. \tag{5.5}$$

Its power spectrum is shown in Figure 5.3. This filter has the desired property of returning a zero in regions of constant intensity and in regions of constantly varying intensity (e.g., ramps), but a nonzero value at their onset or termination. Laplacian filters tend to be very sensitive to image noise. Often a low-pass Gaussian filter is first used to smooth the image before the Laplacian filter is applied. This is equivalent to calculating the Laplacian of the Gaussian itself, and then using the result to derive a high-pass filter. Recall that the normalized Gauss function in two dimensions is given by

$$\frac{1}{2\pi\sigma^2}\exp\left(-\frac{1}{2\sigma^2}\left(x_1^2 + x_2^2\right)\right),$$

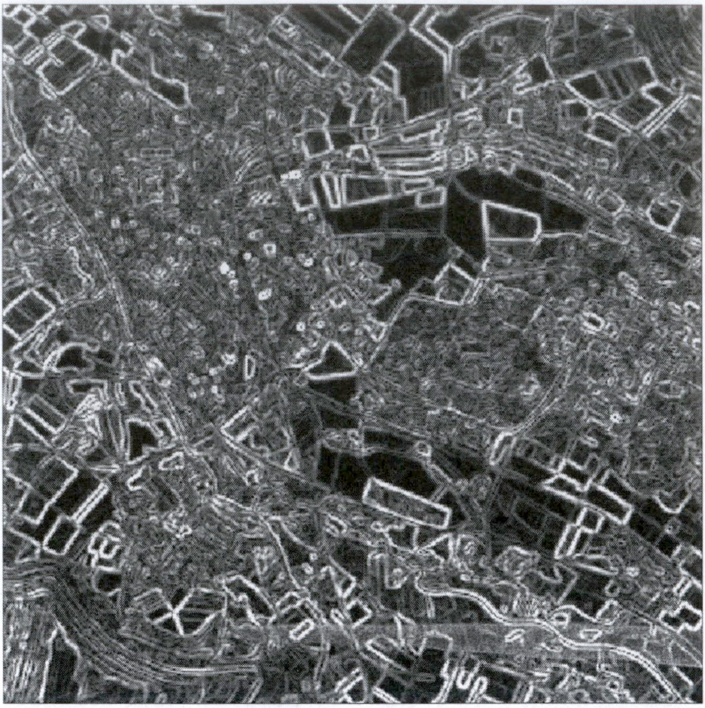

FIGURE 5.2
Sobel edge detection on the 3N band of the ASTER image of Figure 1.1.

where the parameter σ determines its extent. Taking the second partial derivatives gives the *Laplacian of Gaussian* (LoG) filter:

$$\frac{1}{2\pi\sigma^6}(x_1^2 + x_2^2 - 2\sigma^2)\exp\left(-\frac{1}{2\sigma^2}(x_1^2 + x_2^2)\right). \tag{5.6}$$

Listing 5.2 illustrates the creation and use of a LoG filter together with the determination of sign change to generate thin edges or contours from a grayscale image. The filtering is carried out, after appropriate padding, in the frequency domain (lines 24–27), and the zero crossings in the horizontal and vertical directions are determined from the products of the image with a copy of itself shifted by one pixel to the right and upward, respectively (lines 29 and 30). Sign changes correspond to negative values in the product arrays, and these define the contours. The contour pixel intensities are set equal to the magnitude of the local gradient as determined by a Sobel filter (lines 32–35). This allows subsequent thresholding to identify the more significant contours. The program generates the surface plot of the two-dimensional LoG filter, shown in Figure 5.4; the filtered image is displayed in Figure 5.5. The "spaghetti" effect is characteristic.

FIGURE 5.3
Power spectrum image of the Laplacian filter, Equation 5.5. Low frequencies are at the center of the image.

5.2.2 Invariant Moments

Moments and functions of moments are employed extensively in image classification, target identification, and image scene analysis (Prokop and Reeves, 1992). A feature, such as a closed contour extracted from the image in Figure 5.5, can be described in terms of its *geometric moments*. Let S denote the set of pixels belonging to the feature. The (discrete) geometric moment of order p, q of the feature is defined by (see, e.g., Haberächer, 1995; Gonzalez and Woods, 2002)

$$m_{pq} = \sum_{i,j \in S} g(i,j) i^p j^q, \quad p, q = 0, 1, 2 \ldots. \tag{5.7}$$

Thus, m_{00} is the total intensity of the feature or, in a binary representation in which $g(i,j) = 1$ if $(i,j) \in S$ and $g(i,j) = 0$ otherwise, m_{00} is the number of pixels. The center of gravity, $(\bar{x}_1, \bar{x}_2)$, of the feature is

$$\bar{x}_1 = \frac{m_{10}}{m_{00}}, \quad \bar{x}_2 = \frac{m_{01}}{m_{00}}. \tag{5.8}$$

Listing 5.2
Illustrating LoG filtering with sign change detection.

```
 1  PRO EX5_2
 2  ; create a LoG filter (and plot it)
 3      sigma = 2.0
 4      filter = fltarr(16,16)
 5      FOR i=0L,15 DO FOR j=0L,15 DO filter[i,j] = $
 6      (1/(2*!pi*sigma^6))*((i-8)^2+(j-8)^2-2*sigma^2) $
 7       *exp(-((i-8)^2+(j-8)^2)/(2*sigma^2))
 8      thisDevice =!D.Name
 9      set_plot, 'PS'
10      Device, Filename='fig5_4.eps',xsize=3,ysize=3, $
11                /encapsulated
12      shade_surf,filter
13      device,/close_file
14      set_plot,thisDevice
15  ; get an image band and pad it
16      envi_select, title='Choose_multispectral_band', $
17                fid=fid, dims=dims,pos=pos, /band_only
18      num_cols = dims[2]-dims[1]+1
19      num_rows = dims[4]-dims[3]+1
20      image = fltarr(num_cols+16,num_rows+16)
21      image[0:num_cols-1,0:num_rows-1] = $
22          envi_get_data(fid=fid,dims=dims,pos=0)
23  ; pad the filter as well
24      filt = image*0
25      filt[0:15,0:15] = filter
26  ; perform filtering in frequency domain
27      image= float(fft(fft(image)*fft(filt),1))
28  ; get zero-crossing positions
29      indices = where( (image*shift(image,1,0) LT 0) $
30                OR (image*shift(image,0,1) LT 0) )
31  ; perform Sobel filter for edge strengths
32      edge_strengths = sobel(image)
33  ; create the contour image and return it to ENVI
34      image = image*0
35      image[indices]=edge_strengths[indices]
36      envi_enter_data, image
37  END
```

The translation-invariant *centralized moments*, μ_{pq}, are obtained by shifting the origin to the center of gravity,

$$\mu_{pq} = \sum_{i,j \in S} g(i - \bar{x}_1, j - \bar{x}_2)(i - \bar{x}_1)^p (j - \bar{x}_2)^q, \qquad (5.9)$$

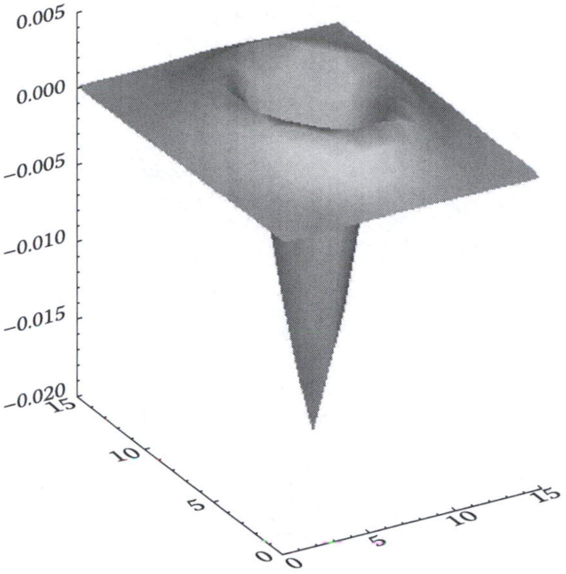

FIGURE 5.4
Laplacian of Gaussian filter on a 16×16 grid, with $\sigma = 2$.

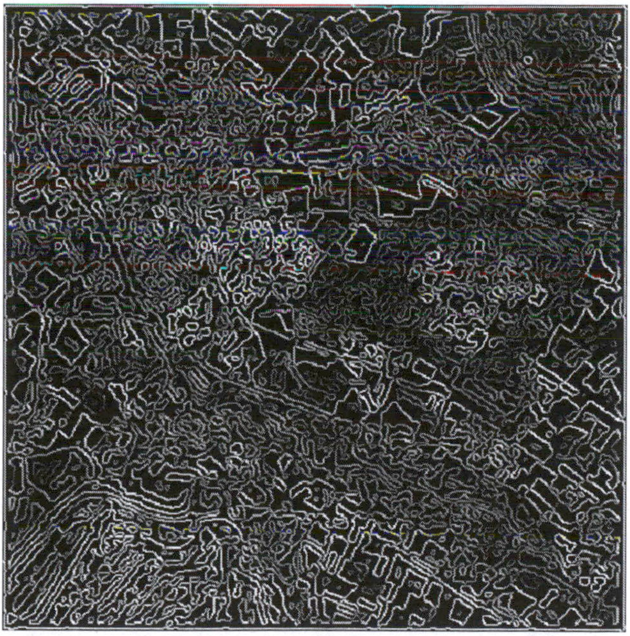

FIGURE 5.5
Image contours from a spatial subset of the 3N band of the Jülich ASTER image calculated with the program in Listing 5.2.

and the *normalized centralized moments*, η_{pq}, by the following normalization (see Exercise 5):

$$\eta_{pq} = \frac{1}{\mu_{00}^{(p+q)/2+1}} \mu_{pq}. \tag{5.10}$$

The normalized centralized moments are, apart from effects of digital quantization, invariant under both translations and scale changes. For example, in a binary representation, the normalized centralized moment, η_{20}, is

$$\eta_{20} = \frac{1}{\mu_{00}^2} \mu_{20} = \frac{1}{n^2} \sum_{i,j \in S} (i - \bar{x}_1)^2 = \frac{1}{n} \sum_{i \in S} (i - \bar{x}_1)^2,$$

where $n = |S|$, the cardinality of S. This is just the variance of the feature in the x_1 direction.

Finally, we can define *Hu moments*, which are functions of the normalized centralized moments of orders $p + q \leq 3$, and which are invariant under rotations, see Hu (1962) or Gonzalez and Woods (2002). There are seven such moments in all, the first four of which are given by

$$\begin{aligned} h_1 &= \eta_{20} + \eta_{02} \\ h_2 &= (\eta_{20} - \eta_{02})^2 + 4\eta_{11}^2 \\ h_3 &= (\eta_{30} - 3\eta_{12})^2 + (\eta_{03} - 3\eta_{21})^2 \\ h_4 &= (\eta_{30} + \eta_{12})^2 + (\eta_{03} + \eta_{21})^2. \end{aligned} \tag{5.11}$$

To illustrate their rotational invariance, consider rotations of the coordinate axes through an angle θ with their origin at the center of gravity of a feature. A point, (i, j), transforms according to

$$\begin{pmatrix} i' \\ j' \end{pmatrix} = \begin{pmatrix} \cos\theta & \sin\theta \\ -\sin\theta & \cos\theta \end{pmatrix} \begin{pmatrix} i \\ j \end{pmatrix} = \boldsymbol{A} \begin{pmatrix} i \\ j \end{pmatrix}.$$

Then, again in a binary image, the first invariant moment in the rotated coordinate system is

$$\begin{aligned} h_1 &= \frac{1}{n^2} \sum_{i',j' \in S} (i'^2 + j'^2) = \frac{1}{n^2} \sum_{i',j' \in S} (i', j') \begin{pmatrix} i' \\ j' \end{pmatrix} \\ &= \frac{1}{n^2} \sum_{i,j \in S} (i, j) \boldsymbol{A}^\top \boldsymbol{A} \begin{pmatrix} i \\ j \end{pmatrix} = \frac{1}{n^2} \sum_{i,j \in S} (i^2 + j^2), \end{aligned}$$

since $\boldsymbol{A}^\top \boldsymbol{A} = \boldsymbol{I}$.

The Hu moments are strictly invariant only in the limit of continuous shapes. The discrete nature of image features introduces errors,

Listing 5.3
Illustrating the Hu invariant moments.

```
PRO Ex5_3
; Airplane
   A = float([[0,0,0,0,0,1,0,0,0,0,0], $
              [0,0,0,0,1,1,1,0,0,0,0], $
              [0,0,0,0,1,1,1,0,0,0,0], $
              [0,0,0,1,1,1,1,1,0,0,0], $
              [0,0,1,1,0,1,0,1,1,0,0], $
              [0,1,1,0,0,1,0,0,1,1,0], $
              [1,0,0,0,0,1,0,0,0,0,1], $
              [0,0,0,0,0,1,0,0,0,0,0], $
              [0,0,0,0,1,1,1,0,0,0,0], $
              [0,0,0,0,0,1,0,0,0,0,0]])
   Im = fltarr(200,200)
   A =  rebin(A,55,50)
   Im[25:79,75:124]=A
; Rotate through 60 deg, magnify by 1.5 and translate
   Im1 = shift(rot(Im,60,1.5,cubic=-0.5),90,-60)
; Rotate through 180 deg, shrink by 0.5 and translate
   Im2 = shift(rot(Im,180,0.5,cubic=-0.5),-90,-60)
; Plot them
   tvscl,Im+Im1+Im2
   thisDevice =!D.Name
   set_plot, 'PS'
   Device,Filename='d:\temp\fig5_6.eps',xsize=6,$
      ysize=6,/encapsulated
   tvscl,Im+Im1+Im2
   device,/close_file
   set_plot, thisDevice
; Invariant moments
   PRINT, hu_moments(Im,/log), format='(7F8.3)'
   PRINT, hu_moments(Im1,/log), format='(7F8.3)'
   PRINT, hu_moments(Im2,/log), format='(7F8.3)'
END
```

which become severe, especially for higher-order moments. Liao and Pawlak (1996) give correction formulae for the geometric moments, Equation 5.7, which reduce the error. This forms the basis of the IDL function `HU_MOMENTS()` described in Appendix C.2.20 for calculating the seven invariant Hu moments. The program in Listing 5.3 generates the features shown in Figure 5.6, and then calls `HU_MOMENTS()` to calculate the logarithms of the invariant moments. The result is as follows:

```
ENVI> ex5_3
  -1.061  -7.363  -6.765  -6.241 -12.752  -9.934 -14.800
  -1.067  -6.819  -6.873  -6.211 -12.754  -9.621 -15.509
  -1.060  -7.355  -6.793  -6.246 -12.771  -9.939 -15.056
```

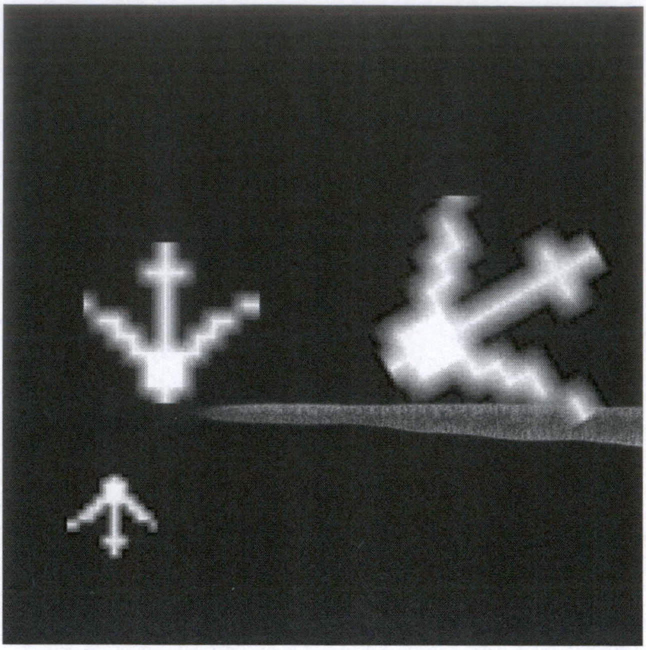

FIGURE 5.6
An "aircraft" feature translated, scaled, and rotated, see Listing 5.3.

The geometric moments are projections of pixel intensities onto the monomials $f_{pq}(i,j) = i^p j^q$. The continuous equivalents, $f_{pq}(\boldsymbol{x}) = x_1^p x_2^q$, are not orthogonal, that is,

$$\int f_{pq}(\boldsymbol{x}) f_{p'q'}(\boldsymbol{x}) d\boldsymbol{x} \neq 0, \quad p,q \neq p'q',$$

so that geometric moments and their invariant derivatives carry redundant information. Alternative moment systems can be defined in terms of orthogonal Legendre or Zernike polynomials, which are particularly relevant for image reconstruction, see Teague (1980).

5.3 Panchromatic Sharpening

The change detection and classification algorithms that we will cover in the next chapters, exploit both the spatial and the spectral information provided by remote sensing imagery. Many common satellite platforms (e.g., LANDSAT 7 ETM+, SPOT, IKONOS, and QuickBird) supply coregistered panchromatic images with considerably higher ground resolutions than that of multispectral bands. However, without additional processing, the application of spectral change detection or classification methods is restricted in the first instance to the poorer spatial resolution of the multispectral data. Conventional image fusion techniques, such as the well-known HSV transformation discussed below, can be used to sharpen the spectral components. However, the result of mixing-in of the panchromatic image is often to "dilute" significantly the spectral information and, in the case of classification for instance, to reduce class separability in the multidimensional feature space. Another disadvantage of the HSV transformation is that one is restricted to using three of the available spectral channels. In the following text, after outlining the HSV method, we consider a number of alternative (and better) fusion techniques.

5.3.1 HSV Fusion

In computers with 24 bit graphics (referred to as "true color" on Windows systems), three bands of a multispectral image can be displayed with 8 bits for each of the additive primary colors, red, green, and blue. There are $2^{24} \approx 16$ million possible colors. The monitor displays the bands as an RGB color composite image that, depending on the choice of spectral bands and histogram functions, may or may not appear to be natural (see, for instance, Figure 1.1). The color of a given pixel in the image can be represented as a vector or a point in a Cartesian coordinate system (the RGB color cube), as illustrated in Figure 5.7.

An alternative color representation is in terms of *hue, saturation*, and *value* (HSV). Value (sometimes called *intensity*) can be thought of as the distance along an axis equidistant from the three orthogonal primary color axes in the RGB representation (the main diagonal of the RGB cube shown as the dashed line in Figure 5.7). Since all points on the axis have equal R, G, and B values, they appear as shades of gray. Hue refers to the actual color and is defined as an angle on a plane perpendicular to the value axis as measured from a reference line pointing in the direction of the red vertex of the RGB cube. Saturation is the "amount" of color present, and is represented by the radius of a circle described in the hue plane. A commonly used method for the fusion of three low-resolution multispectral bands with a higher resolution panchromatic image is to resample the former to the panchromatic resolution, transform them from RGB to HSV coordinates, replace the V component with the grayscale panchromatic image, eventually after performing some kind of histogram matching or normalization of the two, and then transform the bands back to the RGB space.

This procedure is referred to as *HSV panchromatic image sharpening*, and is illustrated in Listing 5.4. The code also illustrates the use of ENVI batch procedures in an IDL program to perform standard manipulations, in this case, histogram stretching and HSV sharpening. The ENVI main menu command

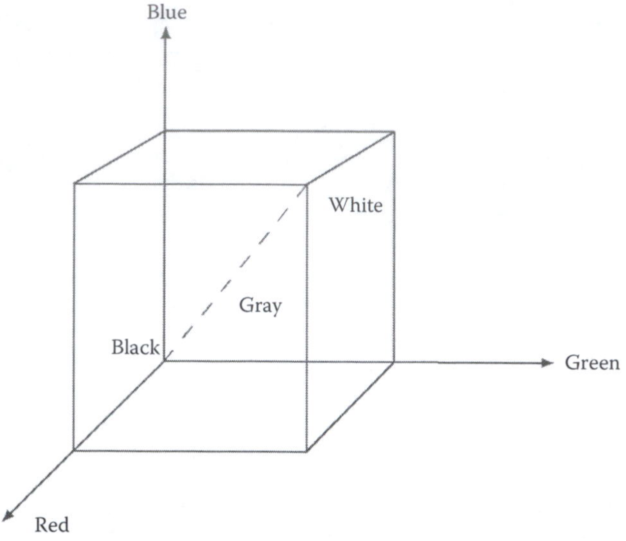

FIGURE 5.7
The RGB color cube.

`Transform/Image Sharpening/HSV` offers similar functionality; however, it does not work with 11 bit data, such as IKONOS and QuickBird imagery. Apparently, the routines that are invoked from the main menu command expect byte-encoded pixel intensities. This drawback is remedied in Listing 5.4. (The `EVENT` parameter in the procedure declaration allows the program to be called from the ENVI menu system.)

5.3.2 Brovey Fusion

In *Brovey* or *color normalized* fusion (Vrabel, 1996), each resampled multispectral pixel is multiplied by the ratio of the corresponding panchromatic pixel intensity to the sum of all of the multispectral intensities. The corrected pixel intensities, $\bar{g}_k(i,j)$, in the kth fused multispectral channel are given by

$$\bar{g}_k(i,j) = g_k(i,j) \cdot \frac{g_p(i,j)}{\sum_{k'} g_{k'}(i,j)}, \tag{5.12}$$

where $g_k(i,j)$ is the (resampled) pixel intensity in the kth spectral band and $g_p(i,j)$ is the corresponding pixel intensity in the panchromatic image. This technique assumes that the spectral range spanned by the panchromatic image is essentially the same as that covered by the multispectral bands. This is seldom the case. Moreover, to avoid bias, the intensities used should be the radiances at the satellite sensors, implying use of the sensors' calibration. The ENVI environment also offers Brovey fusion in its main menu.

Listing 5.4
HSV panchromatic sharpening with ENVI batch procedures.

```
PRO HSV, event

; get MS image
envi_select, title='Select␣3-band␣MS␣input␣file', $
                     fid=fid1, dims=dims1, pos=pos1
IF (fid1 EQ -1) OR (n_elements(pos1) NE 3) THEN RETURN

; get PAN image
envi_select, title='Select␣panchromatic␣image', $
             fid=fid2, pos=pos2, dims=dims2, /band_only
IF (fid2 EQ -1) THEN RETURN

; make batch files available
envi_check_save, /transform

; linear stretch the images and convert to byte format
envi_doit,'stretch_doit',fid=fid1,dims=dims1,pos=pos1,$
  method=1, r_fid=r_fid1, out_min=0, out_max=255, $
  range_by=0, i_min=0, i_max=100, out_dt=1, /in_memory

envi_doit,'stretch_doit',fid=fid2,dims=dims2,pos=pos2,$
  method=1, r_fid=r_fid2, out_min=0, out_max=255, $
  range_by=0, i_min=0, i_max=100, out_dt=1, /in_memory

envi_file_query, r_fid2, ns=f_ns, nl=f_nl
f_dims = [-1l, 0, f_ns-1, 0, f_nl-1]

; HSV pan-sharpening
envi_doit, 'sharpen_doit', $
  fid=[r_fid1,r_fid1,r_fid1],pos=[0,1,2],f_fid=r_fid2,$
  f_dims=f_dims,f_pos=[0],method=0,interp=0,/in_memory

; remove temporary files
envi_file_mng, id=r_fid1, /remove
envi_file_mng, id=r_fid2, /remove

END
```

5.3.3 PCA Fusion

Panchromatic sharpening using principal components analysis (Welch and Ahlers, 1987) is similar to the HSV method. After the PCA transformation, the first principal component is replaced by the panchromatic image, again after some kind of normalization, and the transformation is inverted (see Figure 5.8). Image sharpening using PCA and the closely related

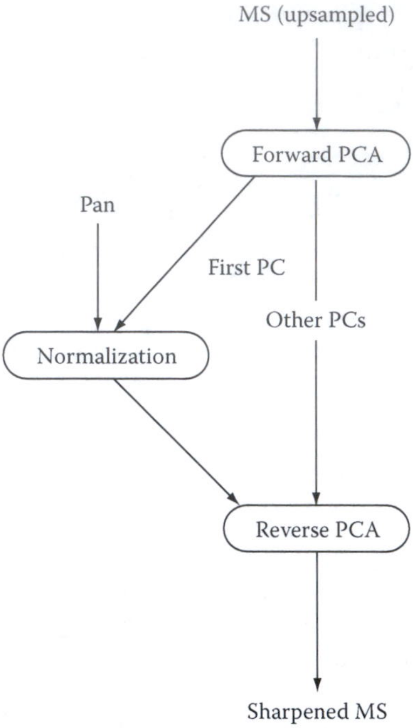

FIGURE 5.8
Panchromatic fusion with the principal components transformation.

Gram–Schmidt transformation are likewise available from the ENVI main menu.

5.3.4 DWT Fusion

The discrete wavelet transform of a two-dimensional image was shown in Chapter 4 to be equivalent to an iterative application of the high- and low-pass filter bank of Figure 4.9 (see also Figures 4.10 and 4.11). The DWT filter bank can be used in an elegant way (Ranchin and Wald, 2000) to achieve panchromatic sharpening when the panchromatic and multispectral spatial resolutions differ by exactly a factor of 2^n for integer n. This is often the case, for example, in SPOT, LANDSAT 7 ETM+, IKONOS, QuickBird, and ASTER imagery.

A DWT fusion procedure for IKONOS imagery for instance, in which the resolutions of the panchromatic and the four multispectral bands differ exactly by a factor of 4, is as follows:

- The panchromatic image, represented by $f_n(i,j)$, is filtered twice to give a degraded image, $f_{n-2}(i,j)$, plus wavelet coefficients at two scales (see Figures 4.9 and 4.11).
- Both the low-pass filtered portion of the panchromatic image as well as the four multispectral bands are filtered once again. (The low-pass

portions of all five bands now have 8m pixel resolutions in the case of IKONOS data.)

- The high-frequency components, C^z_{n-3}, $z = H, V, D$, are sampled to estimate radiometric normalization coefficients, a^z and b^z, for each multispectral band:

$$
\begin{aligned}
a^z &= \hat{\sigma}^z_{ms}/\hat{\sigma}^z_{pan} \\
b^z &= \hat{m}^z_{ms} - a^z \hat{m}^z_{pan},
\end{aligned}
\tag{5.13}
$$

where $\hat{m}^z$ and $\hat{\sigma}^z$ denote estimated mean and standard deviation, respectively.

- These coefficients are then used to normalize the wavelet coefficients for the panchromatic image to those of each multispectral band at the $n-2$ and $n-1$ scales:

$$
C^z_k(i,j) \rightarrow a^z C^z_k(i,j) + b^z, \quad z = H, V, D,\ k = n-2, n-1. \tag{5.14}
$$

- Finally, the degraded portion of the panchromatic image, $f_{n-2}(i,j)$, is replaced with each of the four original multispectral bands in turn, and the filter bank is inverted.

We thus obtain "what would be seen if the multispectral sensors had the resolution of the panchromatic sensor" (Ranchin and Wald, 2000). An ENVI/IDL extension for panchromatic sharpening with the DWT is described in Appendix C.2.2. Figure 5.9 shows an example of the DWT pan-sharpening of IKONOS imagery.

5.3.5 À *Trous* Fusion

The spectral fidelity of the pan-sharpened images obtained with the discrete wavelet transform is excellent, as will be shown in the following text. However, the DWT is not *shift invariant* (Bradley, 2003), which means that small spatial displacements in the input array can cause major variations in the wavelet coefficients at their various scales. This has no effect on perfect reconstruction if one simply inverts the transformation. However, a small misalignment can occur when the multispectral bands are "injected" into the panchromatic image pyramid, as described in Section 5.3.4. This sometimes leads to spatial artifacts (blurring, shadowing, and staircase effect) in the sharpened product (Yocky, 1996). These effects are visible in Figure 5.9 upon close inspection.

As an alternative to the DWT, the *à trous wavelet transform* (ATWT) has been proposed for image sharpening (Aiazzi et al., 2002). The ATWT is a multiresolution decomposition defined formally by a low-pass filter, $H = \{h(0), h(1), \ldots\}$, and a high-pass filter, $G = \delta - H$, where δ denotes an all-pass

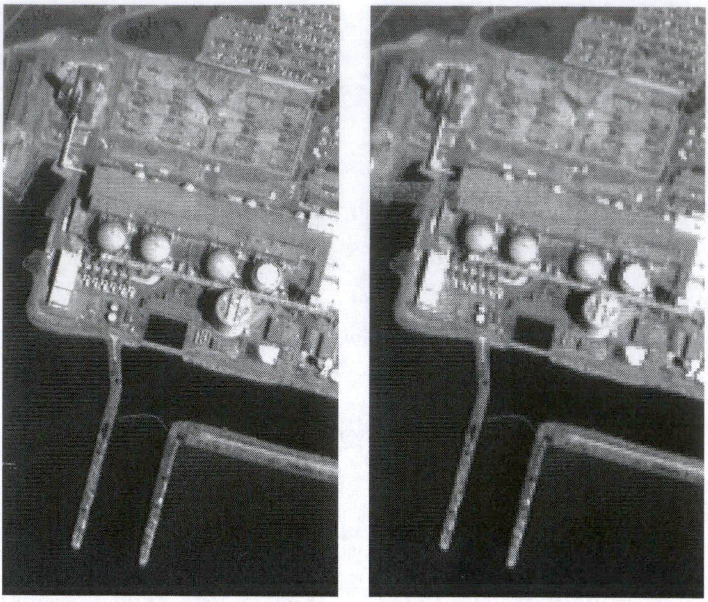

FIGURE 5.9
Panchromatic sharpening of an IKONOS multispectral image with the DWT filter bank. Left: panchromatically sharpened multispectral band of an image over a nuclear power reactor complex in Canada. Right: Original band upsampled by a factor of 4.

filter. The high-frequency part is then just the difference between the original image and the low-pass filtered image. Not surprisingly, this transformation does not allow perfect reconstruction if the output is downsampled. Therefore, downsampling is not performed at all. Rather, at the kth iteration of the low-pass filter, 2^{k-1} zeroes are inserted between the elements of H. This means that every other pixel is interpolated (averaged) on the first iteration,

$$H = \{h(0), 0, h(1), 0, \ldots\},$$

while on the second iteration,

$$H = \{h(0), 0, 0, h(1), 0, 0, \ldots\},$$

etc. (hence the name *à trous*, equivalent to "with holes"). The low-pass filter is usually chosen to be symmetric (unlike the Daubechies wavelet filters for example). The cubic B-spline filter turns out to be a good choice (Núnez et al., 1999), $H = \{1/16, 1/4, 3/8, 1/4, 1/16\}$ (see Chapter 3, Exercise 9).

Figure 5.10 outlines the scheme implemented in the ENVI/IDL extension for ATWT panchromatic sharpening, described in Appendix C.2.3, assuming a difference in spatial resolution of a factor of two: The MS band is resampled to match the dimensions of the high-resolution band. The

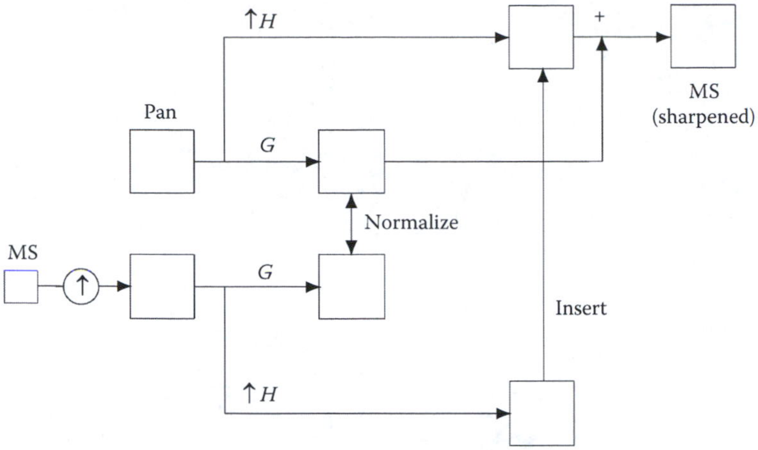

FIGURE 5.10
À trous image sharpening scheme for a multispectral to panchromatic resolution ratio of two. The symbol ↑ H denotes the upsampled low-pass filter.

à trous transformation is applied to both bands (columns and rows are filtered with the upsampled cubic spline filter, with the difference from the original image determining the high-pass result). The high-frequency component of the panchromatic image is normalized to that of the MS image in a similar way as for DWT sharpening (Equations 5.13 and 5.14). Then the smoothed panchromatic component is replaced by the filtered MS image and the transformation is inverted.

The transformation is highly redundant and requires considerably more computer storage to implement. However, when used for image sharpening, it is much less sensitive to misalignment between the multispectral and panchromatic images. A comparison with the DWT method is made in Figure 5.11.

5.3.6 Quality Index

Several quantitative measures have been suggested for determining the degree to which the spectral properties of the multispectral bands have been degraded in the pan-sharpening process (Tsai, 2004). Wang and Bovik (2002) suggest a particularly simple and intuitive measure of spectral fidelity between two image bands with pixel intensities represented by f and g:

$$Q = \frac{\sigma_{fg}}{\sigma_f \sigma_g} \cdot \frac{2\bar{f}\bar{g}}{\bar{f}^2 + \bar{g}^2} \cdot \frac{2\sigma_f \sigma_g}{\sigma_f^2 + \sigma_g^2} = \frac{4\sigma_{fg}\bar{f}\bar{g}}{(\bar{f}^2 + \bar{g}^2)(\sigma_f^2 + \sigma_g^2)}, \tag{5.15}$$

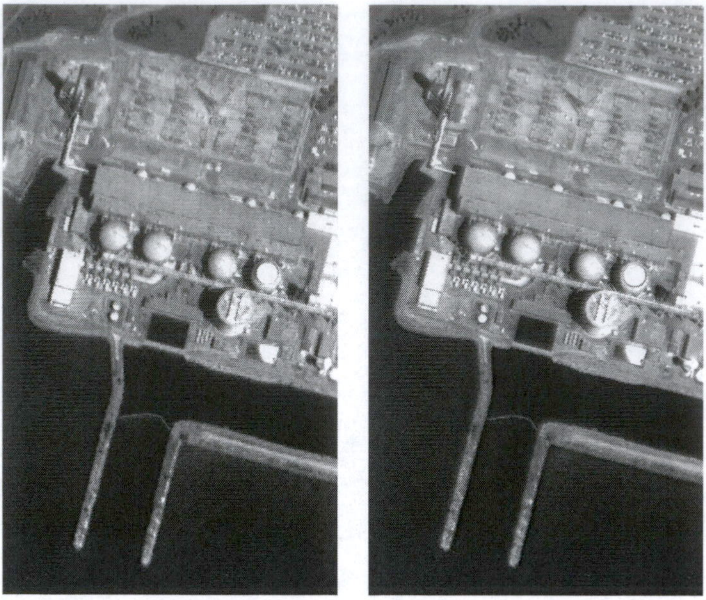

FIGURE 5.11
Same as Figure 5.9, except that the right-hand side now shows panchromatic sharpening with the *à trous* wavelet transform.

Listing 5.5
A routine for the Wang–Bovik quality index.

```
FUNCTION QI, x, y
   Sxy = correlate(x,y,/covariance)
   xm  = mean(x)
   ym  = mean(y)
   Sxx = variance(x)
   Syy = variance(y)
   RETURN, 4*Sxy*xm*ym/((Sxx+Syy)*(xm^2+ym^2))
END
```

where $\bar{f}$ ($\bar{g}$) and σ_f^2 (σ_g^2) are the mean and the variance of band f (g), and σ_{fg} is the covariance of the two bands (see Listing 5.5). The first factor in Equation 5.15 is seen to be the correlation coefficient between the two images, with values in $[-1, 1]$; the second factor compares their average brightness, with values in $[0, 1]$; and the third factor compares their contrasts, also in $[0, 1]$. Thus, perfect spectral correspondence would give a value $Q = 1$.*

* It should be remarked that this index is "marginal" in the sense that the original and the sharpened images are compared band-wise. A more stringent approach would be to test simultaneously for equal mean vectors and covariance matrices (see Anderson, 2003).

TABLE 5.1

Quality Indices for Three Panchromatic Sharpening Methods

MS Band	HSV	ATWT	DWT
1	0.313	0.612	0.811
2	0.217	0.661	0.828
3	0.312	0.641	0.815
4	—	0.674	0.796

Since image quality is normally not spatially invariant, it is usual to compute Q in, say, M sliding windows and then average over all such windows:

$$Q = \frac{1}{M}\sum_{j=1}^{M} Q_j.$$

An ENVI/IDL extension for determining the quality index for pan-sharpened images is given in Appendix C.2.4. Table 5.1 gives the results for the HSV, the DWT, and the ATWT fusion methods for the images in Figure 5.9. The discrete wavelet transform performs best.

5.4 Topographic Correction

Satellite images are two-dimensional representations of the three-dimensional Earth surface. The inclusion of the third dimension—the elevation—is required for terrain modeling and accurate georeferencing.

5.4.1 Rotation, Scaling, and Translation

Transformations of spatial coordinates in three dimensions that involve only rotations, scalings, and translations can be conveniently represented by a 4×4 transformation matrix, $\boldsymbol{A}$,

$$v^* = \boldsymbol{A}v, \tag{5.16}$$

where v is the column vector containing the original coordinates,

$$v = (X, Y, Z, 1)^\top,$$

and v^* contains the transformed coordinates,

$$v^* = (X^*, Y^*, Z^*, 1)^\top.$$

For example, the translation

$$X^* = X + X_0$$
$$Y^* = Y + Y_0$$
$$Z^* = Z + Z_0$$

corresponds to the transformation matrix

$$T = \begin{pmatrix} 1 & 0 & 0 & X_0 \\ 0 & 1 & 0 & Y_0 \\ 0 & 0 & 1 & Z_0 \\ 0 & 0 & 0 & 1 \end{pmatrix},$$

a uniform scaling by 50% to

$$S = \begin{pmatrix} 1/2 & 0 & 0 & 0 \\ 0 & 1/2 & 0 & 0 \\ 0 & 0 & 1/2 & 0 \\ 0 & 0 & 0 & 1 \end{pmatrix},$$

and a simple rotation, θ, about the z-axis to

$$\boldsymbol{R}_\theta = \begin{pmatrix} \cos\theta & \sin\theta & 0 & 0 \\ -\sin\theta & \cos\theta & 0 & 0 \\ 0 & 0 & 1 & 0 \\ 0 & 0 & 0 & 1 \end{pmatrix}.$$

The combined rotation, scaling, translation (RST) transformation is in this case,

$$\boldsymbol{v}^* = (\boldsymbol{R}_\theta \boldsymbol{S}\boldsymbol{T})\boldsymbol{v} = \boldsymbol{A}\boldsymbol{v}. \tag{5.17}$$

The inverse transformation is given by $\boldsymbol{A}^{-1} = \boldsymbol{T}^{-1}\boldsymbol{S}^{-1}\boldsymbol{R}_\theta^{-1}$.

5.4.2 Imaging Transformations

An imaging (or perspective) transformation projects 3D points onto a plane. It is used to describe the formation of a camera image and, unlike the RST transformation, is nonlinear since it involves division by coordinate values.

In Figure 5.12, a camera coordinate system (x, y, z) is shown, aligned with a *world coordinate system* (X, Y, Z) describing the terrain to be imaged. The camera focal length is λ. From simple geometry (similar triangles) we

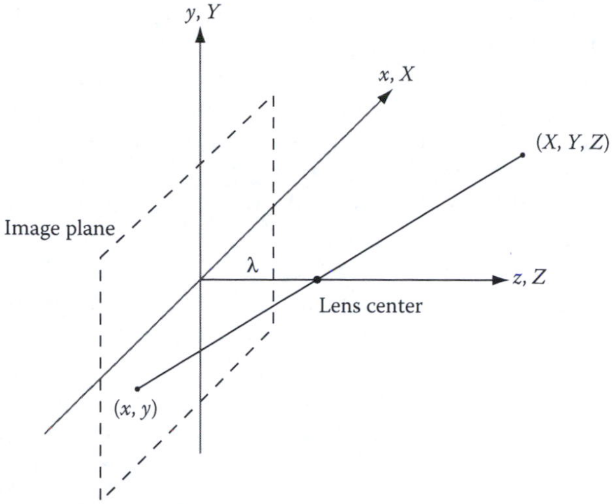

FIGURE 5.12
Basic imaging process. Camera coordinates are (x, y, z) and coincide with world coordinates (X, Y, Z). The focal length is λ.

obtain expressions for the image plane coordinates in terms of the world coordinates:

$$x = \frac{\lambda X}{\lambda - Z}$$
$$y = \frac{\lambda Y}{\lambda - Z}. \tag{5.18}$$

Solving for the X and Y world coordinates,

$$X = \frac{x}{\lambda}(\lambda - Z)$$
$$Y = \frac{y}{\lambda}(\lambda - Z). \tag{5.19}$$

Thus, in order to extract the geographical coordinates (X, Y) of a point on the earth's surface from its image coordinates (x, y), we require knowledge of the elevation Z. Correcting for the elevation in this way constitutes the process of *orthorectification*, resulting in an image without perspective distortion—every point appears as if viewed directly from above.

5.4.3 Camera Models and RFM Approximations

Equations 5.19 are very simple because they assume that the world and the image coordinates coincide. But in order to apply them, one has first to transform the world coordinate system to the satellite coordinate system. This can be done in a straightforward way with the rotation and translation transformations introduced above. However, it requires an accurate knowledge of the altitude, the orientation, and the detailed properties of the satellite imaging system at the time of image acquisition (or, more precisely, *during* the acquisition, since the latter is never instantaneous). The resulting nonlinear equations that relate image and world coordinates, referred to as the *collinearity equations*, constitute the *camera model* for that particular image.

The direct use of a camera model for image processing is complicated as it requires extremely exact, sometimes proprietary information about the sensor system and its orbit. An alternative exists if the image provider also supplies a so-called *rational function model* (RFM) with the imagery. The RFM approximates the camera model for each acquisition in terms of ratios of polynomials, (see, e.g., Tao and Hu, 2001). The RFMs have the form

$$
\begin{aligned}
r' &= f(X', Y', Z') = \frac{a(X', Y', Z')}{b(X', Y', Z')} \\
c' &= g(X', Y', Z') = \frac{c(X', Y', Z')}{d(X', Y', Z')},
\end{aligned}
\tag{5.20}
$$

where c' and r' are the column and row (x, y) coordinates in the image plane relative to an origin (c_0, r_0) and scaled by a factor, c_s, respectively to r_s, that is,

$$
c' = \frac{c - c_0}{c_s}, \quad r' = \frac{r - r_0}{r_s}.
$$

Similarly, X', Y', and Z' are relative, scaled world coordinates:

$$
X' = \frac{X - X_0}{X_s}, \quad Y' = \frac{Y - Y_0}{Y_s}, \quad Z' = \frac{Z - Z_0}{Z_s}.
$$

The polynomials a, b, c, and d in Equation 5.14 are typically to the third order in the world coordinates, for example,

$$
\begin{aligned}
a(X, Y, Z) = {} & a_0 + a_1X + a_2Y + a_3Z + a_4XY + a_5XZ + a_6YZ + a_7X^2 + a_8Y^2 \\
& + a_9Z^2 + a_{10}XYZ + a_{11}X^3 + a_{12}XY^2 + a_{13}XZ^2 + a_{14}X^2Y + a_{15}Y^3 \\
& + a_{16}YZ^2 + a_{17}X^2Z + a_{18}Y^2Z + a_{19}Z^3.
\end{aligned}
$$

The advantage of using *ratios* of polynomials is that they are less subject the to interpolation error than simple polynomials.

For a given acquisition, the provider will fit an RFM to his camera model with a least squares fitting procedure using a two- and a three-dimensional grid of points covering the image respectively to world spaces. The RFM is capable of representing the camera model extremely well and can be used as a replacement for it. Both Space Imaging Corp. and DigitalGlobe Inc. provide RFMs with their so-called *ortho-ready* high-resolution imagery (IKONOS, QuickBird, and WorldView-1 platforms).

To illustrate a simple use of RFM data, consider a vertical structure in a high-resolution image, such as a chimney or building facade. Suppose that we determine the image coordinates of the bottom and the top of the structure to be (r_b, c_b) and (r_t, c_t), respectively. Then, from Equations 5.20,

$$\begin{aligned} r'_b &= f(X', Y', Z'_b) \\ c'_b &= g(X', Y', Z'_b) \\ r'_t &= f(X', Y', Z'_t) \\ c'_t &= g(X', Y', Z'_t), \end{aligned} \tag{5.21}$$

since the (X', Y') coordinates must be the same. This would appear to constitute a set of four equations in four unknowns, X', Y', Z'_b, and Z'_t; however, the solution is unstable because of the close similarity of Z'_t to Z'_b. Nevertheless the object height, $Z'_t - Z'_b$, can be obtained by the following procedure:

1. Get (r_b, c_b) and (r_t, c_t) from the image and convert these to scaled values (r'_b, c'_b) and (r'_t, c'_t).
2. Solve the first two of Equations 5.21, with Newton's method (Press et al., 2002) for instance, for X' and Y' with Z'_b set equal to the average elevation in the scene, that is, Z_0/Z_s, if no digital elevation model (DEM) is available, otherwise to the true, properly scaled elevation.
3. For a range of Z'_t values increasing from Z'_b to some maximum value well exceeding the expected height, calculate (r'_t, c'_t) from the last two of Equations 5.21. Choose for Z'_t the value that gives the closest agreement to the (r_t, c_t) values read in.

Quite generally, the RFM can be used as an alternative for providing end users with the necessary information to perform their own photogrammetric processing. An ENVI/IDL extension for object height determination from RFM data is given in Appendix C.2.5. It is implemented as a GUI (Graphics User Interface) making use of IDL widgets. A screenshot of its use with a QuickBird image is shown in Figure 5.13.

5.4.4 Stereo Imaging and Digital Elevation Models

The missing elevation information, Z, in Equation 5.18 or Equation 5.19 can be obtained with stereoscopic imaging techniques. Figure 5.14 depicts two cameras viewing the same world point, w, from two positions. The separation B of the lens centers is the *baseline*. The objective is to find the coordinates

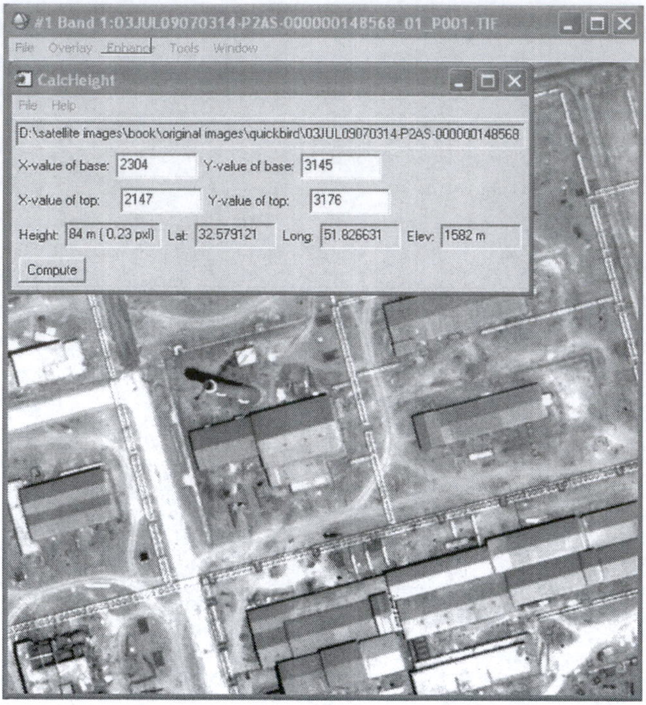

FIGURE 5.13
Determining the height of a vertical structure (the chimney, center left) in a QuickBird panchromatic image with the help of an RFM.

of the point w if its image points have coordinates (x_1, y_1) and (x_2, y_2). We assume that the cameras are identical and that their image coordinate systems are perfectly aligned, differing only in the location of their origins. The Z coordinate of w is the same for both coordinate systems.

In Figure 5.14, the coordinate system of the first camera is shown as coinciding with the world coordinate system. Therefore, from the first of Equations 5.19,

$$X_1 = \frac{x_1}{\lambda}(\lambda - Z).$$

Alternatively, if the second camera is brought to the origin of the world coordinate system,

$$X_2 = \frac{x_2}{\lambda}(\lambda - Z).$$

But, from Figure 5.14,

$$X_2 = X_1 + B,$$

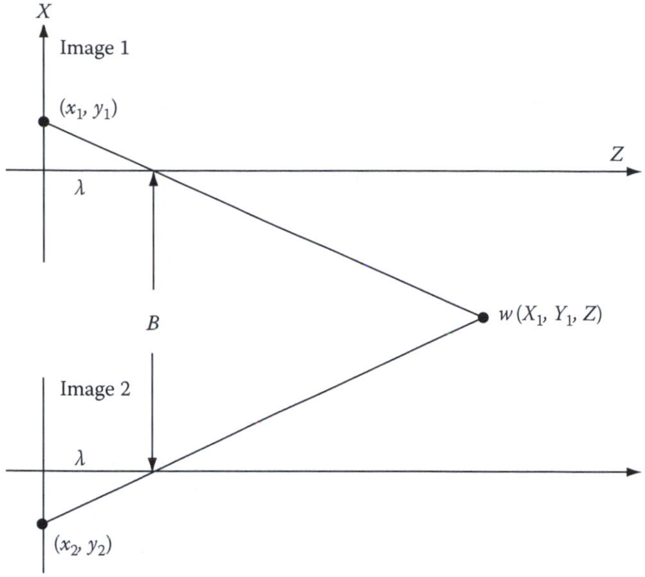

FIGURE 5.14
The stereo imaging process. The coordinates of the point w are relative to a world coordinate system coinciding with the upper camera.

where B is the baseline. We have from the above three equations,

$$Z = \lambda - \frac{\lambda B}{x_2 - x_1}. \tag{5.22}$$

Thus, if the displacement of the image coordinates of the point w, namely, $x_2 - x_1$, can be determined, then the Z coordinate can be calculated. The task is then to find two corresponding points in different images of the same scene. This is usually accomplished by spatial correlation techniques and is closely related to the problem of image–image registration, discussed later in this chapter.

Because the stereo images must be correlated, best results are obtained if they are acquired within a very short time of each other, preferably "along track" if a single platform is used (see Figure 5.15). This figure shows the orientation and the imaging geometry of the VNIR 3N and 3B cameras on the ASTER platform for acquiring a stereo full scene. The satellite travels at a speed of 6.7 km/s at a height of 705 km. A $60 \times 60\,\text{km}^2$ full scene is scanned in 9 s. Then, 55 s later, the same scene is scanned by the back-looking camera, corresponding to a baseline of $B = 370\,\text{km}$. The along-track geometry means that the *epipolar lines* of the stereo pair are parallel, that is, the displacements due to the viewing angle are only along a common direction, in this case,

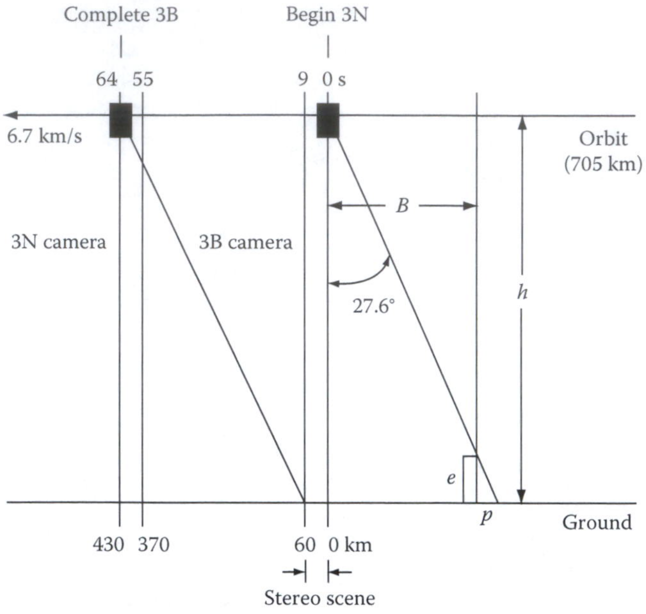

FIGURE 5.15
ASTER along-track stereo acquisition geometry. Parallax p can be related to elevation e by similar triangles: $e/p = (h - e)/B \approx h/B$. (Adapted from Lang, H.R. and Welch, R., Algorithm theoretical basis document for ASTER digital elevation models, Technical Report, Jet Propulsion Laboratory, Pasadena, CA and University of Georgia, Athens, GA, 1999.)

the y axis, in the imaging planes of the two cameras. Therefore, the spatial correlation algorithm used to match points can be one dimensional. If carried out on a pixel-for-pixel basis, one obtains a DEM having approximately the same resolution as that of the stereo imagery.

As an example, Figures 5.16 and 5.17 show an ASTER along-track stereo pair. The IDL program in Listing 5.6 calculates a very rudimentary DEM making use of the routines `CORREL_IMAGES()` and `CORRMAT_ANALYZE` from the IDL Astronomy User's Library* to calculate the local cross correlations for the two images. For each pixel in the nadir image, a 15 × 15 window is moved along a 15 × 51 window (called the *epipolar segment*) in the back-looking image centered at the corresponding position (lines 31–40), allowing for a maximum disparity of ±18 pixels. The point of maximum correlation determines the parallax or disparity, p. This is related to the elevation e of the pixel by

$$e = p \cdot \frac{h}{B} \times 15\text{m},$$

where h is the height of the sensor and B is the baseline (see Figure 5.15).

* Available at www.astro.washington.edu/deutsch/idl/htmlhelp/slibrary01.html

FIGURE 5.16
ASTER 3N band (nadir camera) over a hilly region in North Korea.

Figure 5.18 shows the result. Clearly, there are problems due to correlation errors; however, the relative elevations are approximately correct when compared to the DEM determined with the ENVI commercial add-on `AsterDTM` (Sulsoft, 2003) (see Figure 5.19). This much more sophisticated approach uses image pyramids to accumulate disparities at increasing scales (see also Quam, 1987).

In generating a DEM in this way, either the complete camera model or an RFM can be referred to, but usually neither is sufficient for determining the absolute elevations. Most often, additional ground reference points within the image with known elevations are also required for an absolute calibration. Orthorectification of the image (correction of displacement errors due to parallax and off-nadir viewing angles) is carried out on the basis of the DEM and consists of referring the (X, Y, Z) coordinates of each pixel to the (X, Y) coordinates of a given map projection.

5.4.5 Slope and Aspect

Topographic modeling, or terrain analysis, involves the processing of elevation data provided by a DEM. Specifically, we consider here the generation of *slope images*, which give the steepness of the terrain at each pixel, and *aspect*

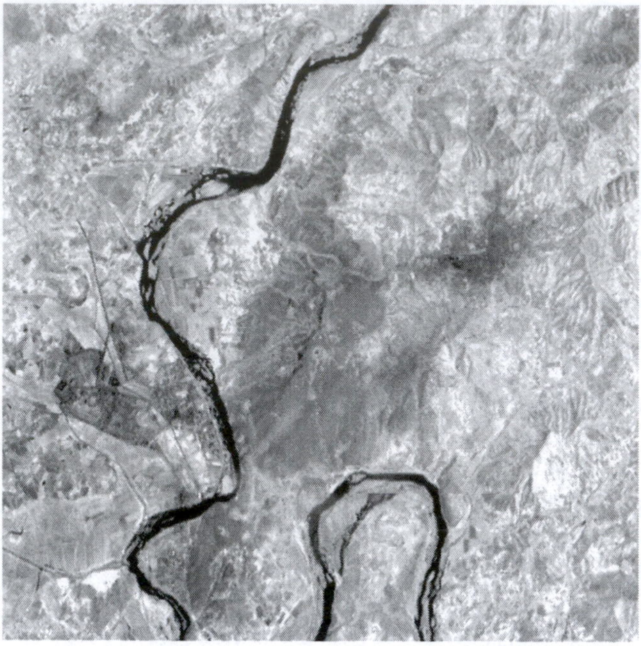

FIGURE 5.17
ASTER 3B band (back-looking camera) registered to Figure 5.16 by a first-order polynomial transformation (see Section 5.5).

images, which give the direction relative to the north of a vector normal to the landscape at each pixel.

A 3×3 pixel window can be used to determine both the slope and the aspect as follows (see Figure 5.20). Define

$$\Delta x_1 = c - a \quad \Delta y_1 = a - g$$
$$\Delta x_2 = f - d \quad \Delta y_2 = b - h$$
$$\Delta x_3 = i - g \quad \Delta y_3 = c - i$$

and

$$\Delta x = (\Delta x_1 + \Delta x_2 + \Delta x_3)/3$$
$$\Delta y = (\Delta y_1 + \Delta y_2 + \Delta y_3)/3.$$

Then the slope angle in radians at the central pixel position is given by (Exercise 10)

Listing 5.6

Calculation of a DEM with cross correlation.

```
PRO ex5_4

height = 705.0
base = 370.0
pixel_size = 15.0

envi_select, title='Choose␣nadir␣image', $
   fid=fid1, dims=dims1, pos=pos1, /band_only
IF (fid1 EQ -1) THEN RETURN
envi_select, title='Choose␣back-looking␣image', $
   fid=fid2, dims=dims2, pos=pos2, /band_only
IF (fid2 EQ -1) THEN RETURN
im1 = envi_get_data(fid=fid1,dims=dims1,pos=pos1)
im2 = envi_get_data(fid=fid2,dims=dims2,pos=pos2)

n_cols = dims1[2]-dims1[1]+1
n_rows = dims1[4]-dims1[3]+1
parallax = fltarr(n_cols,n_rows)

progressbar = Obj_New('progressbar',Color='blue', $
 Text='0', title='Cross␣correlation,␣column␣...', $
                 xsize=250,ysize=20)
progressbar->start
FOR i=7L,n_cols-8 DO  BEGIN
   IF progressbar->CheckCancel() THEN BEGIN
      envi_enter_data,pixel_size*parallax*(height/base)
      progressbar->Destroy
      RETURN
   ENDIF
   progressbar->Update,(i*100)/n_cols,text=strtrim(i,2)
   FOR j=25L,n_rows-26 DO BEGIN
      cim = correl_images(im1[i-7:i+7,j-7:j+7], $
             im2[i-7:i+7,j-25:j+25],xoffset_b=0, $
             yoffset_b=-25,xshift=0,yshift=25)
      corrmat_analyze,cim,xoff,yoff,maxcorr,edge, $
             plateau,magnification=2
      IF (yoff LT -5) OR (yoff GT 18) THEN $
        parallax[i,j]=parallax[i,j-1] ELSE $
        parallax[i,j]=2*yoff
   ENDFOR
ENDFOR
progressbar->destroy
envi_enter_data, pixel_size*parallax*(height/base)

END
```

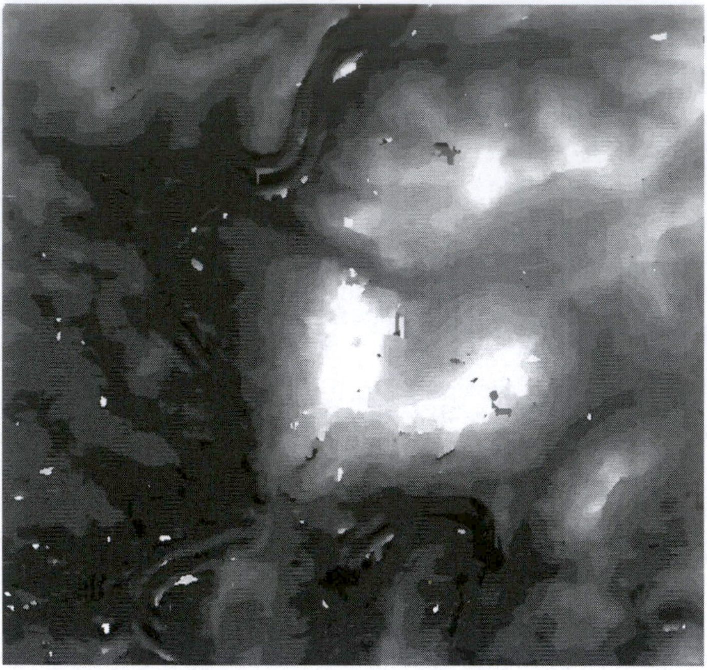

FIGURE 5.18
A rudimentary DEM from the ASTER stereo pair of Figures 5.16 and 5.17 using the program in Listing 5.5.

$$\theta_p = \tan^{-1}\left(\frac{\sqrt{(\Delta x)^2 + (\Delta y)^2}}{2 \cdot \mathrm{GSD}}\right), \tag{5.23}$$

whereas the aspect in radians measured clockwise from the north is

$$\phi_o = \tan^{-1}\left(\frac{\Delta x}{\Delta y}\right). \tag{5.24}$$

Slope/aspect determinations from a DEM are available in the ENVI main menu under `Topographic/Topographic Modeling` (see Figures 5.21 and 5.22).

5.4.6 Illumination Correction

Topographic modeling can be used to correct images for the effects of the local solar illumination. The local illumination depends not only upon the sun's position (elevation and azimuth) but also upon the slope and the aspect

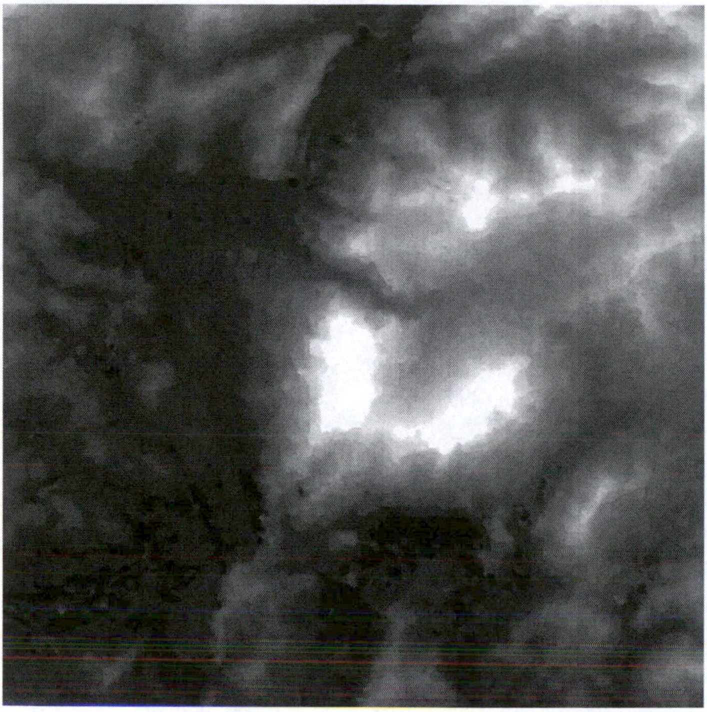

FIGURE 5.19
A DEM generated with the commercial product AsterDTM. (From Sulsoft, Aster DTM 2.0 installation and user's guide, Technical Report, Sulsoft Ltd, Porto Alegre, Brazil, 2003.)

a	*b*	*c*
d	*e*	*f*
g	*h*	*i*

FIGURE 5.20
Pixel elevations in an 8-neighborhood. The letters represent elevations in meters.

of the terrain being illuminated. Figure 5.23 shows the angles involved. The quantity to be calculated is the local solar incidence angle, γ_i, which determines the local irradiance. From trigonometry we can calculate the relation

$$\cos\gamma_i = \cos\theta_p\cos\theta_z + \sin\theta_p\sin\theta_z\cos(\phi_a - \phi_o). \tag{5.25}$$

An example of a $\cos\gamma_i$ image in hilly terrain is shown in Figure 5.24.

Image classification and change detection algorithms, the subjects of the next chapters, will achieve better accuracies if variable image properties extrinsic to the actual surface reflectance are first removed. For a Lambertian surface, the reflected radiance L_H from a horizontal surface toward a sensor (ignoring all atmospheric effects) is given by Equation 1.1,

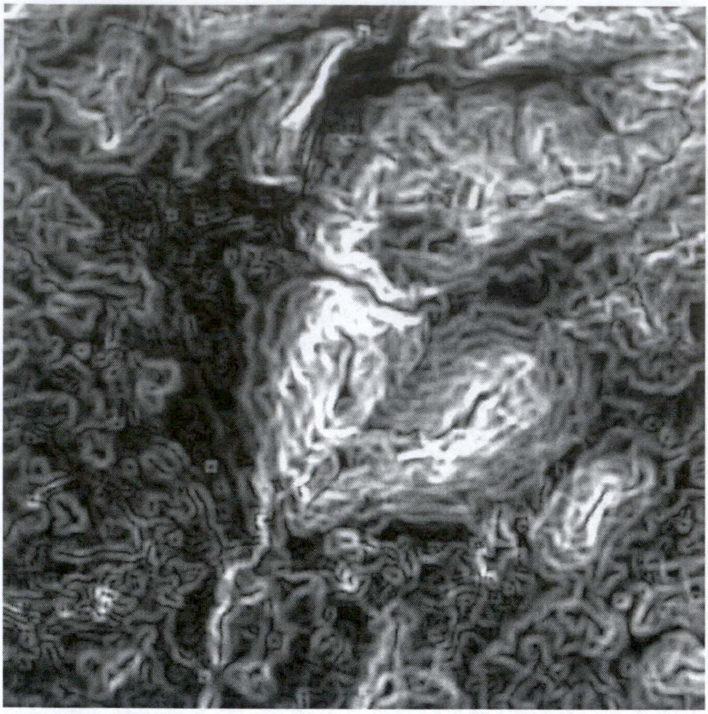

FIGURE 5.21
Slope image calculated with the DEM of Figure 5.19.

which we write in the simplified form

$$L_H = E \cdot \cos\theta_z \cdot R. \tag{5.26}$$

For a surface in rough terrain, the reflected radiance L_T is similarly

$$L_T = E \cdot \cos\gamma_i \cdot R, \tag{5.27}$$

thus giving the standard *cosine correction* relating the observed radiance L_T to that which would have been observed had the terrain been flat, namely,

$$L_H = L_T \frac{\cos\theta_z}{\cos\gamma_i}. \tag{5.28}$$

The Lambertian assumption is in general a poor approximation, the actual reflectance being governed by a *bidirectional reflectance distribution function* (BRDF), which describes the dependence of reflectance on both

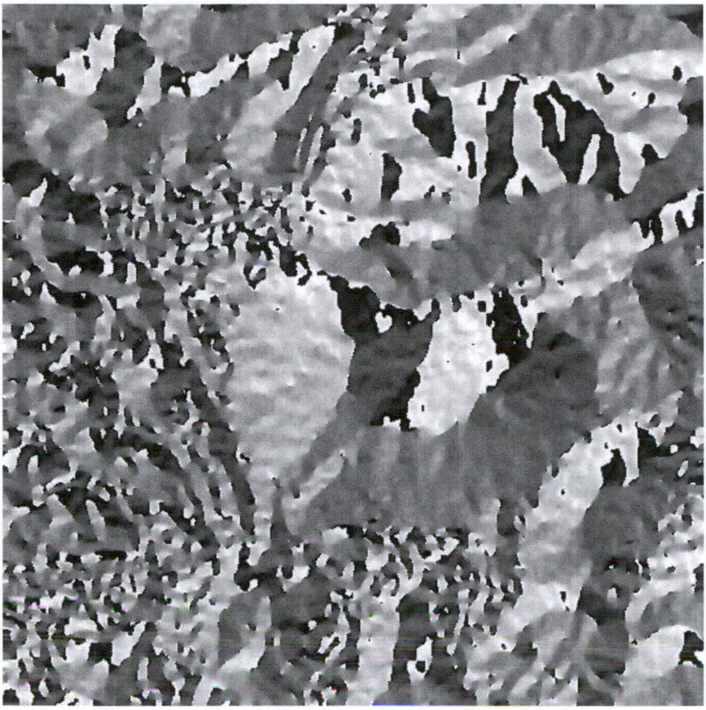

FIGURE 5.22
Aspect image calculated with the DEM of Figure 5.19.

illumination and viewing angles as well as on wavelength (Beisl, 2001). Particularly for the large range of incident angles involved with rough terrain, the cosine correction will over- or underestimate the extremes and lead to unwanted artifacts in the corrected imagery.

An example of an approach that takes a better account of BRDF effects is the semiempirical cosine correction (C-correction) method, suggested by Teillet et al. (1982). We replace Equations 5.26 and 5.27 by

$$L_H = m \cdot \cos\theta_z + b, \quad L_T = m \cdot \cos\gamma_i + b.$$

The parameters m and b can be estimated from a linear regression of observed radiance L_T vs. $\cos\gamma_i$ for a particular image band.* Then, instead of Equation 5.28, one uses

$$L_H = L_T \left(\frac{\cos\theta_z + b/m}{\cos\gamma_i + b/m} \right) \tag{5.29}$$

* The regression may be carried out separately for different land cover categories in order to take into account the variation of BRDF effects with the land cover.

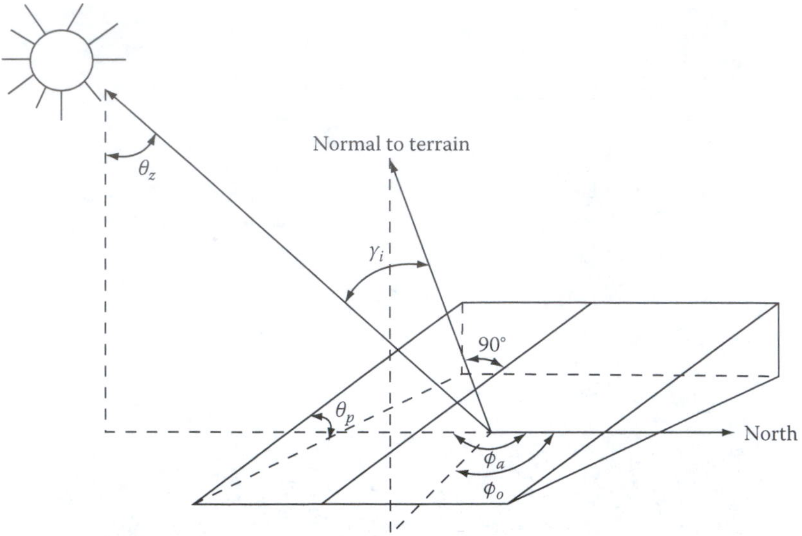

FIGURE 5.23
Angles involved in the computation of local solar incidence: θ_z = solar zenith angle, ϕ_a = solar azimuth, θ_p = slope, ϕ_o = aspect, and γ_i = local solar incidence angle. (Adapted from Riano, D. et al., *IEEE Trans. Geosci. Remote Sens.*, 41, 1056, 2003.)

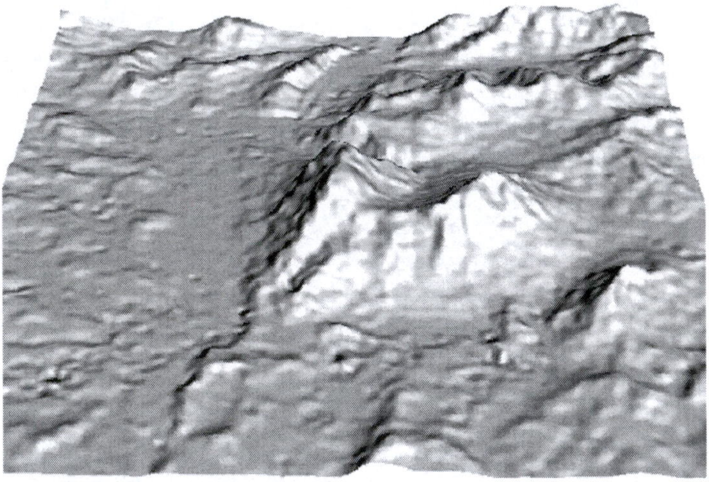

FIGURE 5.24
Cosine of solar incidence angle, γ_i, stretched across the DEM of Figure 5.19; solar zenith angle $\theta_z = 32.3°$ and solar azimuth $\phi_a = 151.3°$. The 3D effect was generated from the ENVI main menu with the command `Topographic/3D SurfaceView`; elevation exaggeration factor = 2.

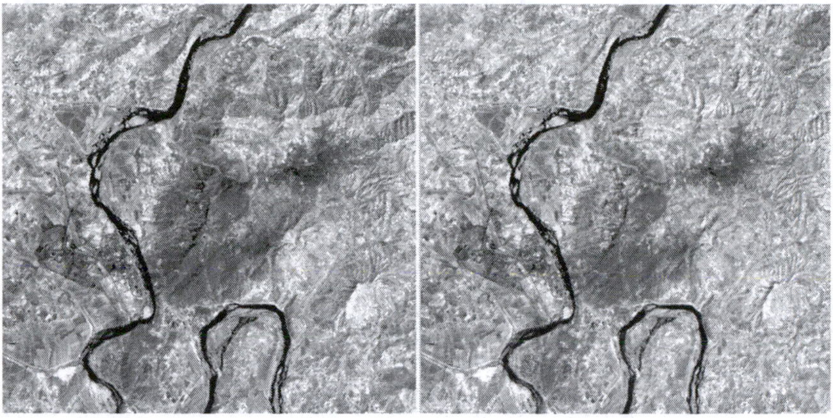

FIGURE 5.25
Original image (left) and image after application of the C-correction method (right) for the scene of Figure 5.16. Hillsides sloped away from the sun (see Figure 5.24) are generally brighter after correction.

as a correction formula. An ENVI/IDL extension for illumination correction with the C-correction approximation, including land cover masking, is given in Appendix C.2.6. An example of its application is shown in Figure 5.25.

5.5 Image–Image Registration

Image registration, either to another image or to a map, is a fundamental task in remote sensing data processing. It is required for georeferencing, stereo imaging, accurate change detection, and indeed for any kind of multitemporal image analysis. A tedious task associated with manual coregistration in the past has been the setting of tie-points or, as they are often called, ground control points (GCPs), since, in general, it was necessary to resort to manual entry. Fortunately, there now exist many reliable automatic or semiautomatic procedures for locating tie-points. These can be divided roughly into four classes (Reddy and Chatterji, 1996):

1. Algorithms that use pixel intensity values directly, such as correlation methods or methods that maximize mutual information
2. Frequency- or wavelet-domain methods that use, for example, the fast Fourier transform
3. Feature-based methods that use low-level features, such as shapes, edges, and corners
4. Algorithms that use high-level features and the relations between them (object-based methods)

In ENVI Version 4.2, tie-point generation using both correlation methods (referred to in the ENVI documentation as "area-based") as well as feature-based methods was introduced. We will consider in the following text of frequency-domain and feature-based algorithms that can be used to complement ENVI's built-in registration capability and that illustrate some of the principles involved.

5.5.1 Frequency-Domain Registration

Consider two $c \times c$ grayscale images, $g_1(i,j)$ and $g_2(i,j)$, where g_2 is offset relative to g_1 by an integer number of pixels:

$$g_2(i,j) = g_1(i',j') = g_1(i-i_0, j-j_0).$$

Taking the Fourier transform, we have

$$\hat{g}_2(k,\ell) = \frac{1}{c^2}\sum_{ij} g_1(i-i_0, j-j_0)e^{-\mathrm{i}2\pi(ik+j\ell)/c},$$

or, with a change of indices to $i'j'$,

$$\begin{aligned}\hat{g}_2(k,\ell) &= \frac{1}{c^2}\sum_{i'j'} g_1(i',j')e^{-\mathrm{i}2\pi(i'k+j'\ell)/c}e^{-\mathrm{i}2\pi(i_0k+j_0\ell)/c}\\ &= \hat{g}_1(k,\ell)e^{-\mathrm{i}2\pi(i_0k+j_0\ell)/c}.\end{aligned}$$

This is the Fourier translation property that we saw in Chapter 3 (see Equation 3.11). Therefore, we can write

$$\hat{g}_2(k,\ell)\hat{g}_1^*(k,\ell) = |\hat{g}_1(k,\ell))|^2 e^{-\mathrm{i}2\pi(i_0k+j_0\ell)/c},$$

where $\hat{g}_1^*$ is the complex conjugate of $\hat{g}_1$, and hence

$$\frac{\hat{g}_2(k,\ell)\hat{g}_1^*(k,\ell)}{|\hat{g}_2(k,\ell)\hat{g}_1^*(k,\ell)|} = e^{-\mathrm{i}2\pi(i_0k+j_0\ell)/c}. \tag{5.30}$$

The inverse transform of the right-hand side of Equation 5.30 exhibits a delta function (spike) at the coordinates (i_0, j_0). Thus, if two otherwise identical (or closely similar) images are offset by an integer number of pixels, the offset can be found by taking their Fourier transforms, computing the ratio on the left-hand side of Equation 5.30 (the so-called *cross-power spectrum*) and then taking the inverse transform of the result. The position of the maximum value in the inverse transform gives the offset values of i_0 and j_0. The IDL program in Listing 5.7 illustrates the procedure; the result is shown in

Listing 5.7
Image matching by phase correlation.

```
PRO ex5_5

; read a JPEG image, cut out two 512x512 pixel arrays
filename = Dialog_Pickfile(Filter='*.jpg',/READ)

IF filename EQ '' THEN RETURN ELSE BEGIN
   Read_JPeG,filename,image
   g1 = image[0,10:521,10:521]
   g2 = image[0,0:511,0:511]

; perform Fourier transforms
   f1 = fft(g1, /double)
   f2 = fft(g2, /double)

; Determine the offset
   g = fft(f2*conj(f1)/abs(f1*conj(f1)),$
        /inverse,/double)
   pos = where(g EQ max(g))
   PRINT,'Offset='+strtrim(pos MOD 512) $
        +strtrim(pos/512)

; output as EPS file
   thisDevice =!D.Name
   set_plot, 'PS'
   Device, Filename='c:\temp\fig5_25.eps', $
          xsize=4,ysize=4,/inches,/Encapsulated
   shade_surf,g[0,0:50,0:50]
   device,/close_file
   set_plot, thisDevice
ENDELSE

END
```

Figure 5.26. Images that differ not only by an offset but also by a rigid rotation and/or a change of scale can in principle be registered similarly (Reddy and Chatterji, 1996). Subpixel registration is also possible with a refinement of the above method (Shekarforoush et al., 1995) (see the routine PHASE_CORR in Appendix C.2.4).

5.5.2 Feature Matching

Various techniques for an automatic determination of tie-points based on low-level features have been suggested in the literature. We next discuss and implement one such method, namely, the modification of a

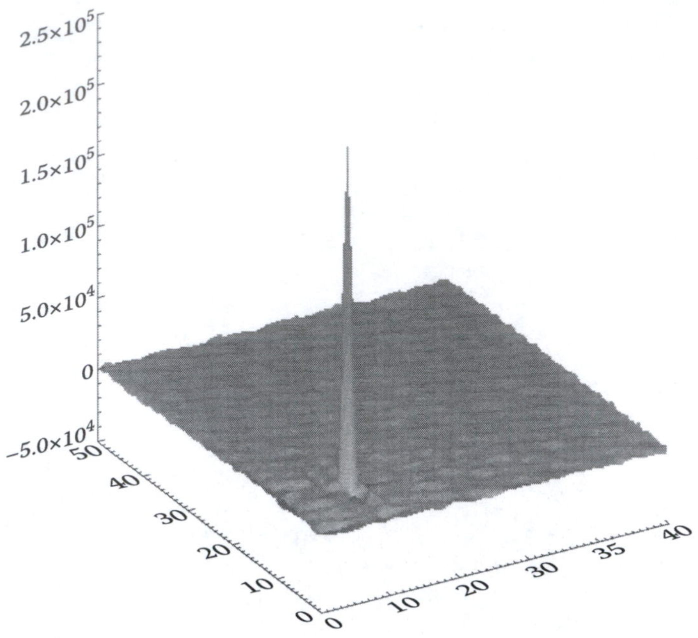

FIGURE 5.26
Phase correlation of two identical images shifted relative to one another by 10 pixels; image generated by the program shown in Listing 5.6.

contour-matching procedure proposed by Li et al. (1995). It functions especially well in bitemporal scenes in which vegetation changes do not dominate, and can of course be augmented by other automatic feature-matching methods or by manual selection. The required steps are shown in Figure 5.27 and are described below.

5.5.2.1 High-Pass Filtering

The first step involves the application of a Laplacian of Gaussian filter to both images in the manner discussed in Section 5.2.1. After determining the

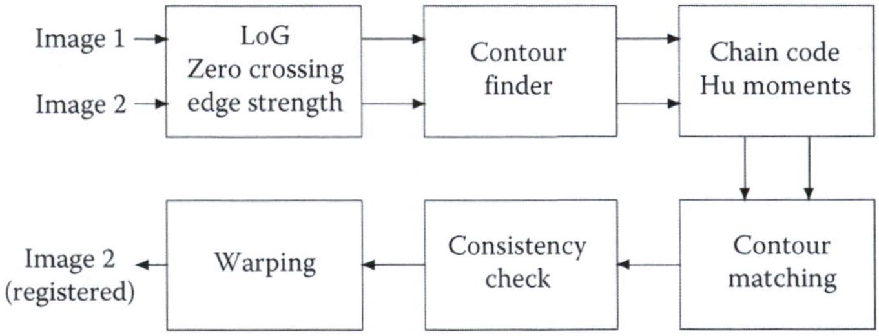

FIGURE 5.27
Image–image registration with contour matching.

contours by examining zero crossings of the LoG-filtered image, the contour strengths are encoded in the pixel intensities. Strengths are taken to be proportional to the magnitude of the gradient at the zero crossing determined by a Sobel filter, as illustrated in the program shown in Listing 5.2.

5.5.2.2 Closed Contours

In the next step, all closed contours with strengths above some given threshold are determined by tracing the contours. Pixels that have been visited during tracing are set to zero so that they will not be visited again. A typical result is shown in Figure 5.28.

5.5.2.3 Chain Codes and Moments

For subsequent matching purposes, all significant closed contours found in the preceding step are *chain encoded*. Any curve or contour can be represented by an integer sequence $\{a_1, a_2 \ldots a_i \ldots\}$, $a_i \in \{0, 1, 2, 3, 4, 5, 6, 7\}$, depending on

FIGURE 5.28
Closed contours derived from the 3N band of an ASTER image over Nevada, acquired in July 2003.

the relative position of the current pixel with respect to the previous pixel in the curve. A shift to the east is coded as 0, to the northeast as 1, and so on. This simple code has the drawback that some contours produce a wraparound. For example, the line in the direction -22.5° has the chain code $\{707070\ldots\}$. Li et al. (1995) suggest the smoothing operation

$$\{a_1 a_2 \ldots a_n\} \rightarrow \{b_1 b_2 \ldots b_n\},$$

where $b_1 = a_1$ and $b_i = q_i$. The integer q_i satisfies $(q_i - a_i) \bmod 8 = 0$ and $|q_i - b_{i-1}| \rightarrow \min$, $i = 2,3\ldots n$.* They also suggest applying the smoothing filter $\{0.1, 0.2, 0.4, 0.2, 0.1\}$ to the result. After both processing steps, two chain codes can be easily compared by "sliding" one over the other and determining their maximum correlation.

The closed contours are further characterized by determining their first four Hu moments, $h_1 \ldots h_4$ (Equations 5.11).

5.5.2.4 Contour Matching

Each significant contour in one image is first matched with contours in the second image according to their invariant moments. This is done by setting a threshold on the allowed differences, for instance, one standard deviation. If one or more matches are found, the best candidate for a tie-point is then chosen to be that matched contour in the second image for which the chain code correlation with the contour in the first image is maximum. If the maximum correlation is less than some threshold, for example, 0.9, then the match is rejected. The tie-point coordinates are taken to be the centers of gravity, $(\bar{x}_1, \bar{x}_2)$, of the matched contour pairs (see Equations 5.8).

5.5.2.5 Consistency Check

The contour-matching procedure invariably generates some false tie-points, so a further processing step is required. In Li et al. (1995), the fact that distances are preserved under a rigid transformation, is used. Let $\overline{A_1 A_2}$ represent the distance between two points A_1 and A_2 in an image. For two sets of m matched contour centers, $\{A_i \mid i = 1\ldots m\}$ and $\{B_i \mid i = 1\ldots m\}$, in images 1 and 2, the ratios

$$\overline{A_i A_j}/\overline{B_i B_j}, \quad i = 1\ldots m,\ j = i+1\ldots m,$$

* This is rather cryptic, so here is an example: For the wraparound sequence $\{707070\ldots\}$, we have $a_1 = b_1 = 7$ and $a_2 = 0$. Therefore, we must choose $q_2 = 8$, since this satisfies $(q_2 - a_2) \bmod 8 = 0$ and $|q_2 - b_1| = 1$. (For the alternatives $q_2 = 0, 16, 24\ldots$, the difference $|q_2 - b_1|$ is larger.) Continuing the same argument leads to the new sequence $\{787878\ldots\}$ with no wraparound.

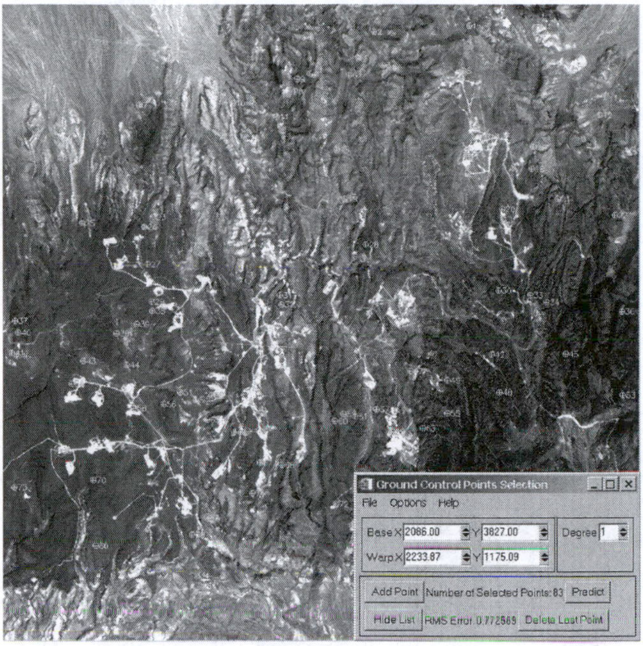

FIGURE 5.29
(See color insert following page 114.) 85 tie-points obtained by matching the contours of Figure 5.28 with those obtained from a similar image acquired in June 2001. The RMS error is 0.77 pixel for first-order polynomial warping (see text).

are calculated. These should form a cluster, so that indices associated with ratios scattered away from the cluster center can be rejected as false matches.

5.5.2.6 Implementation in IDL

An ENVI/IDL extension for tie-point determination using the above techniques is described in Appendix C.2.7. In this implementation, use of the scale and the rotational invariance of the Hu moments is in fact not made, because the chain codes as defined above are only translationally invariant.* The program's intended application is for an accurate subpixel image–image registration of multitemporal scenes that are acquired with the same sensor (or that have been aligned and resampled to the same GSD), for instance as a preliminary to change detection. Figures 5.28 and 5.29 illustrate the program's application to bitemporal ASTER data. For comparison, Figure 5.30 shows the tie-points generated by ENVI's automatic feature-based matching

* Rotational invariance can be achieved for instance by differencing the chain codes, scale invariance by resampling them (see Gonzalez and Woods, 2002, Chapter 11).

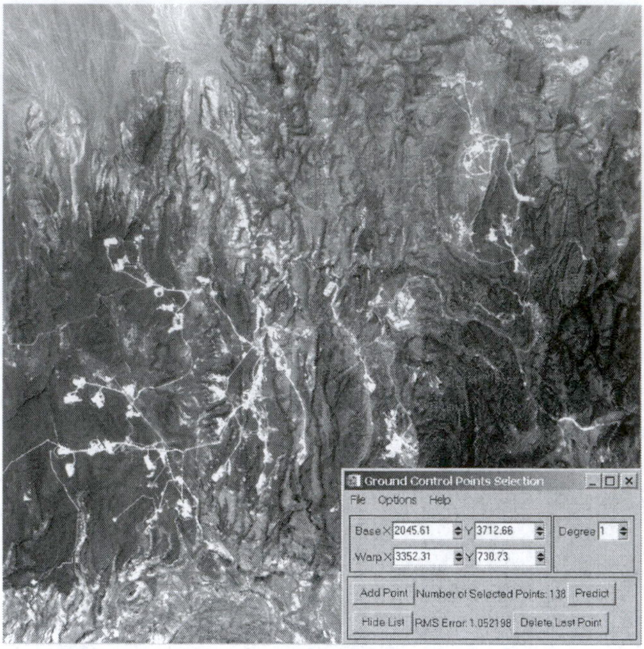

FIGURE 5.30
(See color insert following page 114.) 138 tie-points obtained with ENVI's feature-based matching procedure applied to the July 2003 and June 2001 ASTER scenes. The RMS error is 1.05 pixel for first-order polynomial warping.

algorithm. Although ENVI finds considerably more reference points, they are less uniformly distributed than the matched contours.

5.5.3 Resampling and Warping

If we represent with (x_1, x_2) the coordinates of a point in image 1 and with (u_1, u_2) the corresponding point in image 2, then, for example, a second-order polynomial map of image 2 to image 1 is given by

$$u_1 = a_0 + a_1x_1 + a_2x_2 + a_3x_1x_2 + a_4x_1^2 + a_5x_2^2$$
$$u_2 = b_0 + b_1x_1 + b_2x_2 + b_3x_1x_2 + b_4x_1^2 + b_5x_2^2.$$

Since there are 12 unknown coefficients, at least six tie-point pairs are needed to determine the map (each pair generates two equations). If more than six pairs are available, the coefficients can be found by least squares fitting. This has the advantage that an RMS error for the mapping can be estimated, as shown in Figures 5.29 and 5.30. Similar considerations apply for lower- or higher-order polynomial maps.

Having determined the map coefficients, image 2 can be registered to image 1 by resampling. *Nearest-neighbor resampling* simply chooses the actual pixel in image 2 that has its center nearest to the calculated coordinates (u_1, u_2), and transfers it to the location (x_1, x_2). This is the preferred technique for classification or change detection, since the registered image consists of the original pixel brightnesses, simply rearranged in position to give a correct image geometry. Other commonly used resampling methods are *bilinear interpolation* and *cubic convolution interpolation* (see, e.g., Jensen, 2005, for a good explanation). These methods interpolate, and therefore mix, the spectral intensities of neighboring pixels.

5.6 Exercises

1. Design a lookup table for byte-encoded data to perform 2% linear saturation. (2% of the dark and bright pixels saturate to 0 and 255, respectively.)
2. The *decorrelation stretch* generates a more color-intensive RGB composite image of highly correlated spectral bands than is obtained by simple linear stretching of the individual bands (Richards and Jia, 2006). Using ENVI batch procedures, write a routine to implement it:

   (a) Do a principal components transformation of three selected image bands.

   (b) Then do a linear histogram stretch of the principal components.

   (c) Finally, invert the transformation and place the result in ENVI's available bands list.

   (*Note:* This is merely an exercise in batch programming, as ENVI provides the decorrelation stretch transformation in its main menu.)
3. The *Roberts operator* or *Roberts filter* is available in the ENVI menu system and approximates intensity gradients in diagonal directions:

$$\nabla_1(i,j) = [g(i,j) - g(i+1,j+1)]$$
$$\nabla_2(i,j) = [g(i+1,j) - g(i,j+1)].$$

   Modify the program in Listing 5.1 to calculate its power spectrum.
4. An edge detector introduced by to Smith and Brady (1997) called SUSAN (*smallest univalue segment assimilating nucleus*) employs a circular mask, typically approximated by 37 pixels, that is,

```
  000
 00000
0000000
0000000
0000000
 00000
  000
```

Let r be any pixel under the mask, $g(r)$ be its intensity, and r_0 be the central pixel. Define the function

$$c(r, r_0) = \begin{cases} 1 & \text{if } |g(r) - g(r_0)| \le t \\ 0 & \text{if } |g(r) - g(r_0)| > t, \end{cases}$$

where t is a threshold. Associate with r_0 the sum

$$n(r_0) = \sum_r c(r, r_0).$$

If the mask covers a region of a sufficiently low contrast, $n(r_0) = n_{max} = 37$. As the mask moves toward an intensity "edge" having any orientation in an image, the quantity $n(r_0)$ will decrease, reaching a minimum as the center crosses the edge. Accordingly, an edge strength can be defined as

$$E(r_0) = \begin{cases} g - n(r_0) & \text{if } n(r_0) < h \\ 0 & \text{otherwise.} \end{cases}$$

The parameter h is chosen (from experience) as $0.75 \cdot n_{max}$.

(a) A convenient way to calculate $c(r, r_0)$ is to use the continuous approximation

$$c(r, r_0) = e^{-(g(r) - g(r_0))/t)^6}. \tag{5.31}$$

Write an IDL procedure to plot this function for $g(r_0) = 127$ and for $g(r) = 0 \ldots 255$.

(b) Write an IDL program to implement the SUSAN edge detector for arbitrary grayscale images. *Hint*: Create a lookup table to evaluate Equation 5.31:

```
LUT   = fltarr(256)
FOR i=0,255 DO LUT[i] = exp(-(i/t)^6)
```

5. One can approximate the centralized moments of a feature (Equation 5.9) by the integral

$$\mu_{pq} = \iint (x - x_x)^p (y - y_c)^q f(x,y)dx\, dy,$$

where the integration is over the whole image and where $f(x,y) = 1$ if the point (x,y) lies on the feature and $f(x,y)=0$ otherwise. Use this approximation to prove that the normalized centralized moments, η_{pq}, given in Equation 5.10 are invariant under scaling transformations of the form

$$\begin{pmatrix} x_1' \\ x_2' \end{pmatrix} = \begin{pmatrix} \alpha & 0 \\ 0 & \alpha \end{pmatrix} \begin{pmatrix} x_1 \\ x_2 \end{pmatrix}.$$

6. Wavelet noise reduction (Gonzalez and Woods, 2002).
   (a) Apply the discrete wavelet transformation to reduce the noise in a multispectral image by modifying the example program in Listing 4.3 to perform the following steps:

   - Select a multispectral image in ENVI and determine the number of columns, rows, and spectral bands
   - Create a band sequential (BSQ) array of the same dimensions for the output
   - For each band
     - Read the band into a new `DWT` object instance
     - Compress once
     - For each of the three quadrants containing the detailed wavelet coefficients
       - Extract the coefficients with the class method `GET_QUADRANT()`
       - Determine their mean and standard deviation
       - Zero all coefficients with the absolute value relative to the mean smaller than three standard deviations
       - Inject them back into the transformed image with the class method `INJECT`
     - Expand back
     - Store the modified band in the output array
     - Destroy the object instance
   - Return the output array to ENVI

   Note: The coefficients are extracted as one-dimensional arrays. When injecting them back they must be reformed to two-dimensional arrays. The correct dimensions for doing this are returned from the `DWT` object with the methods `GET_NUM_COLS()`

and `GET_NUM_ROWS()` after the compression. (See also the IDL function `WV_DENOISE` for similar functionality.)

(b) Test your program with a noisy three-band image, for example, with the last three components of the MNF transformation of a LANDSAT TM image. Use the example program in Listing 3.5 to determine the noise covariance matrix before and after carrying through the above procedure.

7. Show that the means and standard deviations of the renormalized panchromatic wavelet coefficients, C_k^z, in Equation 5.14 are equal to those of the multispectral bands.
8. Write an ENVI/IDL extension to perform additive *à trous* fusion (see Núnez et al., 1999).
9. Use the Wang-Bovik quality index routine (Appendix C.2.4) to assess the spectral fidelity of a pan-sharpened image using the complete range of methods offered by ENVI and by the extensions described in this chapter: HSV, Brovey, Gram-Schmidt, PCA, DWT, and ATWT.
10. From the definition of the gradient (Equation 5.3) show that the terrain slope angle, θ_p, can be approximated from a DEM by Equation 5.23.

6

Supervised Classification: Part 1

Land cover classification of remote sensing imagery is a task that falls into the general category of *pattern recognition*. Pattern recognition problems, in turn, are usually approached by developing appropriate *machine learning* algorithms. Broadly speaking, machine learning involves tasks for which there is no known direct method to compute a desired output from a set of inputs. The strategy adopted is for the computer to "learn" from a set of representative examples.

In the case of supervised classification, the task can often be seen as one of modeling probability distributions. On the basis of representative data for K land cover classes presumed to be present in a scene, the *a posteriori* probabilities for class k conditional on observation g, $\Pr(k \mid g)$, $k = 1 \ldots K$, are "learned" or approximated. This is usually called the *training phase* of the classification procedure. Then these probabilities are used to classify all of the pixels in the image, a step referred to as the *generalization phase*.

In this chapter, we will consider three representative models for supervised classification that involve this sort of posterior probability estimation: a *parametric model* (the Bayes maximum likelihood classifier), a *nonparametric model* (Gaussian kernel classification), and a *semiparametric* or *mixture model* (the feed-forward neural network). For the nonparametric case, ENVI provides little functionality, and its neural network classifier uses a decidedly suboptimal training algorithm. Therefore, both here and in Appendix B, we shall make some considerable effort to develop our own classification routines. Finally, the *support vector machine* (SVM) classifier, available in ENVI since Version 4.3, will be discussed in detail. SVMs are also nonparametric in the sense that they make direct use of a subset of the labeled training data (the support vectors) to effect a partitioning of the feature space, however, unlike the three aforementioned classifiers, without reference to the statistical distributions of the training data.

To illustrate the various algorithms developed here and in the following chapters on image classification, we will work with the ASTER scene shown in Figure 6.1 (see also Figure 1.1). In this data set, the six SWIR bands have been sharpened to the 15m ground resolution of the three VNIR bands with the *à trous* wavelet fusion method of Section 5.3.5 (the Wang-Bovik quality index ranged from 0.72 to 0.88 over the six bands), and a principal components analysis of the stacked nine-band image has been performed. All classification examples in the text, both supervised and unsupervised, will be carried out with subsets of the principal components.

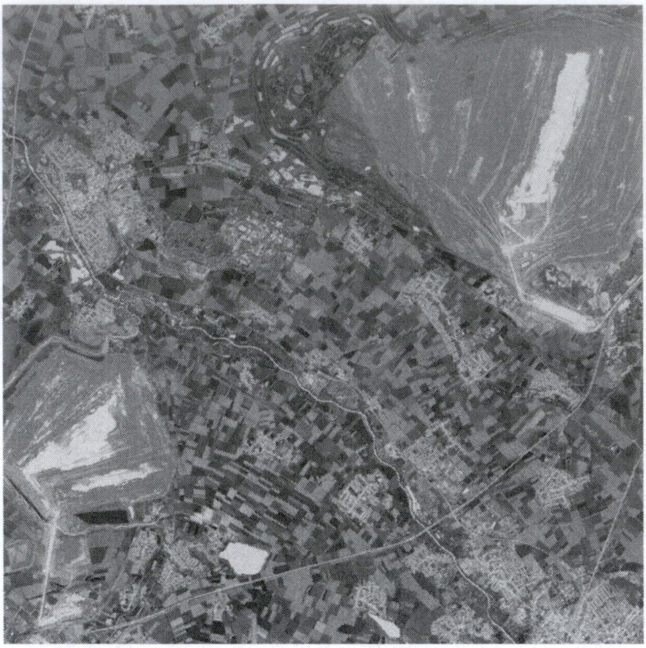

FIGURE 6.1
(See color insert following page 114.) An RGB color composite (1000 × 1000 pixels, linear 2% saturation stretch) of the first three principal components, 1 (red), 2 (green), and 3 (blue), of the nine nonthermal bands of the ASTER scene acquired over Jülich, Germany, on May 1, 2007.

6.1 Maximum *a Posteriori* Probability

The basis for most of the classifiers that we consider in this chapter is a decision rule based on the *a posteriori* probabilities, $\Pr(k \mid g)$, so this rule will be our starting point.

Let us begin by defining a *loss function*, $L(k, g)$, that measures the cost of associating the observation g with the class k. Let λ_{kj} be the loss incurred if g in fact belongs to class k, but is classified as belonging to class j. It can reasonably be assumed that

$$\lambda_{kj} \begin{cases} =0 & \text{if } k=j \\ >0 & \text{otherwise,} \end{cases} \qquad k,j = 1\ldots K, \tag{6.1}$$

that is, correct classifications do not incur losses while misclassifications do. The loss function can then be expressed as a sum over the individual losses, weighted according to their probabilities of occurrence, $\Pr(j \mid g)$:

$$L(k, g) = \sum_{j=1}^{K} \lambda_{kj} \Pr(j \mid g). \tag{6.2}$$

Without further specifying λ_{kj}, a loss-minimizing decision rule for classification may be defined (ignoring the possibility of ties) as

$$g \text{ is in class } k \text{ provided } L(k, g) \le L(j, g) \text{ for all } j = 1 \ldots K. \tag{6.3}$$

So far we have been quite general. Now, suppose that the losses are independent of the kind of misclassification that occurs (for instance, the classification of a "forest" pixel into the class "meadow" is just as costly as classifying it as "urban area," etc.). Then we can write

$$\lambda_{kj} = 1 - \delta_{kj}, \tag{6.4}$$

where $\delta_{kj} = 1$ for $k = j$ and 0 otherwise. Thus, any given misclassification ($j \ne k$) has a unit cost, and a correct classification ($j = k$) costs nothing, as before. We then obtain from Equation 6.2

$$L(k, g) = \sum_{j=1}^{K} \Pr(j \mid g) - \Pr(k \mid g) = 1 - \Pr(k \mid g),\ k = 1 \ldots K, \tag{6.5}$$

and from Equation 6.3 the following decision rule:

$$g \text{ is in class } k \text{ provided } \Pr(k \mid g) \ge \Pr(j \mid g) \text{ for all } j = 1 \ldots K. \tag{6.6}$$

In other words, we assign each new observation to the class with the highest *a posteriori* probability. As indicated in the introduction, our main task will therefore be to determine the posterior probabilities, $\Pr(k \mid g)$.

6.2 Training Data and Separability

The choice of training data is arguably the most difficult and critical part of the supervised classification process. The standard procedure is to select areas within a scene that are representative of each class of interest. In the ENVI environment, the areas are entered as *regions of interest* (ROIs), from which the training observations are selected. Some fraction of the representative data may be retained for later accuracy assessment. These comprise the so-called *test data* and are withheld from the training phase in order not to bias the subsequent evaluation. We will refer to the set of labeled training data as *training pairs* or *training examples* and write it in the form

$$\mathcal{T} = \{g(\nu), \ell(\nu)\}, \quad \nu = 1 \ldots m, \tag{6.7}$$

FIGURE 6.2
Ground reference data for four land cover categories, photographed on May 1, 2007. (a) Cereal grain; (b) grassland; (c) rapeseed; and (d) sugar beets.

where m is the number of observations and

$$\ell(\nu) \in \mathcal{K} = \{1 \ldots K\} \tag{6.8}$$

is the class label of observation, $\mathbf{g}(\nu)$.

As an illustration, ground reference data were collected on the same day as the acquisition of the ASTER image in Figure 6.1. Figure 6.2 shows photographs of 4 of the 10 land cover categories used for classification. The others were water, suburban settlement, urban area/industrial park, herbiferous forest, coniferous forest, and open cast mining. In all, 30 ROIs were identified in the scene as representatives of the 10 classes, involving 7173 pixels. They are shown in Figure 6.3. Two-thirds of these, sampled uniformly across the ROIs, were used for training and the remainder reserved for testing, that is, estimating the generalization error on new observations.

The *degree of separability* of the training observations will give some indication of the prospects for success of the classification procedure and can help in deciding how the data should be processed prior to classification. A very commonly used separability measure may be derived by considering

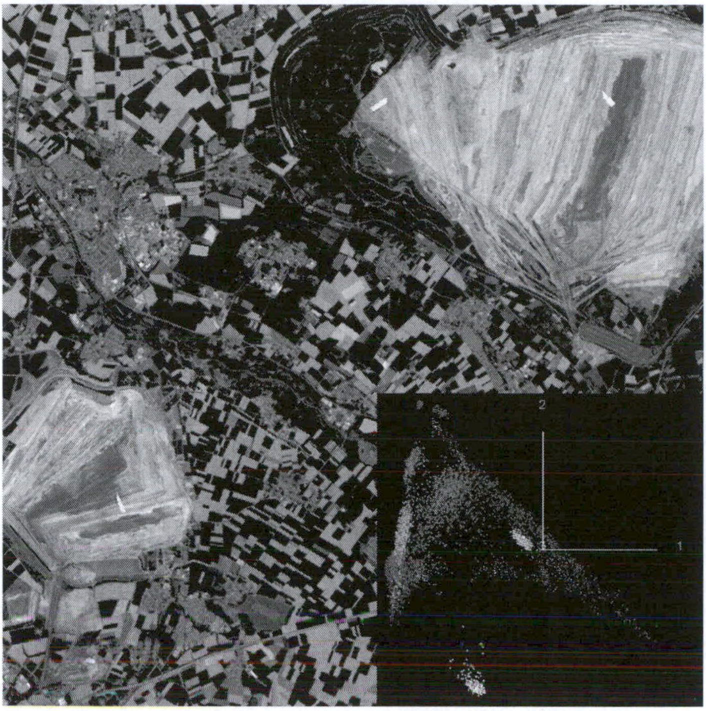

FIGURE 6.3
(See color insert following page 114.) ROIs for supervised classification. The inset shows the training observations projected onto the plane of the first two principal axes.

the *Bayes error*. Suppose that there are just two classes involved, $\mathcal{K} = \{1, 2\}$. If we apply the decision rule (Equation 6.6) for some pixel intensity vector $\boldsymbol{g}$, we must assign the class as that having the maximum *a posteriori* probability. Therefore, the probability $r(\boldsymbol{g})$ of incorrectly classifying the pixel is given by

$$r(\boldsymbol{g}) = \min[\ \Pr(1 \mid \boldsymbol{g}), \Pr(2 \mid \boldsymbol{g})\ ].$$

The Bayes error, ϵ, is defined to be the average value of $r(\boldsymbol{g})$, which we can calculate as the integral of $r(\boldsymbol{g})$ times the probability density $p(\boldsymbol{g})$, taken over all of the observations $\boldsymbol{g}$:

$$\begin{aligned}\epsilon = \int r(\boldsymbol{g})p(\boldsymbol{g})d\boldsymbol{g} &= \int \min[\ \Pr(1 \mid \boldsymbol{g}), \Pr(2 \mid \boldsymbol{g})\ ]p(\boldsymbol{g})d\boldsymbol{g} \\ &= \int \min[\ p(\boldsymbol{g} \mid 1)\Pr(1), p(\boldsymbol{g} \mid 2)\Pr(2)\ ]d\boldsymbol{g}. \qquad (6.9)\end{aligned}$$

Bayes' theorem (Equation 2.57) was invoked in the last equality. The Bayes error may be used as a measure of the separability of the two classes: the smaller the error, the better the separability.

Calculating the Bayes error is in general difficult, but we can at least get an approximate upper bound on it as follows (Fukunaga, 1990). First note that, for any $a, b \geq 0$,

$$\min[\, a, b\,] \leq a^s b^{1-s}, \quad 0 \leq s \leq 1.$$

For example, if $a < b$, then the inequality can be written as

$$a \leq a \left(\frac{b}{a}\right)^{1-s},$$

which is clearly true. Applying this inequality to Equation 6.9, we get the *Chernoff bound*, ϵ_u, on the Bayes error,

$$\epsilon \leq \epsilon_u = \Pr(1)^s \Pr(2)^{1-s} \int p(g \mid 1)^s p(g \mid 2)^{1-s} dg. \tag{6.10}$$

The least upper bound is then determined by minimizing ϵ_u with respect to s. If $p(g \mid 1)$ and $p(g \mid 2)$ are multivariate normal distributions with equal covariance matrices $\boldsymbol{\Sigma}_1 = \boldsymbol{\Sigma}_2$, then it can be shown that the minimum in fact occurs at $s = 1/2$. Approximating the minimum as $s = 1/2$ also for the case where $\boldsymbol{\Sigma}_1 \neq \boldsymbol{\Sigma}_2$ leads to the (somewhat less tight) *Bhattacharyya bound*, ϵ_B,

$$\epsilon \leq \epsilon_B = \sqrt{\Pr(1)\Pr(2)} \int \sqrt{p(g \mid 1) p(g \mid 2)}\, dg. \tag{6.11}$$

This integral can be evaluated explicitly (Exercise 1). The result is

$$\epsilon_B = \sqrt{\Pr(1)\Pr(2)}\, e^{-B},$$

where B is the *Bhattacharyya distance*, given by

$$B = \frac{1}{8}(\boldsymbol{\mu}_2 - \boldsymbol{\mu}_1)^\top \left[\frac{\boldsymbol{\Sigma}_1 + \boldsymbol{\Sigma}_2}{2}\right]^{-1} (\boldsymbol{\mu}_2 - \boldsymbol{\mu}_1) + \frac{1}{2}\log\left(\frac{|\boldsymbol{\Sigma}_1 + \boldsymbol{\Sigma}_2|/2}{\sqrt{|\boldsymbol{\Sigma}_1||\boldsymbol{\Sigma}_2|}}\right). \tag{6.12}$$

Large values of B imply small upper limits on the Bayes error and hence good separability. The first term in B is a squared average Mahalanobis distance (see Section 6.3) and expresses the class separability due to the dissimilarity of the class means.* The second term measures the difference between the covariance matrices of the two classes. It vanishes when $\boldsymbol{\Sigma}_1 = \boldsymbol{\Sigma}_2$.

The Bhattacharyya distance as a measure of separability has the disadvantage that it continues to grow even after the classes have become so well separated that any classification procedure could distinguish them perfectly. The *Jeffries-Matusita distance* measures separability of two classes on a more convenient scale $[0 - 2]$ in terms of B:

$$J = 2(1 - e^{-B}). \tag{6.13}$$

* This term is proportional to the maximum value of the Fisher linear discriminant, as can be seen by substituting Equation 3.80 into Equation 3.79.

TABLE 6.1

Lowest 10 Paired Class Separabilities for the First Four Principal Components of the ASTER Scene

Class 1	Class 2	J-M Distance
Grain [Red]	Grassland [Red2]	1.42
Settlement [Magenta]	Industry [Maroon]	1.51
Grassland [Red2]	Herbiferous [Green]	1.68
Settlement [Magenta]	Herbiferous [Green]	1.91
Settlement [Magenta]	Grassland [Red2]	1.94
Industry [Maroon]	Coniferous [Sea Green]	1.94
Coniferous [Sea Green]	Herbiferous [Green]	1.95
Sugar beet [Cyan]	Mining [White]	1.98
Grain [Red]	Herbiferous [Green]	1.99
Industry[Maroon]	Herbiferous [Green]	1.99

As B continues to grow, the measure saturates at the value 2. The factor 2 comes from the fact that the Jeffries-Matusita distance can be derived independently as the average distance between two density functions (see Richards and Jia, 2006, and Exercise 1). The ENVI menu command

```
Basic Tools/Region OF Interest/Compute ROI Separability
```

calculates Jeffries-Matusita distances between all pairs of classes defined in a given set of ROIs by estimating the class means and covariance matrices from the pixels contained within them and then using Equations 6.12 and 6.13. Some examples are shown in Table 6.1 for the training data of Figure 6.3.

6.3 Maximum Likelihood Classification

Consider once again Bayes' theorem, expressed in the form of Equation 2.60:

$$\Pr(k \mid g) = \frac{p(g \mid k)\Pr(k)}{p(g)}, \tag{6.14}$$

where $\Pr(k)$, $k = 1 \ldots K$, are prior probabilities, $p(g \mid k)$ is a class-specific probability density function, and where $p(g)$ is given by

$$p(g) = \sum_{j=1}^{K} p(g \mid j)\Pr(j).$$

Since $p(g)$ is independent of k, we can write the decision rule (Equation 6.6) as

$$g \text{ is in class } k \text{ provided } p(g \mid k)\Pr(k) \geq p(g \mid j)\Pr(j) \text{ for all } j = 1 \ldots K. \tag{6.15}$$

Now suppose that the observations from class k are sampled from a multivariate normal distribution. Then the density functions are given by

$$p(g \mid k) = \frac{1}{(2\pi)^{N/2}|\boldsymbol{\Sigma}_k|^{1/2}} \exp\left(-\frac{1}{2}(g - \boldsymbol{\mu}_k)^\top \boldsymbol{\Sigma}_k^{-1}(g - \boldsymbol{\mu}_k)\right). \tag{6.16}$$

Taking the logarithm of Equation 6.16 gives

$$\log\big(p(g \mid k)\big) = -\frac{N}{2}\log(2\pi) - \frac{1}{2}\log|\boldsymbol{\Sigma}_k| - \frac{1}{2}(g - \boldsymbol{\mu}_k)^\top \boldsymbol{\Sigma}_k^{-1}(g - \boldsymbol{\mu}_k).$$

The first term may be ignored, as it too is independent of k. Together with Equation 6.15 and the definition of the *discriminant function*

$$d_k(g) = \log(\Pr(k)) - \frac{1}{2}\log|\boldsymbol{\Sigma}_k| - \frac{1}{2}(g - \boldsymbol{\mu}_k)^\top \boldsymbol{\Sigma}_k^{-1}(g - \boldsymbol{\mu}_k), \tag{6.17}$$

we obtain the *maximum likelihood classifier*:

$$g \text{ is in class } k \text{ provided } d_k(g) \geq d_j(g) \text{ for all } j = 1 \ldots K. \tag{6.18}$$

There may be no information about the prior class probabilities, $\Pr(k)$, in which case they can be set equal and ignored in the classification. Then the factor 1/2 in Equation 6.17 can be dropped as well, and the discriminant becomes

$$d_k(g) = -\log|\boldsymbol{\Sigma}_k| - (g - \boldsymbol{\mu}_k)^\top \boldsymbol{\Sigma}_k^{-1}(g - \boldsymbol{\mu}_k). \tag{6.19}$$

The second term in Equation 6.19 is the square of the so-called *Mahalanobis distance*

$$\sqrt{(g - \boldsymbol{\mu}_k)^\top \boldsymbol{\Sigma}_k^{-1}(g - \boldsymbol{\mu}_k)}.$$

The contours of constant multivariate probability density in Equation 6.16 are hyperellipsoids of constant Mahalanobis distance to the mean $\boldsymbol{\mu}_k$.

The moments $\boldsymbol{\mu}_k$ and $\boldsymbol{\Sigma}_k$, which appear in the discriminant functions, may be estimated from the training data using the maximum likelihood parameter estimates (see Section 2.5):

$$\begin{aligned} \boldsymbol{\mu}_k \approx \boldsymbol{m}_k &= \frac{1}{m_k}\sum_{\{\nu \mid \ell(\nu)=k\}} g(\nu) \\ \boldsymbol{\Sigma}_k \approx \boldsymbol{C}_k &= \frac{1}{m_k}\sum_{\{\nu \mid \ell(\nu)=k\}} (g(\nu) - \boldsymbol{\mu}_k)(g(\nu) - \boldsymbol{\mu}_k)^\top, \end{aligned} \tag{6.20}$$

where m_k is the number of training pixels with class label k.

Having estimated the parameters from the training data, the generalization phase consists simply of applying the rule of Equation 6.18 to all of the pixels in the image. Because of the small number of parameters to be estimated, maximum likelihood classification is extremely fast. Its weakness lies in the restrictiveness of the assumption that all observations are drawn from multivariate normal probability distributions. Computational efficiency eventually achieved at the cost of generality is a characteristic of *parametric classification models*, to which category the maximum likelihood classifier belongs (Bishop, 1995).

6.3.1 ENVI's Maximum Likelihood Classifier

Note that applying the rule of Equation 6.18 will place any observation into one of the K classes no matter how small its maximum discriminant function turns out to be. If it is thought that some classes may have been overlooked, or if no training data were available for one or two known classes, then it might be desirable to assume that observations with small maximum discriminant functions belong to one of these inaccessible classes. Then they should perhaps not be classified at all. If this is desired, it may be achieved simply by setting a threshold on the maximum discriminant, $d_k(g)$, marking the observations lying below the threshold as "unclassified." This possibility is provided with ENVI's maximum likelihood classification algorithm. The algorithm is called from the ENVI main menu with

```
Classification/Supervised/Maximum Likelihood
```

To facilitate setting thresholds for unclassified observations, the ENVI maximum likelihood classifier optionally generates a *rule image* consisting of the discriminant functions (Equation 6.17) for each class and training observation, whereby the prior probabilities, $\Pr(k)$, are set equal to one another. If g is normally distributed, the squared Mahalanobis distance term in Equation 6.17 is chi-square distributed with N degrees of freedom, where N is the dimensionality of the observations. The histogram of a rule image band for class k will therefore resemble a chi-square distribution, reflected about zero due to the minus sign, and shifted due to the $\log(\Pr(k))$ and $\frac{1}{2}\log|\boldsymbol{\Sigma}_k|$ terms. Thresholds can be set in a post-classification session with the *rule classifier*, accessible from the ENVI main menu under

```
Classification/Post Classification/Rule Classifier
```

and a new classified image can be generated.

For further postprocessing of classification results, we shall later require the posterior class membership probabilities, $\Pr(k \mid g)$. The IDL routine in Listing 6.1 converts the ENVI maximum likelihood rule images to membership probabilities (byte-scaled to save storage space) under the assumption that there is no "unclassified" class (the membership probabilities sum to unity).

Listing 6.1

Calculating class membership probabilities from rule images.

```
21 PRO rule_convert
22
23 envi_select, title='Choose_rule_image', $
24     fid=fid, dims=dims,pos=pos,/no_dims,/no_spec
25 IF (fid EQ -1) THEN BEGIN
26     PRINT, 'cancelled'
27     RETURN
28 ENDIF
29
30 map_info = envi_get_map_info(fid=fid)
31
32 num_cols = dims[2]-dims[1]+1
33 num_rows = dims[4]-dims[3]+1
34 num_classes = n_elements(pos)
35 num_pixels = num_cols*num_rows
36
37 ; get MaxLike rule image
38 rule_image = dblarr(num_pixels, num_classes)
39 FOR i=0,num_classes-1 DO rule_image[*,i] = $
40     envi_get_data(fid=fid,dims=dims,pos=pos[i])
41
42 ; exponentiate and normalize
43 prob_image = exp(rule_image)
44 den = total(prob_image,2,/double)
45 FOR i=0,num_classes-1 DO  prob_image[*,i]= $
46     prob_image[*,i]/den
47
48 ; write to memeory
49 envi_enter_data, $
50    reform(bytscl(prob_image,min=0.0,max=1.0),num_cols, $
51     num_rows,num_classes),map_info=map_info
52
53 END
```

6.3.2 Modified Maximum Likelihood Classifier

In order to carry out an unbiased assessment of the accuracy of supervised classification methods, ENVI's built-in evaluation procedures allow comparison with the so-called *ground truth* ROIs or images containing areas of labeled data not used during the training phase. We will prefer a somewhat different evaluation philosophy, arguing that, if other representative training areas are indeed available for evaluation, then they should also be used to train the classifier. For evaluation purposes, some portion of the pixels in *all* of the training areas can be held back, but such test data should be selected from the pool of available labeled observations. This point of view assumes that all training/test areas are equally representative of their respective classes, but if that were not the case then there would be no justification to use them at all.

A wrapper for the ENVI maximum likelihood classifier called `MAXLIKE_RUN`, which separates ROI training area pixels into a training data set and a test data set in the ratio 2:1, is described in Appendix C.2.8. It makes use of the ENVI batch procedure `CLASS_DOIT` to perform the actual classification of both the test data and the full image, and can serve as an ENVI/IDL extension. Test results can optionally be saved to a file in a format consistent with that used by the other classification routines to be described in the remainder of this chapter. Classification evaluation using the test results will be discussed in Chapter 7.

An excerpt from `MAXLIKE_RUN` is shown in Listing 6.2. The labeled observations are stored in the data matrix format in the $N \times m$ array variable `Gs`,

Listing 6.2

Excerpt from the program module MAXLIKE_RUN.PRO.

```
; split into training and test data
   seed = 12345L
   num_test = m/3

; sampling with replacement
   test_indices = randomu(seed,num_test,/long) MOD m
   test_indices =  test_indices[sort(test_indices)]
   train_indices=difference(lindgen(m),test_indices)
   Gs_test = Gs[*,test_indices]
   Ls_test = Ls[*,test_indices]
   m = n_elements(train_indices)
   Gs = Gs[*,train_indices]
   Ls = Ls[*,train_indices]

; train the classifier
   mn =  fltarr(num_bands,K)
   cov = fltarr(num_bands,num_bands,K)
   class_names = strarr(K+1)
   class_names[0]='unclassified'
   void = max(transpose(Ls),labels,dimension=2)
   labels = byte((labels/m))
   FOR i=0,K-1 DO BEGIN
      class_names[i+1]='class'+string(i+1)
      indices = where(labels EQ i,count)
      IF count GT 1 THEN BEGIN
         GGs = Gs[*,indices]
         FOR j=0,num_bands-1 DO mn[j,i] = mean(GGs[j,*])
         cov[*,*,i] = correlate(GGs,/covariance)
      ENDIF
   ENDFOR
```

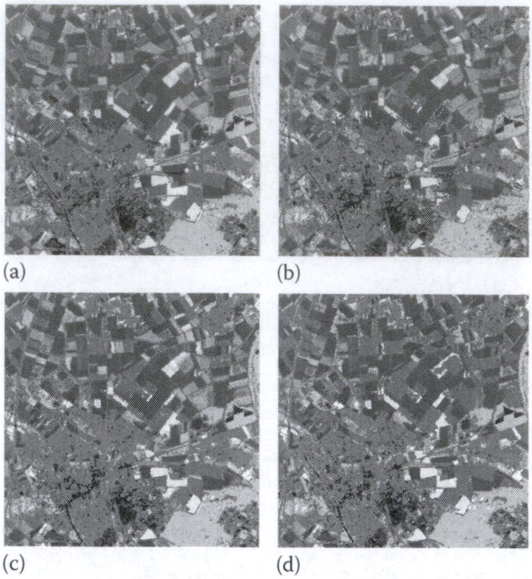

FIGURE 6.4
(See color insert following page 114.) Supervised classification of a portion of the Jülich ASTER scene: (a) maximum likelihood; (b) Gaussian kernel; (c) neural network; (d) SVM. Five land cover categories (water: blue, settlement: magenta, herbifierous forest: green, rapeseed: yellow, cereal grain: red) are superimposed on VNIR band 3.

and the corresponding labels in the $K \times m$ array variable `Ls`. Here, m is the number of ground reference observations, N is their dimensionality, and K is the number of classes. (The individual labels are in fact K-element arrays with zeroes everywhere except at the position of the class. For example, if there are five classes, class 2 corresponds to the label array $[0, 1, 0, 0, 0]$. This convention will turn out to be convenient when we come to consider neural network classifiers.) In lines 118–125, the training set is split, with a random sample (with replacement) of 1/3 of the observations held back in the variables `GS_TEST, LS_TEST`. In lines 139 and 140, the means and the covariance matrices of the individual classes are estimated from the remaining training pixels. They serve as an input to the `CLASS_DOIT` batch procedure (not shown).

An example of a classification with `MAXLIKE_RUN` is shown in Figure 6.4a. The input data consisted of the first four principal components of the Jülich ASTER scene of Figure 6.1 together with the training data of Figure 6.3.

6.4 Gaussian Kernel Classification

Nonparametric classification models may estimate the class-specific probability densities, $p(g \mid k)$, as in the preceding section, from a set of training data. However, unlike the maximum likelihood classifier, no strong prior assumptions about the nature of the densities are made. In the *Parzen window* approach to nonparametric classification (Duda et al., 2001), each training observation $g(\nu)$, $\nu = 1 \ldots m$, is used as the center of a local kernel function. The probability density for class k at a point, g, is taken to be the average of the kernel functions for the training data in that class, evaluated at g. For example, using a Gaussian kernel, the probability density for the kth class is estimated as

$$p(g \mid k) \approx \frac{1}{m_k} \sum_{\{\nu \mid \ell(\nu)=k\}} \frac{1}{\sqrt{2\pi}\sigma} \exp\left(-\frac{\|g - g(\nu)\|^2}{2\sigma^2}\right). \tag{6.21}$$

The quantity σ is a smoothing parameter, which has been chosen in this case to be class independent. Since the Gaussian functions are normalized, we have

$$\int_{-\infty}^{\infty} p(g \mid k) dg = \frac{1}{m_k} \sum_{\{\nu \mid \ell(\nu)=k\}} 1 = 1,$$

as required of a probability density. Under fairly general conditions, the right-hand side of Equation 6.21 can be shown to converge to $p(g \mid k)$ as the number of training observations tends to infinity. If we set the prior probabilities in the decision rule (Equation 6.6) equal as before, then an observation, g, will be assigned to class k when

$$p(g \mid k) \geq p(g \mid j), \quad j = 1 \ldots K.$$

Training the Gaussian kernel classifier involves searching for an optimal value of the smoothing parameter, σ. Too large a value will wash out the class dependency, too small a value will lead to poor generalization on new data. Training can be effected very conveniently by minimizing the misclassification rate with respect to σ. When presenting an observation vector, $g(\nu)$, to the classifier during the training phase, the contribution to the probability density at the point $g(\nu)$ from class k is, from Equation 6.21 and apart from a constant factor, given by

$$\begin{aligned} p(g(\nu) \mid k) &= \frac{1}{m_k} \sum_{\{\nu' \mid \ell(\nu')=k\}} \exp\left(-\frac{\|g(\nu) - g(\nu')\|^2}{2\sigma^2}\right) \\ &= \frac{1}{m_k} \sum_{\{\nu' \mid \ell(\nu')=k\}} (\mathcal{K})_{\nu\nu'}, \end{aligned} \tag{6.22}$$

where $\mathcal{K}$ is an $m \times m$ Gaussian kernel matrix (see Equation 4.25). It is advisable to delete the contribution of $g(\nu)$ itself to the sum in the above equation in order to avoid biasing the classification in favor of the training observation's own label, a bias that would otherwise arise due to the appearance of

Listing 6.3

Excerpt from the program module GPUKERNEL_RUN.PRO.

```
FUNCTION Output, sigma, Hs, symm=symm
   COMMON examples, Gs, Gs_gpu, ells, K, m
   IF n_elements(symm) EQ 0 THEN symm = 0
   result = fltarr(n_elements(Hs[0,*]),K)
   IF gpuDetectDeviceMode() EQ 1 THEN BEGIN
      IF symm THEN Hs_gpu = Gs_gpu $
              ELSE Hs_gpu = gpuputarr(Hs)
      Kappa_gpu = gpugausskernel_matrix(Gs_gpu,$
                            Hs_gpu,gma=0.5/sigma^2)
      IF NOT symm THEN gpufree,Hs_gpu
      Kappa = gpugetarr(Kappa_gpu)
      gpufree,Kappa_gpu
   END $
   ELSE Kappa=gausskernel_matrix(Gs,Hs,gma=0.5/sigma^2)
   IF symm THEN $
      Kappa[indgen(m),indgen(m)] = 0.0
   FOR j=0,K-1 DO BEGIN
      Kpa = Kappa
      idx = where(ells NE j, ncomplement=nj)
      Kpa[*,idx] = 0
      result[*,j] = total(Kpa,2)/nj
   ENDFOR
   RETURN, result
END

FUNCTION Theta, sigma
   COMMON examples, Gs, Gs_gpu, ells, K, m
   _ = max(output(sigma,Gs,/symm),labels,dimension=2)
   _ = where(labels/m NE ells, count)
   result = float(count)/m
   oplot,[alog10(sigma)],[result],psym=5,color=0
   RETURN, result
END
```

a zero in the exponent and a dominating contribution to $p(\mathbf{g}(\nu) \mid k)$ (see Masters, 1995). This amounts to zeroing the diagonal of $\mathcal{K}$ before performing the sum in Equation 6.22.

In Appendix C.2.9, an ENVI/IDL extension GPUKERNEL_RUN is described, which implements these ideas. Listings 6.3 and 6.4 show excerpts. The function OUTPUT(SIGMA,HS) in Listing 6.3 calculates the arrays of class probability densities for all pixels' vectors in the data matrix HS using the current value of the smoothing parameter SIGMA. The common block variable GS contains the training pixels, also in the data matrix format. Their labels are stored in the m-dimensional array ELLS. In the training phase, the keyword SYMM is set, indicating that HS and GS are in fact identical. The diagonal of

Listing 6.4
Excerpt from the program module GPUKERNEL_RUN.PRO.

```
; set training class labels in common block
   _ = max(transpose(Ls),ells,dimension=2)
   ells = ells/m

; set GPU array for training data if possible
   IF gpuDetectDeviceMode() EQ 1 THEN $
                    Gs_gpu = gpuputarr(Gs)

   window,12,xsize=600,ysize=400, $
     title='Bracketing,␣please␣wait␣...'
   wset, 12
   widget_control, /hourglass

; bracket the minimum
   s1=0.1  & s2=10.0
   minF_bracket, s1,s2,s3, FUNC_NAME="Theta"

   window,12,xsize=600,ysize=400, $
     title='Iteration␣...'
   PLOT,[alog10(s1),alog10(s3)],[theta(s1),theta(s3)],$
      psym=5,color=0,background='FFFFFF'XL, $
      title = 'Gaussian␣Kernel␣Classification',$
      xtitle = 'log(sigma)', $
      ytitle = 'theta'

; hunt it down
   minF_parabolic, s1,s2,s3, sigma, theta_min, $
                 FUNC_NAME='Theta'
```

the symmetric kernel matrix KAPPA is then set to zero (line 37). The sums in Equation 6.22 are calculated in the FOR-loop in lines 38–43. As was the case for the kernel PCA program (Section 4.4.2), this ENVI extension makes use of the GPULib interface to CUDA, if available, in order to accelerate calculation.

The minimization of the misclassification rate with respect to σ takes place in two steps (Listing 6.4): First, the minimum is bracketed by a call to procedure MINF_BRACK with initial estimates $0.1 \leq \sigma \leq 10$, and then, it is approximated iteratively using Brent's parabolic interpolation method, procedure MINF_PARABOLIC. The two procedures are described in Press et al. (2002); the code was taken from the IDL Astronomy User's Library (see Section 5.4.4). The function THETA(SIGMA) (Listing 6.3) passed to both procedures, calculates the misclassification rate with a call to OUTPUT(SIGMA,GS,/SYMM). Figure 6.5 shows a plot of the misclassification rate vs. σ during the training phase.

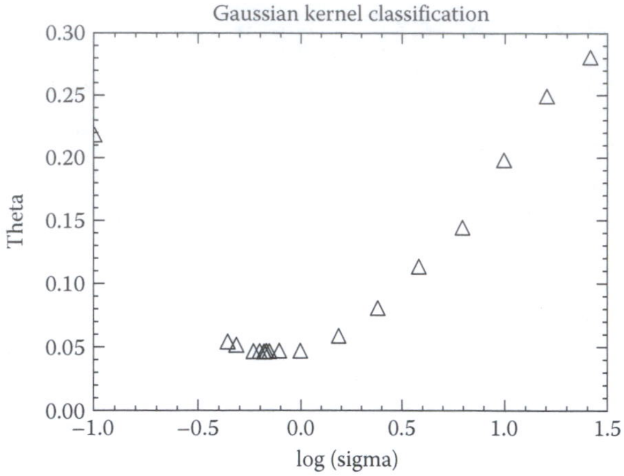

FIGURE 6.5
Training the Gaussian kernel classifier with parabolic interpolation.

The classification of the entire image proceeds row-by-row using ENVI's tiling facility with repeated calls to `OUTPUT(SIGMA,HS)`, where `SIGMA` is fixed to its optimum value and `HS` is a row of image pixel vectors in the data matrix format. A classification result, again for the Jülich ASTER scene, is shown in Figure 6.4b. Similarly to the modified maximum likelihood classifier, the labeled observations can be split into training and test pixels with the latter held back for later accuracy evaluation.

The Gaussian kernel classifier, like many other nonparametric methods, suffers from the drawback of requiring all training data points to be used in the generalization phase (a so-called *memory-based* classifier). The evaluation is very slow if the number of training points is large, which should be the case if a reasonable approximation of the class-specific densities is to be achieved. The routine `KERNEL_RUN` is in fact unacceptably slow for training data sets exceeding a few thousand pixels, even with CUDA enabled. Moreover, the required number of training samples grows exponentially with the dimensionality N of the data.* Quite generally, the complexity of the calculation is determined by the amount of training data, not by the difficulty of the classification problem itself, an undesirable state of affairs.

6.5 Neural Networks

Neural networks belong to the category of semiparametric models for probability density estimation, a category that lies somewhere between the parametric and nonparametric extremes (Bishop, 1995). They make no strong

* The so-called *curse of dimensionality* (Bellman, 1961).

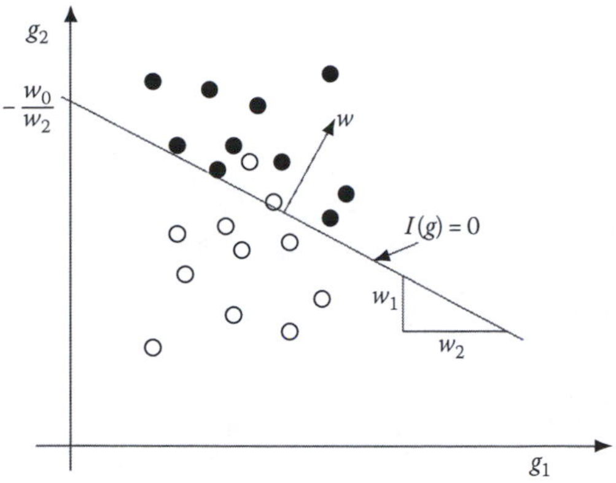

FIGURE 6.6
A linear discriminant for two classes. The vector $w = (w_1, w_2)^\top$ is normal to the separating line in the direction of class $k = 1$, shown as black dots.

assumptions about the form of the probability distributions and can be adjusted flexibly to the complexity of the system that they are being used to model. They, therefore, provide an attractive compromise.

To motivate their use for classification, let us consider two classes, $k = 1$ and $k = 2$, for two-dimensional observations, $g = (g_1, g_2)^\top$. We can write the maximum likelihood decision rule, Equation 6.18 with Equation 6.17, in terms of a new discriminant function,

$$I(g) = d_1(g) - d_2(g),$$

and say that

$$g \text{ is class } \begin{cases} 1 & \text{if } I(g) \geq 0 \\ 2 & \text{if } I(g) < 0. \end{cases}$$

The discriminant $I(g)$ is a rather complicated quadratic function of g. The *simplest* discriminant that could conceivably decide between the two classes is a linear function of the form*

$$I(g) = w_0 + w_1 g_1 + w_2 g_2, \tag{6.23}$$

where w_0, w_1, and w_2 are parameters. The decision boundary occurs for $I(g) = 0$, that is, for

$$g_2 = -\frac{w_1}{w_2} g_1 - \frac{w_0}{w_2},$$

as depicted in Figure 6.6.

* A linear decision boundary will arise in a maximum likelihood classifier if the covariance matrices for the two classes are identical (see Exercise 2).

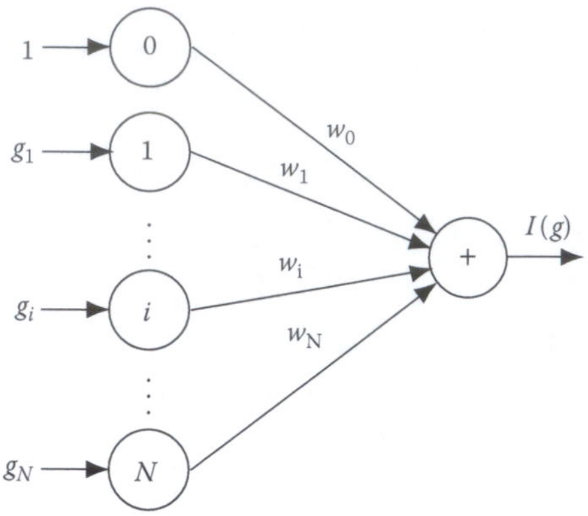

FIGURE 6.7
An artificial neuron representing Equation 6.24. The first input is always unity and is called the bias.

Extending the discussion now to N-dimensional observations, we can work with the discriminant

$$I(\boldsymbol{g}) = w_0 + w_1 g_1 + \cdots + w_N g_N = \boldsymbol{w}^\top \boldsymbol{g} + w_0. \tag{6.24}$$

In this higher-dimensional feature space, the decision boundary, $I(\boldsymbol{g}) = 0$, generalizes to an *oriented hyperplane*. Equation 6.24 can be represented schematically as an *artificial neuron* or a *perceptron*, as shown in Figure 6.7, along with some additional jargon. Thus, the "input signals," $g_1 \ldots g_N$, are multiplied with "synaptic weights" $w_1 \ldots w_N$ and the results are summed in a "neuron" to produce "the output signal," $I(\boldsymbol{g})$. The w_0 term is treated by introducing a "bias" input of unity, which is multiplied by w_0 and included in the summation.

In keeping with the biological analogy, the output $I(\boldsymbol{g})$ may be modified by a so-called *sigmoid* (= S-shaped) "activation function," for example, by the *logistic* function

$$f(\boldsymbol{g}) = \frac{1}{1 + e^{-I(\boldsymbol{g})}}.$$

$I(\boldsymbol{g})$ is then referred to as the *activation* of the neuron inducing the output signal, $f(\boldsymbol{g})$. The IDL code

```
thisDevice =!D.Name
set_plot, 'PS'
device, Filename='c:\temp\fig6_8.eps', $
   xsize=15,ysize=10,/Encapsulated
x=(findgen(100)-50)/10
PLOT, x,1/(1+exp(-x))
device,/close_file
set_plot,thisDevice
```

generates the plot of the logistic function shown in Figure 6.8. This modification of the discriminant has the advantage that the output signal saturates

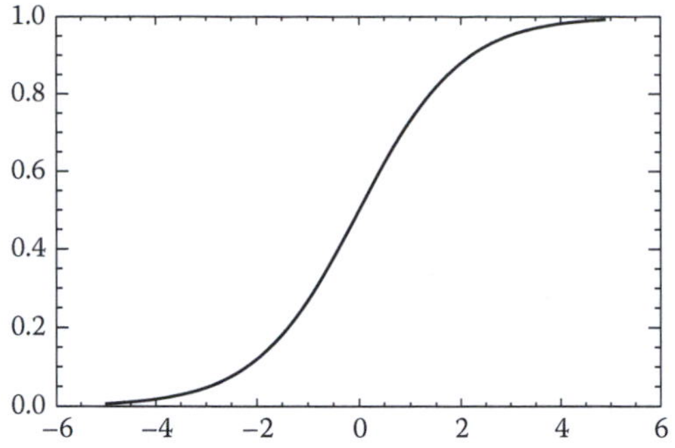

FIGURE 6.8
The logistic activation function.

at values 0 and 1 for large negative and positive inputs, respectively. However, Bishop (1995) suggests that there is also a good statistical justification for using it. Suppose that the two classes are normally distributed with $\boldsymbol{\Sigma}_1 = \boldsymbol{\Sigma}_2 = \boldsymbol{I}$. Then,

$$p(\boldsymbol{g} \mid k) = \frac{1}{2\pi} \exp\left(\frac{-\|\boldsymbol{g} - \boldsymbol{\mu}_k\|^2}{2}\right),$$

for $k = 1, 2$, and we have with Bayes' theorem

$$\begin{aligned} \Pr(1 \mid \boldsymbol{g}) &= \frac{p(\boldsymbol{g} \mid 1)\Pr(1)}{p(\boldsymbol{g} \mid 1)\Pr(1) + p(\boldsymbol{g} \mid 2)\Pr(2)} \\ &= \frac{1}{1 + p(\boldsymbol{g} \mid 2)\Pr(2)/(p(\boldsymbol{g} \mid 1)\Pr(1))} \\ &= \frac{1}{1 + \exp(-\frac{1}{2}[\|\boldsymbol{g} - \boldsymbol{\mu}_2\|^2 - \|\boldsymbol{g} - \boldsymbol{\mu}_1\|^2])(\Pr(2)/\Pr(1))}. \end{aligned}$$

With the substitution

$$e^{-a} = \Pr(2)/\Pr(1),$$

we get

$$\begin{aligned} \Pr(1 \mid \boldsymbol{g}) &= \frac{1}{1 + \exp(-\frac{1}{2}[\|\boldsymbol{g} - \boldsymbol{\mu}_2\|^2 - \|\boldsymbol{g} - \boldsymbol{\mu}_1\|^2] - a)} \\ &= \frac{1}{1 + \exp(-\boldsymbol{w}^\top \boldsymbol{g} - w_0)} \\ &= \frac{1}{1 + e^{-I(\boldsymbol{g})}} = f(\boldsymbol{g}). \end{aligned}$$

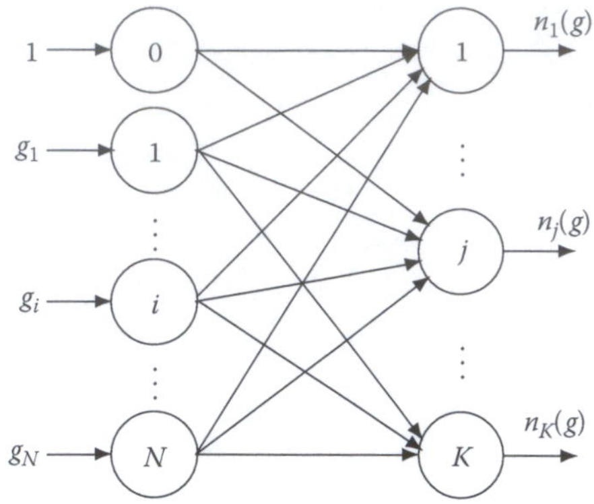

FIGURE 6.9
A single-layer neural network.

In the second equality above, we have made the additional substitutions

$$\boldsymbol{w} = \boldsymbol{\mu}_1 - \boldsymbol{\mu}_2$$
$$w_0 = -\frac{1}{2}\|\boldsymbol{\mu}_1\|^2 + \frac{1}{2}\|\boldsymbol{\mu}_2\|^2 + a.$$

Thus, we expect that the output signal, $f(\boldsymbol{g})$, of the neuron will not only discriminate between the two classes, *but also that it will approximate the posterior class membership probability* $\Pr(1 \mid \boldsymbol{g})$.

The extension of linear discriminants from two to K classes is straightforward, and leads to the *single-layer neural network* of Figure 6.9. There are K neurons (the circles on the right), each of which calculates its own discriminant

$$n_j(\boldsymbol{g}) = f(I_j(\boldsymbol{g})),\ j = 1 \ldots K.$$

The observation $\boldsymbol{g}$ is assigned to the class whose neuron produces the maximum output signal, i.e.,

$$k = \arg\max_j n_j(\boldsymbol{g}).$$

Each neuron is associated with a synaptic weight vector $\boldsymbol{w}_j$, which we from now on will understand to include the bias weight w_0. Thus, for the jth neuron,

$$\boldsymbol{w}_j = (w_{0j}, w_{1j} \ldots w_{Nj})^\top,$$

and, for the whole network,

$$\mathbf{W} = (\boldsymbol{w}_1, \boldsymbol{w}_2 \ldots \boldsymbol{w}_K) = \begin{pmatrix} w_{01} & w_{02} & \cdots & w_{0K} \\ w_{11} & w_{12} & \cdots & w_{1K} \\ \vdots & \vdots & \ddots & \vdots \\ w_{N1} & w_{N2} & \cdots & w_{NK} \end{pmatrix}, \tag{6.25}$$

which we shall call the *synaptic weight matrix* for the neuron layer.

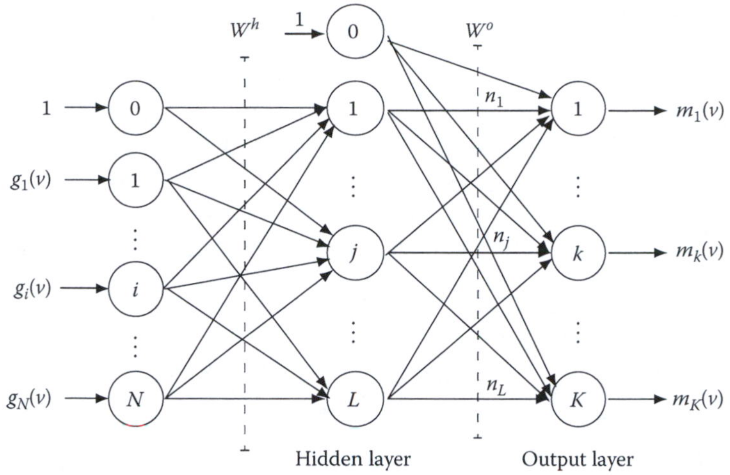

FIGURE 6.10
A two-layer feed-forward neural network with L hidden neurons for the classification of N-dimensional data into K classes. The argument ν identifies a training example.

6.5.1 Neural Network Classifier

Single-layer networks turn out to be rather limited in the kinds of classification tasks that they can handle. In fact, only so-called *linearly separable* problems, problems in which classes of training observations can be separated by hyperplanes, are fully solvable (see Exercise 3). On the other hand, networks with just one additional layer of processing neurons can approximate any given decision boundary arbitrarily closely (Bishop, 1995) provided that the first-layer (or hidden-layer) outputs are nonlinear.* Accordingly, we shall develop a classifier based on the two-layer feed-forward architecture† shown in Figure 6.10.

For the νth training pixel, the input to the network is the $(N + 1)$-component (biased) observation vector:

$$g(\nu) = (1, g_1(\nu) \dots g_N(\nu))^\top.$$

This input is distributed simultaneously to all of the L neurons in the hidden layer of neurons. These in turn determine an $(L + 1)$-component vector of

* The input data are transformed by the hidden layer into a higher-dimensional function space in which the problem becomes linearly separable (see Müller et al., 2001). This is the strategy used for designing *SVMs* (Belousov et al., 2002) (see Section 6.6).

† The adjective *feed-forward* merely serves to differentiate this network structure from other networks having feedback connections.

intermediate outputs (adding in the bias input for the next layer),

$$n(\nu) = (1, n_1(\nu) \dots n_L(\nu))^\top,$$

in which $n_j(\nu)$ is shorthand for

$$f(I_j^h(g(\nu))), \quad j = 1 \dots L.$$

In this expression, the activation I_j^h of the hidden neurons is given by

$$I_j^h(g(\nu)) = w_j^{h\top} g(\nu),$$

where the vector w_j^h is the weight vector for the jth neuron in the hidden layer:

$$w_j^h = (w_{0j}^h, w_{1j}^h \dots w_{Nj}^h)^\top.$$

In terms of a hidden weight matrix, W^h, having the form of Equation 6.25, namely,

$$W^h = (w_1^h, w_2^h \dots w_L^h),$$

we can write all of this more compactly in vector notation as

$$n(\nu) = \begin{pmatrix} 1 \\ f(W^{h\top} g(\nu)) \end{pmatrix}. \tag{6.26}$$

Here, we just have to interpret the logistic function of a vector v, $f(v)$, as a vector of logistic functions of the components of v.

The vector $n(\nu)$ is then fed in the same manner to the *output layer* with its associated output weight matrix,

$$W^o = (w_1^o, w_2^o \dots w_K^o),$$

and the output signal, $m(\nu)$, is calculated similarly as for Equation 6.26, as

$$m(\nu) = f(W^{o\top} n(\nu)). \tag{6.27}$$

However, this last equation is not quite satisfactory. According to our previous considerations, we would like to interpret the network outputs as class membership probabilities. This means we must ensure that

$$0 \le m_k(\nu) \le 1, \quad k = 1 \dots K,$$

and, furthermore, that

$$\sum_{k=1}^{K} m_k(\nu) = 1.$$

The logistic function, f, satisfies the first condition, but there is no reason why the second condition should be met. It can be enforced, however, by using a modified logistic activation function for the output neurons, called *softmax* (Bridle, 1990). The softmax function is defined as

$$m_k(\nu) = \frac{e^{I_k^o(\boldsymbol{n}(\nu))}}{e^{I_1^o(\boldsymbol{n}(\nu))} + e^{I_2^o(\boldsymbol{n}(\nu))} + \cdots + e^{I_M^o(\boldsymbol{n}(\nu))}}, \tag{6.28}$$

where

$$I_k^o(\boldsymbol{n}(\nu)) = \boldsymbol{w}_k^{o\top}\boldsymbol{n}(\nu), \quad k = 1\ldots K, \tag{6.29}$$

and it clearly guarantees that the output signals sum to unity.

Equations 6.26, 6.28, and 6.29 now provide a complete mathematical representation of the neural network classifier shown in Figure 6.10. It turns out to be a very useful classifier indeed. To quote Bishop (1995):

> [...] [two-layer, feed-forward] networks can approximate arbitrarily well any functional continuous mapping from one finite dimensional space to another, provided the number [L] of hidden units is sufficiently large. [...] An important corollary of this result is, that in the context of a classification problem, networks with sigmoidal nonlinearities and two layers of weights can approximate any decision boundary to arbitrary accuracy. [...] More generally, the capability of such networks to approximate general smooth functions allows them to model posterior probabilities of class membership.

The IDL object class `FFN`, an excerpt of which is given in Listing 6.5, mirrors the network architecture of Figure 6.10. It will form the basis for the implementation of the backpropagation training algorithm developed below, and also for the more efficient training algorithms described in Appendix B. A summary of all of the program modules involved in the ENVI/IDL extension for neural network classification is given in Appendix C.2.10.

6.5.2 Cost Functions

We have not yet considered the correct choice of synaptic weights, that is, how to go about training the neural network classifier. As mentioned in Section 6.3.2, the training data are most conveniently represented as the set of labeled pairs:

$$\mathcal{T} = \{(\boldsymbol{g}(\nu), \boldsymbol{\ell}(\nu)) \mid \nu = 1\ldots m\},$$

where the label

$$\boldsymbol{\ell}(\nu) = (0\ldots 0, 1, 0\ldots 0)^\top$$

Listing 6.5

Excerpt from the object class FFN.

```
FUNCTION FFN::Init, Gs, Ls, L
; network architecture
   self.LL = L
   self.np = n_elements(Gs[*,0])
   self.NN = n_elements(Gs[0,*])
   self.KK = n_elements(Ls[0,*])
; biased output vector from hidden layer
   self.N = ptr_new(dblarr(L+1))
; biased exemplars (column vectors)
   self.Gs = ptr_new([[dblarr(self.np)+1],[Gs]])
   self.Ls = ptr_new(Ls)
; weight matrices
   self.Wh = ptr_new(randomu(seed,L,self.NN+1,/double)$
                                                -0.5)
   self.Wo = ptr_new(randomu(seed,self.KK,L+1,/double)$
                                                -0.5)
   RETURN,1
END

FUNCTION FFN::forwardPass, G
; forward pass through network
   expnt = transpose(*self.Wh)##G
   *self.N = [[1.0],[1/(1+exp(-expnt))]]
; softmax activation for output neurons
   I = transpose(*self.Wo)##*self.N
   A = exp(I-max(I))
   RETURN, A/total(A)
END

FUNCTION FFN::classify, Gs, Probs
; vectorized class membership probabilities
   nx = n_elements(Gs[*,0])
   Ones = dblarr(nx) + 1.0
   expnt = transpose(*self.Wh)##[[Ones],[Gs]]
   N = [[Ones],[1/(1+exp(-expnt))]]
   Io = transpose(*self.Wo)##N
   maxIo = max(Io,dimension=2)
   FOR k=0,self.KK-1 DO Io[*,k]=Io[*,k]-maxIo
   A = exp(Io)
   sum = total(A,2)
   Probs = fltarr(nx,self.KK)
   FOR k=0,self.KK-1 DO Probs[*,k] = A[*,k]/sum
   void = max(probs,labels,dimension=2)
   RETURN, byte(labels/nx+1)
END
```

is a K-dimensional column vector of zeroes, except with the "1" at the kth position to indicate that $\boldsymbol{g}(\nu)$ belongs to class k.

Under certain assumptions about the distribution of the training data, the *quadratic cost function*,

$$E(\boldsymbol{W}^h, \boldsymbol{W}^o) = \frac{1}{2}\sum_{\nu=1}^{m} \|\boldsymbol{\ell}(\nu) - \boldsymbol{m}(\nu)\|^2, \tag{6.30}$$

can be justified as a training criterion for feed-forward networks (Exercise 4). The network weights, $\boldsymbol{W}^h$ and $\boldsymbol{W}^o$, must be adjusted so as to minimize E. This minimization will clearly tend to make the network produce the output signal

$$\boldsymbol{m} = (1, 0 \ldots 0 \ldots 0)^\top$$

whenever it is presented with a training observation, $\boldsymbol{g}(\nu)$, from class $k = 1$, and similarly for the other classes.

This, of course, is what we wish it to do. However, a more appropriate cost function for classification problems can be obtained with the following criterion: Choose the synaptic weights so as to *maximize the probability of observing the training data.* The joint probability for observing the training example $(\boldsymbol{g}(\nu), \boldsymbol{\ell}(\nu))$ is

$$\Pr(\boldsymbol{g}(\nu), \boldsymbol{\ell}(\nu)) = \Pr(\boldsymbol{\ell}(\nu) \mid \boldsymbol{g}(\nu))\Pr(\boldsymbol{g}(\nu)), \tag{6.31}$$

where we have used Equation 2.53. The neural network, as was argued, approximates the posterior class membership probability, $\Pr(\boldsymbol{\ell}(\nu) \mid \boldsymbol{g}(\nu))$. In fact, this probability can be expressed directly in terms of the network output signal in the form

$$\Pr(\boldsymbol{\ell}(\nu) \mid \boldsymbol{g}(\nu)) = \prod_{k=1}^{K} [\, m_k(\boldsymbol{g}(\nu)) \,]^{\ell_k(\nu)}. \tag{6.32}$$

In order to see this, consider the case $\boldsymbol{\ell} = (1, 0 \ldots 0)^\top$. Then, according to Equation 6.32,

$$\Pr((1, 0 \ldots 0)^\top \mid \boldsymbol{g}) = m_1(\boldsymbol{g})^1 \cdot m_2(\boldsymbol{g})^0 \cdots m_K(\boldsymbol{g})^0 = m_1(\boldsymbol{g}),$$

which is the probability that $\boldsymbol{g}$ is in class 1, as desired. Now, substituting Equation 6.32 into Equation 6.31, we therefore wish to maximize

$$\Pr(\boldsymbol{g}(\nu), \boldsymbol{\ell}(\nu)) = \prod_{k=1}^{K} [\, m_k(\boldsymbol{g}(\nu)) \,]^{\ell_k(\nu)}\Pr(\boldsymbol{g}(\nu)).$$

Taking logarithms, dropping terms that are independent of the synaptic weights, and summing over all of the training data, we see that this is equivalent to minimizing the *cross entropy* cost function,

$$E(\mathbf{W}^h, \mathbf{W}^o) = -\sum_{\nu=1}^{m}\sum_{k=1}^{K} \ell_k(\nu) \log[m_k(\mathbf{g}(\nu))], \tag{6.33}$$

with respect to the synaptic weight parameters.

6.5.3 Backpropagation

A minimum of the cost function (Equation 6.33) can be found with various search algorithms. *Backpropagation* is the most well-known and extensively used method, and is described below. It is implemented in the standard ENVI neural network for supervised classification under the main menu command

```
Classification/Supervised/Neural Net
```

Two considerably faster algorithms, the *scaled conjugate gradient* and the *Kalman filter*, are discussed in detail in Appendix B, and ENVI/IDL extensions for supervised classification with a feed-forward neural network trained with these algorithms are given in Appendix C.2.10. The discussions in both appendixes build on the following development and coding of the backpropagation algorithm.

Our starting point is the so-called *local version* of the cost function of Equation 6.33:

$$E(\mathbf{W}^h, \mathbf{W}^o, \nu) = -\sum_{k=1}^{K} \ell_k(\nu) \log[m_k(\mathbf{g}(\nu))], \quad \nu = 1 \dots m.$$

This is just the cost function for a single training example. If we manage to make it smaller at each step of the calculation and cycle, either sequentially or randomly, through the available training pairs, then we are obviously minimizing the overall cost function as well. With the abbreviation $m_k(\nu) = m_k(\mathbf{g}(\nu))$, the local cost function can be written a little more compactly as

$$E(\nu) = -\sum_{k=1}^{K} \ell_k(\nu) \log[m_k(\nu)].$$

Here, the dependence of the cost function on the synaptic weights is also implicit. More compactly still, it can be represented in vector form as an inner product:

$$E(\nu) = -\boldsymbol{\ell}(\nu)^\top \log[\boldsymbol{m}(\nu)]. \tag{6.34}$$

Our problem then is to minimize Equation 6.34 with respect to the synaptic weights, which are the $(N+1) \times L$ elements of the matrix $\boldsymbol{W}^h$ and the $(L+1) \times K$ elements of $\boldsymbol{W}^o$. Let us consider the following algorithm:

Algorithm (Backpropagation)

1. Initialize the synaptic weights with random numbers and set ν equal to a random integer in the interval $[1, m]$.
2. Choose the training pair $(\boldsymbol{g}(\nu), \boldsymbol{\ell}(\nu))$ and determine the output response, $\boldsymbol{m}(\nu)$, of the network.
3. For $k = 1 \dots K$ and $j = 0 \dots L$, replace w^o_{jk} with $w^o_{jk} - \eta \frac{\partial E(\nu)}{\partial w^o_{jk}}$.
4. For $j = 1 \dots L$ and $i = 0 \dots N$, replace w^h_{ij} with $w^h_{ij} - \eta \frac{\partial E(\nu)}{\partial w^h_{ij}}$.
5. If $\sum_\nu E(\nu)$ ceases to change significantly, stop, otherwise set ν equal to a new random integer in $[1, m]$ and go to step 2.

This algorithm jumps randomly through the training data, reducing the local cost function at each step. The reduction is accomplished by changing each synaptic weight, w, by an amount proportional to the negative slope, $-\partial E(\nu)/\partial w$, of the local cost function with respect to that weight parameter, stopping when the overall cost function (Equation 6.33) can no longer be reduced. The constant of proportionality, η, is referred to as the *learning rate* for the network. This algorithm makes use only of the first derivatives of the cost function with respect to the synaptic weight parameters, and belongs to the class of *gradient descent* methods.

To implement the algorithm, the partial derivatives of $E(\nu)$ with respect to the synaptic weights are required. Let us begin with the output neurons, which generate the softmax output signals,

$$m_k(\nu) = \frac{e^{I^o_k(\nu)}}{e^{I^o_1(\nu)} + e^{I^o_2(\nu)} + \cdots + e^{I^o_K(\nu)}}, \tag{6.35}$$

where

$$I^o_k(\nu) = \boldsymbol{w}^{o\top}_k \boldsymbol{n}(\nu).$$

We wish to determine (step 3 of the backpropagation algorithm)

$$\frac{\partial E(\nu)}{\partial w^o_{jk}}, \quad j = 0 \dots L,\ k = 1 \dots K.$$

Recalling the rules for vector differentiation in Chapter 1 and applying the chain rule, we get

$$\frac{\partial E(\nu)}{\partial \boldsymbol{w}^o_k} = \frac{\partial E(\nu)}{\partial I^o_k(\nu)} \frac{\partial I^o_k(\nu)}{\partial \boldsymbol{w}^o_k} = -\delta^o_k(\nu)\boldsymbol{n}(\nu), \quad k = 1 \dots K, \tag{6.36}$$

where we have introduced the quantity $\delta_k^o(\nu)$ given by

$$\delta_k^o(\nu) = -\frac{\partial E(\nu)}{\partial I_k^o(\nu)}. \tag{6.37}$$

This is the negative rate of change of the local cost function with respect to the activation of the kth output neuron.

Again applying the chain rule and using Equations 6.34 and 6.35,

$$\begin{aligned}
-\delta_k^o(\nu) = \frac{\partial E(\nu)}{\partial I_k^o(\nu)} &= \sum_{k'=1}^{K} \frac{\partial E(\nu)}{\partial m_{k'}(\nu)} \frac{\partial m_{k'}(\nu)}{\partial I_k^o(\nu)} \\
&= \sum_{k'=1}^{K} -\frac{\ell_{k'}(\nu)}{m_{k'}(\nu)} \left(\frac{e^{I_k^o(\nu)}\delta_{kk'}}{\sum_{k''=1}^{K} e^{I_{k''}^o(\nu)}} - \frac{e^{I_{k'}^o(\nu)} e^{I_k^o(\nu)}}{\left(\sum_{k''=1}^{K} e^{I_{k''}^o(\nu)}\right)^2} \right).
\end{aligned}$$

Here, $\delta_{kk'}$ is given by

$$\delta_{kk'} = \begin{cases} 0 & \text{if } k \neq k' \\ 1 & \text{if } k = k'. \end{cases}$$

Continuing, making use of Equation 6.35,

$$\begin{aligned}
-\delta_k^o(\nu) &= \sum_{k'=1}^{K} -\frac{\ell_{k'}(\nu)}{m_{k'}(\nu)} m_k(\nu)(\delta_{kk'} - m_{k'}(\nu)) \\
&= -\ell_k(\nu) + m_k(\nu) \sum_{k'=1}^{K} \ell_{k'}(\nu).
\end{aligned}$$

But this last sum over the K components of the label $\boldsymbol{\ell}(\nu)$ is just unity, and, therefore, we have

$$-\delta_k^o(\nu) = -\ell_k(\nu) + m_k(\nu), \quad k = 1 \ldots K,$$

which may be written as the K-component vector,

$$\boldsymbol{\delta}^o(\nu) = \boldsymbol{\ell}(\nu) - \boldsymbol{m}(\nu). \tag{6.38}$$

From Equation 6.36, we can therefore express the third step in the backpropagation algorithm in the form of the matrix equation (see Exercise 6)

$$\boldsymbol{W}^o(\nu + 1) = \boldsymbol{W}^o(\nu) + \eta\, \boldsymbol{n}(\nu)\boldsymbol{\delta}^o(\nu)^\top. \tag{6.39}$$

Here, $\boldsymbol{W}^o(\nu + 1)$ indicates the synaptic weight matrix *after* the update for the νth training pair. Note that the second term on the right-hand side of Equation 6.39 is an outer product, yielding a matrix of dimension $(L+1) \times K$ and so matching the dimension of $\boldsymbol{W}^o(\nu)$.

For the hidden weights, step 4 of the algorithm, we proceed similarly:

$$\frac{\partial E(\nu)}{\partial \boldsymbol{w}_j^h} = \frac{\partial E(\nu)}{\partial I_j^h(\nu)} \frac{\partial I_j^h(\nu)}{\partial \boldsymbol{w}_j^h} = -\delta_j^h(\nu)\boldsymbol{g}(\nu), \quad j = 1 \dots L, \tag{6.40}$$

where $\delta_j^h(\nu)$ is the negative rate of change of the local cost function with respect to the activation of the jth hidden neuron:

$$\delta_j^h(\nu) = -\frac{\partial E(\nu)}{\partial I_j^h(\nu)}.$$

Applying the chain rule again:

$$\begin{aligned} -\delta_j^h(\nu) &= \sum_{k=1}^{K} \frac{\partial E(\nu)}{\partial I_k^o(\nu)} \frac{\partial I_k^o(\nu)}{\partial I_j^h(\nu)} = -\sum_{k=1}^{K} \delta_k^o(\nu) \frac{\partial I_k^o(\nu)}{\partial I_j^h(\nu)} \\ &= -\sum_{k=1}^{K} \delta_k^o(\nu) \frac{\partial \boldsymbol{w}_k^{o\top} \boldsymbol{n}(\nu)}{\partial I_j^h(\nu)} = -\sum_{k=1}^{K} \delta_k^o(\nu) \boldsymbol{w}_k^{o\top} \frac{\partial \boldsymbol{n}(\nu)}{\partial I_j^h(\nu)}. \end{aligned}$$

In the last partial derivative, since $I_j^h(\nu) = \boldsymbol{w}_j^{h\top} \boldsymbol{g}(\nu)$, only the output of the jth hidden neuron is a function of $I_j^h(\nu)$. Therefore,

$$\delta_j^h(\nu) = \sum_{k=1}^{K} \delta_k^o(\nu) w_{jk}^o \frac{\partial n_j(\nu)}{\partial I_j^h(\nu)}. \tag{6.41}$$

Recall that the hidden units use the logistic activation function:

$$n_j(I_j^h) = f(I_j^h) = \frac{1}{1 + e^{-I_j^h}}.$$

This function has a very simple derivative:

$$\frac{\partial n_j(x)}{\partial x} = n_j(x)(1 - n_j(x)).$$

Therefore, we can write Equation 6.41 as

$$\delta_j^h(\nu) = \sum_{k=1}^{K} \delta_k^o(\nu) w_{jk}^o n_j(\nu)(1 - n_j(\nu)), \quad j = 1 \dots L,$$

or, more compactly, as the matrix equation

$$\begin{pmatrix} 0 \\ \boldsymbol{\delta}^h(\nu) \end{pmatrix} = \boldsymbol{n}(\nu) \cdot (\boldsymbol{1} - \boldsymbol{n}(\nu)) \cdot \left(\boldsymbol{W}^o \boldsymbol{\delta}^o(\nu) \right). \tag{6.42}$$

The dot is intended to denote simple component-by-component (the so-called *Hadamard*) multiplication. The equation must be written in this rather awkward way because the expression on the right-hand side has $L+1$ components. This also makes the fact $1 - n_0(\nu) = 0$ explicit. Equation 6.42 is the origin of the term "backpropagation," since it propagates the negative rate of change of the cost function with respect to the output activations, $\boldsymbol{\delta}^o(\nu)$, backward through the network to determine the negative rate of change with respect to the hidden activations, $\boldsymbol{\delta}^h(\nu)$.

Finally, with Equation 6.40, we obtain the update rule for step 4 of the backpropagation algorithm:

$$\boldsymbol{W}^h(\nu+1) = \boldsymbol{W}^h(\nu) + \eta\, \boldsymbol{g}(\nu)\boldsymbol{\delta}^h(\nu)^\top. \tag{6.43}$$

The choice of an appropriate learning rate, η, is problematic: small values imply slow convergence and large values produce oscillation. Some improvement can be achieved with an additional, purely heuristic parameter called *momentum*, which maintains a portion of the preceding weight increments in the current iteration. Equation 6.39 is replaced with

$$\boldsymbol{W}^o(\nu+1) = \boldsymbol{W}^o(\nu) + \Delta^o(\nu) + \alpha\Delta^o(\nu-1), \tag{6.44}$$

where $\Delta^o(\nu) = \eta\, \boldsymbol{n}(\nu)\boldsymbol{\delta}^{o\top}(\nu)$ and α is the momentum parameter. A similar expression replaces Equation 6.43. Typical choices for the backpropagation parameters are $\eta = 0.01$ and $\alpha = 0.5$.

Listing 6.6 is an excerpt from the object class `FFNBP` extending (i.e., inheriting) the class `FFN` to implement the backpropagation algorithm. It shows the code for the procedure method `TRAIN`, which closely parallels the equations developed above.

6.5.4 Overfitting and Generalization

A fundamental and much-discussed dilemma in the application of neural networks (and other learning algorithms) is that of *overfitting*. Reduced to its essentials, the question is: "How many hidden neurons are enough?" The number of neurons in the output layer of the network in Figure 6.10 is determined by the number of training classes. The number in the hidden layer is fully undetermined. If "too few" hidden neurons are chosen (and, thus, too few synaptic weights), there is a danger that the classification will be suboptimal: there will be an insufficient number of adjustable parameters to resolve the class structure of the training data. If, on the other hand, "too many" hidden neurons are selected, and if the training data have a large noise variance, there will be a danger that the network will fit the data all too well, including their detailed random structure. Such a detailed structure is a characteristic of the particular training sample chosen and not of the underlying class distributions that the network is supposed to learn. It is here that one speaks of overfitting. In either case, the capability of the network

Listing 6.6
Excerpt from the object class FFNBP.

```
PRO FFNBP::Train
   iter = 0L & iter100 = 0L
   iterations = 100*self.np
   eta = 0.01 &  alpha = 0.5 ; learn rate & momentum
   progressbar=Obj_New('progressbar', $
    Color='blue',Text='0',$
    title='Training:␣example␣No...',xsize=250,ysize=20)
   progressbar->start
   window,12,xsize=600,ysize=400,title='Cost␣Function'
   wset,12
   inc_o1 = 0 & inc_h1 = 0
   REPEAT BEGIN
      IF progressbar->CheckCancel() THEN BEGIN
         PRINT,'Training␣interrupted'
         progressbar->Destroy & RETURN
      ENDIF
; select example pair at random
      nu = long(self.np*randomu(seed))
      x=(*self.Gs)[nu,*]
      ell=(*self.Ls)[nu,*]
; send it through the network
      m=self->forwardPass(x)
; determine the deltas
      d_o = ell - m
      d_h = (*self.N*(1-*self.N)* $
            (*self.Wo##d_o))[1:self.LL]
; update the synaptic weights
      inc_o = eta*(*self.N##transpose(d_o))
      inc_h = eta*(x##d_h)
      *self.Wo = *self.Wo + inc_o + alpha*inc_o1
      *self.Wh = *self.Wh + inc_h + alpha*inc_h1
      inc_o1 = inc_o & inc_h1 = inc_h
; record cost history
      IF iter MOD 100 EQ 0 THEN BEGIN
         (*self.cost_array)[iter100] = $
                    alog10(self->cost(0))
         iter100 = iter100+1
         progressbar->Update,iter*100/iterations,$
                      text=strtrim(iter,2)
         PLOT,*self.cost_array,xrange=[0,iter100],$
             color=0,background='FFFFFF'XL,$
            xtitle='Iterations/100)',$
            ytitle='log(cross␣entropy)'
      END
      iter=iter+1
   ENDREP UNTIL iter EQ iterations
   progressbar->destroy
END
```

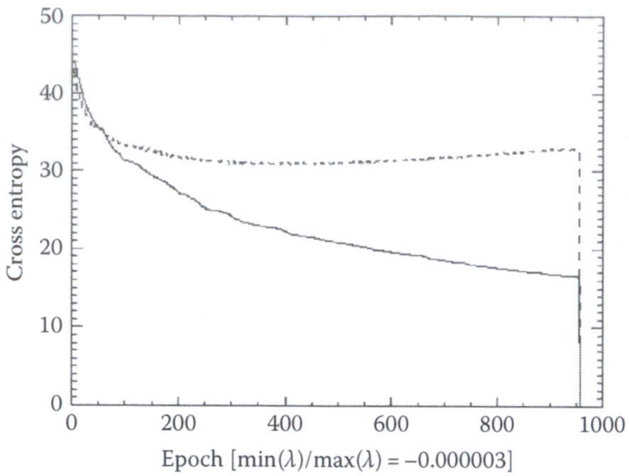

FIGURE 6.11
An example of overfitting during neural network training with the scaled conjugate gradient algorithm. The lower curve shows the cost function calculated with the training data, the upper curve with the validation data. One "epoch" corresponds to a cycle through all of the training examples. The quantity λ is an array of eigenvalues of the Hessian matrix, (see Appendix B).

to generalize to unknown inputs will be impaired. One can find excellent discussions on this subject in Hertz et al. (1991, Chapter 6) and in Bishop (1995, Chapter 9), where regularization techniques are introduced to penalize overfitting. Alternatively, so-called *growth* and *pruning* algorithms can be applied in which the network architecture is optimized during the training procedure.* We shall restrict ourselves here to a solution that presupposes an over-dimensioned network, that is, one with too many hidden weights, as well as the availability of a second data set, which is usually referred to as the *validation* data set.

An option in the ENVI/IDL neural network program of Appendix C.2.10 allows the training data to be split into two, so that half the data are reserved for validation purposes. (If test data were held back, the training data set is then one-third of its original size.) Both halves are still representative of the class distributions and are statistically independent. During the training phase, which is carried out only with the training data, cost functions calculated both with the training data as well as with the validation data are displayed. Overfitting is indicated by a continued decrease in the training cost function accompanied by a gradual increase in the validation cost function. Figure 6.11 shows a fairly typical example, achieved with 12 hidden neurons and 7 classes. The gradual, albeit very slight, increase in the validation cost function is indicative of overfitting: the network is learning the detailed structure of the training data at the cost of its ability to generalize. The algorithm should thus be stopped when the upper curve starts to rise (so-called *early stopping*).

Figure 6.4c shows a classification of the Jülich ASTER scene, as obtained with the neural network classifier. In the next chapter, we shall see how

* A popular growth algorithm is the *cascade correlation neural network* of Fahlman and LeBiere (1990).

to compare the generalization capability of neural networks with that of both the maximum likelihood and Gaussian kernel classifiers, which we developed previously.

6.6 Support Vector Machines

Let us return to the simple linear discriminant function, $I(g)$, for a two-class problem given by Equation 6.24, with the convention that the weight vector does not include the bias term, that is,

$$I(g) = w^\top g(\nu) + w_0,$$

where

$$w = (w_1 \dots w_N)^\top$$

and where the training observations are

$$g(\nu) = (g_1(\nu) \dots g_N(\nu))^\top,$$

with corresponding (this time scalar) labels

$$\ell(\nu) \in \{0, 1\}, \quad \nu = 1 \dots m.$$

A quadratic cost function for training this discriminant on two classes would then be

$$E(w) = \frac{1}{2} \sum_{\nu=1}^{m} (w^\top g(\nu) + w_0 - \ell(\nu))^2. \tag{6.45}$$

This expression is essentially the same as Equation 6.30, when it is written for the case of a single neuron. Training the neuron of Figure 6.7 to discriminate the two classes means finding the weight parameters that minimize the above cost function. This can be done, for example, by using a gradient descent method, or, more efficiently, with the *perceptron algorithm*, as explained in Exercise 3. We shall consider in the following text an alternative to such a cost function approach. The method we describe is reminiscent of the Gaussian kernel method described in Section 6.4 in that the training observations are used also at the classification phase, but, as we shall see, not all of them.

6.6.1 Linearly Separable Classes

It is convenient first of all to relabel the training observations as $\ell(\nu) \in \{+1, -1\}$, rather than $\ell(\nu) \in \{0, 1\}$. With this convention, the product

$$\ell(\nu)(\boldsymbol{w}^\top \boldsymbol{g}(\nu) + w_0), \quad \nu = 1 \dots m,$$

is called the *margin* of the νth training pair, $(\boldsymbol{g}(\nu), \ell(\nu))$, relative to the hyperplane,

$$I(\boldsymbol{g}) = \boldsymbol{w}^\top \boldsymbol{g} + w_0 = 0.$$

The perpendicular distance d_ν of a point, $\boldsymbol{g}(\nu)$, to the hyperplane is (see Figure 6.12) given by

$$d_\nu = \frac{1}{\|\boldsymbol{w}\|}(\boldsymbol{w}^\top \boldsymbol{g}(\nu) + w_0). \tag{6.46}$$

That is, from the figure,

$$d_\nu = \|\boldsymbol{g}(\nu) - \boldsymbol{g}\|,$$

and, since $\boldsymbol{w}$ is perpendicular to the hyperplane,

$$\boldsymbol{w}^\top(\boldsymbol{g}(\nu) - \boldsymbol{g}) = \|\boldsymbol{w}\|\,\|\boldsymbol{g}(\nu) - \boldsymbol{g}\| = \|\boldsymbol{w}\| d_\nu.$$

But the left-hand side of the above equation is just

$$\boldsymbol{w}^\top(\boldsymbol{g}(\nu) - \boldsymbol{g}) = \boldsymbol{w}^\top \boldsymbol{g}(\nu) - \boldsymbol{w}^\top \boldsymbol{g} = \boldsymbol{w}^\top \boldsymbol{g}(\nu) + w_0,$$

from which Equation 6.46 follows. The distance d_ν is understood to be positive for points above the hyperplane and negative for points below it. The quantity

$$\gamma_\nu = \ell(\nu) d_\nu = \frac{1}{\|\boldsymbol{w}\|}\ell(\nu)(\boldsymbol{w}^\top \boldsymbol{g}(\nu) + w_0) \tag{6.47}$$

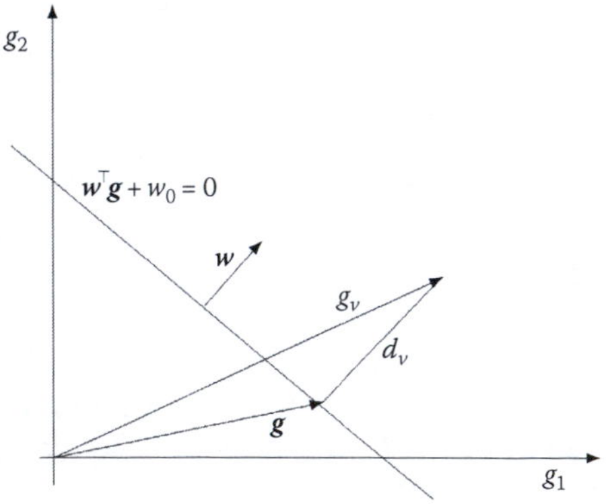

FIGURE 6.12
Distance d_ν to the separating hyperplane.

is called the *geometric margin* for the observation. Observations lying on the hyperplane have zero geometric margins, incorrectly classified observations have negative geometric margins, and correctly classified observations have positive geometric margins.

The *geometric margin of a hyperplane* (relative to a given training set) is defined as the smallest (geometric) margin over the observations in that set, and the *maximal margin hyperplane* is the hyperplane that maximizes the smallest margin, that is, the hyperplane with parameters w, w_0 given by

$$\arg\max_{w,w_0} \left(\frac{1}{\|w\|} \min_{\nu} \left(\ell(\nu)(w^\top g(\nu) + w_0) \right) \right). \tag{6.48}$$

If the training data are linearly separable, then the resulting smallest margin will be positive since all observations are correctly classified. A maximal margin hyperplane is illustrated in Figure 6.13.

6.6.1.1 Primal Formulation

The maxmin problem (Equation 6.48) can be reformulated as follows (Cristianini and Shawe-Taylor, 2000; Bishop, 2006). If we transform the parameters according to

$$w \rightarrow \kappa w, \quad w_0 \rightarrow \kappa w_0$$

for some constant κ, then the distance d_ν will remain unchanged, as is clear from Equation 6.46. Now let us choose κ such that

$$\ell(\nu)(w^\top g(\nu') + w_0) = 1 \tag{6.49}$$

for whichever training observation $g(\nu')$ happens to be closest to the hyperplane. This implies that the following constraints are met:

$$\ell(\nu)I(g(\nu)) = \ell(\nu)(w^\top g(\nu) + w_0) \geq 1, \quad \nu = 1 \ldots m, \tag{6.50}$$

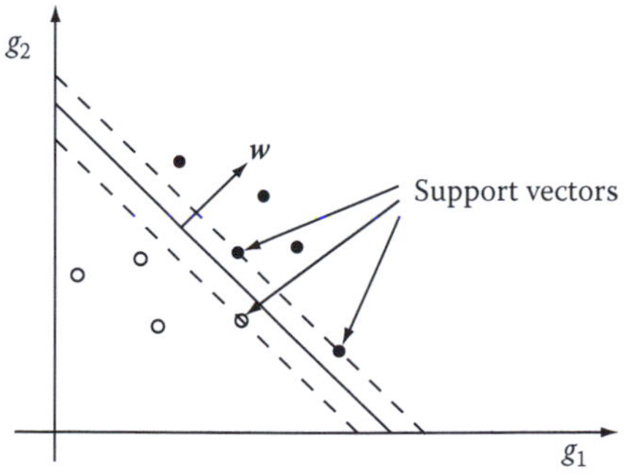

FIGURE 6.13
A maximal margin hyperplane for linearly separable training data, and its support vectors (see text).

and the geometric margin of the hyperplane (the smallest geometric margin over the observations) is, with Equation 6.47, simply $\|w\|^{-1}$. For observations for which equality holds in Equation 6.50, the constraints are called *active*, otherwise *inactive*. Clearly, for any choice of w, w_0, there will always be at least one active constraint. So, the problem expressed in Equation 6.48 is equivalent to maximizing $\|w\|^{-1}$ or, expressed more conveniently, to solving

$$\arg\min_{w} \frac{1}{2}\|w\|^2 \tag{6.51}$$

subject to the constraints of Equation 6.50. These constraints define the *feasible region* for the minimization problem. Taken together, Equations 6.50 and 6.51 constitute the primal formulation of the original maxmin problem (Equation 6.48). The bias parameter, w_0, is determined implicitly by the constraints, as we will see later.

6.6.1.2 Dual Formulation

To solve Equation 6.51, we will apply the Lagrangian formalism for *inequality constraints*, a generalization of the method that we have used extensively up till now and first introduced in Section 1.6. Both Bishop (2006, Appendix E) and Cristianini and Shawe-Taylor (2000, Chapter 5) provide good discussions. We introduce a Lagrange multiplier, α_ν, for each of the inequality constraints, to obtain in the Lagrange function

$$L(\boldsymbol{w}, w_0, \boldsymbol{\alpha}) = \frac{1}{2}\|\boldsymbol{w}\|^2 - \sum_{\nu=1}^{m} \alpha_\nu \big(\ell(\nu)(\boldsymbol{w}^\top \boldsymbol{g}(\nu) + w_0) - 1\big), \tag{6.52}$$

where $\boldsymbol{\alpha} = (\alpha_1 \ldots \alpha_m)^\top \geq \boldsymbol{0}$. Minimization over w and w_0 requires that the respective derivatives be set equal to zero:

$$\frac{\partial L}{\partial \boldsymbol{w}} = \boldsymbol{w} - \sum_{\nu} \ell(\nu)\alpha_\nu \boldsymbol{g}(\nu) = \boldsymbol{0}, \tag{6.53}$$

$$\frac{\partial L}{\partial w_0} = \sum_{\nu} \ell(\nu)\alpha_\nu = 0. \tag{6.54}$$

Therefore, from Equation 6.53,

$$\boldsymbol{w} = \sum_{\nu} \ell(\nu)\alpha_\nu \boldsymbol{g}(\nu).$$

Substituting this back into Equation 6.52 and using Equation 6.54 gives

$$\begin{aligned}L(\boldsymbol{w}, w_0, \boldsymbol{\alpha}) &= \frac{1}{2}\sum_{\nu\nu'} \ell(\nu)\ell(\nu')\alpha_\nu\alpha_{\nu'}(\boldsymbol{g}(\nu)^\top \boldsymbol{g}(\nu')) \\ &\quad - \sum_{\nu\nu'} \ell(\nu)\ell(\nu')\alpha_\nu\alpha_{\nu'}(\boldsymbol{g}(\nu)^\top \boldsymbol{g}(\nu')) + \sum_\nu \alpha_\nu \\ &= \sum_\nu \alpha_\nu - \frac{1}{2}\sum_{\nu\nu'} \ell(\nu)\ell(\nu')\alpha_\nu\alpha_{\nu'}(\boldsymbol{g}(\nu)^\top \boldsymbol{g}(\nu')),\end{aligned} \tag{6.55}$$

in which $\boldsymbol{w}$ and w_0 no longer appear. We thus obtain the *dual formulation* of the minimization problem, Equations 6.50 and 6.51, namely, we maximize

$$\tilde{L}(\boldsymbol{\alpha}) = \sum_\nu \alpha_\nu - \frac{1}{2}\sum_{\nu\nu'} \ell(\nu)\ell(\nu')\alpha_\nu\alpha_{\nu'}(\boldsymbol{g}(\nu)^\top \boldsymbol{g}(\nu')) \tag{6.56}$$

with respect to the *dual variables*, $\boldsymbol{\alpha} = (\alpha_1 \ldots \alpha_m)^\top$, subject to the constraints

$$\begin{aligned}\boldsymbol{\alpha} &\geq \boldsymbol{0}, \\ \sum_\nu \ell(\nu)\alpha_\nu &= 0.\end{aligned} \tag{6.57}$$

We must maximize because, according to the *duality theory*, $\tilde{L}(\boldsymbol{\alpha})$ is a lower limit for $L(\boldsymbol{w}, w_0, \boldsymbol{\alpha})$ and, for our problem,

$$\max_{\boldsymbol{\alpha}} \tilde{L}(\boldsymbol{\alpha}) = \min_{\boldsymbol{w}, w_0} L(\boldsymbol{w}, w_0, \boldsymbol{\alpha})$$

(see, e.g., Cristianini and Shawe-Taylor, 2000, Chapter 5). We shall see how to do this in the sequel; however, let us for now suppose that we have found the solution, $\boldsymbol{\alpha}^*$, which maximizes $\tilde{L}(\boldsymbol{\alpha})$. Then,

$$\boldsymbol{w}^* = \sum_\nu \ell(\nu)\alpha^*_\nu \boldsymbol{g}(\nu)$$

determines the maximal margin hyperplane and it has the geometric margin $\|\boldsymbol{w}^*\|^{-1}$. In order to classify a new observation, $\boldsymbol{g}$, we simply evaluate the sign of

$$I^*(\boldsymbol{g}) = \boldsymbol{w}^{*\top}\boldsymbol{g} + w^*_0 = \sum_\nu \ell(\nu)\alpha^*_\nu(\boldsymbol{g}(\nu)^\top \boldsymbol{g}) + w^*_0, \tag{6.58}$$

that is, we ascertain on which side of the hyperplane the observation lies. (We still need an expression for w^*_0. This is described below.) Note that both the training phase, that is, the solution of the dual problem (Equations 6.56 and 6.57), as well as the generalization phase (Equation 6.58), involve only inner products of the observations $\boldsymbol{g}$.

6.6.1.3 Quadratic Programming and Support Vectors

The optimization problem represented by Equations 6.56 and 6.57 is a *quadratic programming problem*, the objective function $\tilde{L}(\boldsymbol{\alpha})$ being quadratic in the dual variables, α_v. According to the *Karush–Kuhn–Tucker* (KKT) conditions for quadratic programs,* in addition to the constraints of Equations 6.57, the *complementarity condition*

$$\alpha_\nu\big(\ell(\nu)I(g(\nu)) - 1\big) = 0, \quad \nu = 1 \ldots m, \tag{6.59}$$

must be satisfied. Taken together, Equations 6.50, 6.57, and 6.59 are *necessary and sufficient conditions* for a solution $\boldsymbol{\alpha} = \boldsymbol{\alpha}^*$. The complementarity condition says that each of the constraints in Equation 6.50 is either active, that is, $\ell(\nu)I(g(\nu)) = 1$, or is inactive, $\ell(\nu)I(g(\nu)) > 1$, in which case, from Equation 6.59, $\alpha_\nu = 0$.

When classifying a new observation with Equation 6.58, either the training observation, $g(\nu)$, satisfies $\ell(\nu)I^*(g(\nu)) = 1$, that is to say, it has a minimum margin (Equation 6.49), or $\alpha_\nu = 0$ by virtue of the complementarity condition, meaning that it *plays no role in the classification*. The labeled training observations with the minimum margin are called *support vectors* (see Figure 6.13). After the solution of the quadratic programming problem, these are the only observations that can contribute to the classification of new data.

Let us call SV the set of support vectors. Then, from the complementarity condition (Equation 6.59) we have, for $\nu \in SV$,

$$\ell(\nu)\left(\sum_{\nu' \in SV} \ell(\nu')\alpha^*_{\nu'}(g(\nu')^\top g(\nu)) + w_0^*\right) = 1. \tag{6.60}$$

We can therefore write

$$\begin{aligned}\|w^*\|^2 = {w^*}^\top w^* &= \sum_{\nu,\nu'} \ell(\nu)\ell(\nu')\alpha_\nu\alpha_{\nu'}(g(\nu)^\top g(\nu')) \\ &= \sum_{\nu\in SV} \alpha^*_\nu\left(\ell(\nu)\sum_{\nu'\in SV} \ell(\nu')\alpha^*_{\nu'}(g(\nu)^\top g(\nu'))\right),\end{aligned}$$

and, from Equation 6.60,

$$\|w^*\|^2 = \sum_{\nu\in SV} \alpha^*_\nu(1 - \ell(\nu)w_0^*) = \sum_{\nu\in SV} \alpha^*_\nu, \tag{6.61}$$

where in the second equality we have made use of the second constraint in Equation 6.57. Thus, the geometric margin of the maximal hyperplane is

* See again Cristianini and Shawe-Taylor (2000, Chapter 5).

given in terms of the dual variables by

$$\|\boldsymbol{w}^*\|^{-1} = \left(\sum_{\nu \in SV} \alpha_\nu^*\right)^{-1/2}. \tag{6.62}$$

Equation 6.60 can be used to determine w_0^* once the quadratic program has been solved, simply by choosing an arbitrary $\nu \in SV$. Bishop (2006) suggests a numerically more stable procedure: Multiply Equation 6.60 by $\ell(\nu)$ and make use of the fact that $\ell(\nu)^2 = 1$. Then take the average of the equations for all support vectors and solve for w_0^* to get

$$w_0^* = \frac{1}{|SV|} \sum_{\nu \in SV} \left(\ell(\nu) - \sum_{\nu' \in SV} \alpha_{\nu'} \ell(\nu') (\boldsymbol{g}(\nu)^\top \boldsymbol{g}(\nu'))\right). \tag{6.63}$$

6.6.2 Overlapping Classes

With the SVM formalism introduced so far, one can solve two-class classification problems that are linearly separable. Using kernel substitution, which will be treated shortly, nonlinearly separable problems can also be solved. In general, though, one is faced with labeled training data that overlap considerably, so that their complete separation would imply gross overfitting.

To overcome this problem, we introduce the so-called *slack variables*, ξ_ν, associated with each training vector, such that

$$\xi_\nu = \begin{cases} |\ell(\nu) - I(\boldsymbol{g}(\nu))| & \text{if } \boldsymbol{g}(\nu) \text{ is on the wrong side of the margin boundary} \\ 0 & \text{otherwise.} \end{cases}$$

This situation is illustrated in Figure 6.14. The constraints of Equation 6.50 now become

$$\ell(\nu) I(\boldsymbol{g}(\nu)) \geq 1 - \xi_\nu, \quad \nu = 1 \ldots m, \tag{6.64}$$

and are often referred to as *soft margin* constraints.

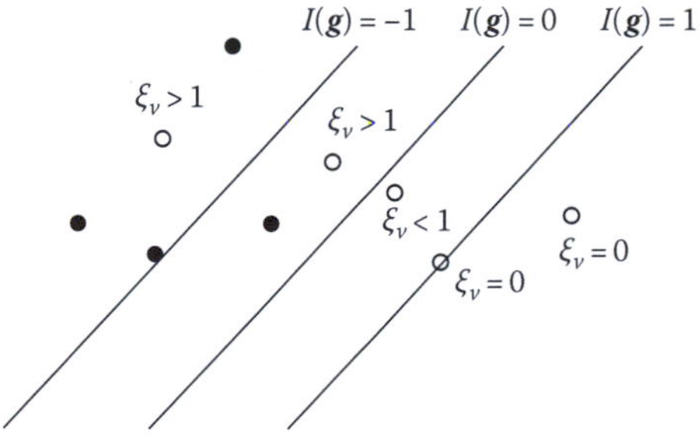

FIGURE 6.14
Slack variables. Observations with values less than 1 are correctly classified.

Our objective is again to maximize the margin, but to penalize points with large values of ξ. We, therefore, modify the objective function of Equation 6.51 by introducing a regularization term:

$$C\sum_{\nu=1}^{m}\xi_\nu + \frac{1}{2}\|w\|^2. \tag{6.65}$$

The parameter C determines the degree of penalization. When minimizing this objective function, in addition to the above inequality constraints we require that

$$\xi_\nu \ge 0, \quad \nu = 1\dots m.$$

The Lagrange function is now given by

$$L(w, w_0, \alpha) = \frac{1}{2}\|w\|^2 + C\sum_{\nu=1}^{m}\xi_\nu - \sum_\nu \alpha_\nu(\ell(\nu)(w^\top g(\nu) + w_0) - 1 + \xi_\nu)$$
$$- \sum_{\nu=1}^{m}\mu_\nu\xi_\nu, \tag{6.66}$$

where $\mu_\nu \ge 0$, $\nu = 1\dots m$, are additional Lagrange multipliers. Setting the derivatives of the Lagrange function with respect to w, w_0 and ξ_ν equal to zero, we get

$$w = \sum_{\nu=1}^{m}\alpha_\nu\ell(\nu)g(\nu) \tag{6.67}$$

$$\sum_{\nu=1}^{m}\alpha_\nu\ell(\nu) = 0 \tag{6.68}$$

$$\alpha_\nu = C - \mu_\nu, \tag{6.69}$$

which leads to the dual form

$$\tilde{L}(\alpha) = \sum_\nu \alpha_\nu - \frac{1}{2}\sum_{\nu\nu'}\ell(\nu)\ell(\nu')\alpha_\nu\alpha_{\nu'}(g(\nu)^\top g(\nu')). \tag{6.70}$$

This is the same as for the separable case (Equation 6.56) except that the constraints are slightly different. From Equation 6.69, $\alpha_\nu = C - \mu_\nu \le C$, so that

$$0 \le \alpha_\nu \le C, \quad \nu = 1\dots m. \tag{6.71}$$

Furthermore, the complementarity conditions now read

$$\alpha_\nu\big(\ell(\nu)I(g(\nu)) - 1 + \xi_\nu\big) = 0$$
$$\mu_\nu\xi_\nu = 0, \quad \nu = 1 \ldots m. \tag{6.72}$$

Equations 6.64, 6.68, 6.71, and 6.72 are again necessary and sufficient conditions for a solution $\boldsymbol{\alpha} = \boldsymbol{\alpha}^*$.

As before, when classifying new data with Equation 6.58, only the support vectors, that is, the training observations for which $\alpha_\nu > 0$, will play a role. If $\alpha_\nu > 0$, there are now two possibilities to be distinguished:

- $0 \leq \alpha_\nu < C$. Then we must have $\mu_\nu > 0$ (Equation 6.69), implying from the second condition in Equation 6.72 that $\xi_\nu = 0$. The training vector thus lies exactly on the margin and corresponds to the support vector for the separable case.
- $\alpha_\nu = C$. The training vector lies inside the margin and is correctly classified ($\xi \leq 1$) or incorrectly classified ($\xi > 1$).

As before, we can use Equation 6.63 to determine w_0^* after solving the quadratic program for w^*, except that the set SV must be restricted to those observations for which $0 < \alpha_\nu < C$, that is, to the support vectors lying on the margin.

6.6.3 Solution with Sequential Minimal Optimization

To train the classifier, we must maximize Equation 6.70 subject to the boundary conditions given by Equations 6.68 and 6.71, which define the feasible region for the problem. We outline here a popular method for doing this, called *sequential minimal optimization* (SMO) (see Cristianini and Shawe-Taylor, 2000). In this algorithm, pairs of Lagrange multipliers $(\alpha_\nu, \alpha_{\nu'})$ are chosen and varied so as to increase the objective function while still satisfying the constraints. (At least two multipliers must be considered in order to guarantee the fulfillment of the equality constraint [Equation 6.68]). The SMO method has the advantage that the maximum can be found analytically at each step, thus leading to fast computation.

Algorithm (Sequential Minimal Optimization)

1. Set $\alpha_\nu^{old} = 0,\ \nu = 1 \ldots m$ (clearly, $\boldsymbol{\alpha}^{old}$ is in the feasible region).
2. Choose a pair of Lagrange multipliers, calling them without loss of generality, α_1 and α_2. Let all of the other multipliers be fixed.
3. Maximize $\tilde{L}(\boldsymbol{\alpha})$ with respect to α_1 and α_2.
4. If the complementarity conditions are satisfied (within some tolerance), stop, else go to Step 2.

In Step 2, heuristic rules must be used to choose appropriate pairs. Step 3 can be carried out analytically by maximizing the quadratic function

$$W(\alpha_1, \alpha_2) = \alpha_1 + \alpha_2 - \frac{1}{2}(g(1)^\top g(1))\alpha_1^2 - \frac{1}{2}(g(2)^\top g(2))\alpha_2^2 - \ell_1\ell_2\alpha_1\alpha_2 - \ell_1\alpha_1 v_1 - \ell_2\alpha_2 v_2 + \text{Const},$$

where

$$v_i = \sum_{\nu=3}^{n} \ell(\nu)\alpha_\nu^{old}(g(\nu)^\top g(i)), \quad i = 1, 2.$$

Maximization is carried out in the feasible region for α_1, α_2. The equality constraint (Equation 6.68) requires

$$\ell_1\alpha_1 + \ell_2\alpha_2 = \ell_1\alpha_1^{old} + \ell_2\alpha_2^{old},$$

or, equivalently,

$$\alpha_2 = -\ell_1\ell_2\alpha_1 + \gamma,$$

where $\gamma = \ell_1\ell_2\alpha_1^{old} + \alpha_2^{old}$, so that (α_1, α_2) lies on a line with slope ± 1, depending on the value of $\ell_1\ell_2$. The inequality constraint (Equation 6.71) defines the endpoints of the line segment that need to be considered, namely, $(\alpha_1 = 0, \alpha_2 = \max(0, \gamma))$ and $(\alpha_1 = C, \alpha_2 = \min(C, -\ell_1\ell_2 C + \gamma))$. Regarding Step 4, the complementarity conditions are as follows (see Equations 6.72):

For all ν,

$$\begin{aligned} &\text{if } \alpha_\nu = 0, \text{ then } \ell(\nu)I(g(\nu)) - 1 \geq 0 \\ &\text{if } \alpha_\nu > 0, \text{ then } \ell(\nu)I(g(\nu)) - 1 = 0 \\ &\text{if } \alpha_\nu = C, \text{ then } \ell(\nu)I(g(\nu)) - 1 \leq 0. \end{aligned}$$

6.6.4 Multiclass SVMs

A distinct advantage of the SVM for remote sensing image classification over neural networks is the unambiguity of the solution. Since one maximizes a quadratic function, the maximum is global and will always be found. There is no possibility of becoming trapped in a local optimum. However SVM classifiers have two disadvantages. They are designed for two-class problems and their outputs, unlike feed-forward neural networks, do not model posterior class membership probabilities in a natural way.

A common way to overcome the two-class restriction is to determine all possible two-class results and then use a voting scheme to decide on the class label (Wu et al., 2004). That is, for K classes, we train $K(K-1)/2$ SVMs on each of the possible pairs $(i, j) \in \mathcal{K} \otimes \mathcal{K}$. For a new observation g and the SVM for (i, j), let

$$\mu_{ij}(g) = \Pr(\ell = i \mid \ell = i \text{ or } j,\ g) = \frac{\Pr(\ell = i \mid g)}{\Pr(\ell = i \text{ or } j \mid g)}. \tag{6.73}$$

The last equality follows from the definition of conditional probability, Equation 2.53, since the joint probability (given g) for i or j and i is just the probability for i. Now suppose that r_{ij} is some rough estimator for μ_{ij}, perhaps simply $r_{ij} = 1$ if $\mu_{ij} > 0.5$ and 0 otherwise. The voting rule is then

$$k = \arg\max_i \left(\sum_{j \neq i}^{K} [[r_{ij} > r_{ji}]] \right), \tag{6.74}$$

where $[[\cdots]]$ is the *indicator function*

$$[[x]] = \begin{cases} 1 & \text{if } x \text{ is true} \\ 0 & \text{if } x \text{ is false.} \end{cases}$$

With regard to the second restiction, a very simple estimate of the posterior class membership probability is the ratio of the number of votes to the total number of classifications,

$$\Pr(k \mid g) \approx \frac{2}{K(K-1)} \sum_{j \neq i}^{K} I\{r_{kj} > r_{jk}\}. \tag{6.75}$$

If r_{ij} is a more realistic estimate, we can proceed as follows (Price et al., 1995):

$$\begin{aligned}
\sum_{j \neq i} \Pr(\ell = i \text{ or } j \mid g) &= \sum_{j \neq i} \big(\Pr(\ell = i \mid g) + \Pr(\ell = j \mid g)\big) \\
&= (K-1)\Pr(\ell = i \mid g) + \sum_{j \neq i} \Pr(\ell = j \mid g) \\
&= (K-2)\Pr(\ell = i \mid g) + \sum_{j} \Pr(\ell = j \mid g) \\
&= (K-2)\Pr(\ell = i \mid g) + 1.
\end{aligned} \tag{6.76}$$

Combining this with Equation 6.73 gives, (see Exercise 8),

$$\Pr(\ell = i \mid g) = \frac{1}{\sum_{j \neq i} \mu_{ij}^{-1} - (K-2)} \tag{6.77}$$

or, substituting the estimate r_{ij}, we estimate the posterior class membership probabilities as

$$\Pr(k \mid g) \approx \frac{1}{\sum_{j \neq k} r_{kj}^{-1} - (K-2)}. \tag{6.78}$$

Because of the substitution, the probabilities will not sum exactly to one, and must be normalized.

6.6.5 Kernel Substitution

Of course the reason why we have replaced the relatively simple optimization problem (Equation 6.51) with the more involved dual formulation is due to the fact that the labeled training observations only enter into the dual formulation in the form of scalar products $g(\nu)^\top g(\nu')$. This allows us once again to use the elegant method of kernelization, introduced in Chapter 4. Then we can apply SVMs to situations in which the classes are not linearly separable.

To give a simple example (Müller et al., 2001), consider the classification problem illustrated in Figure 6.15. While the classes are clearly separable, they cannot be separated with a hyperplane. Now introduce the transformation

$$\phi : \mathbb{R}^2 \mapsto \mathbb{R}^3,$$

such that

$$\phi(g) = \begin{pmatrix} g_1^2 \\ \sqrt{2} g_1 g_2 \\ g_2^2 \end{pmatrix}.$$

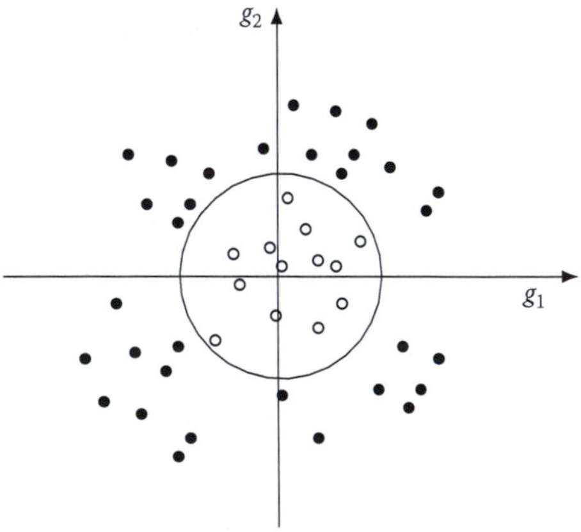

FIGURE 6.15
Two classes that are not linearly separable in the two-dimensional space of observations.

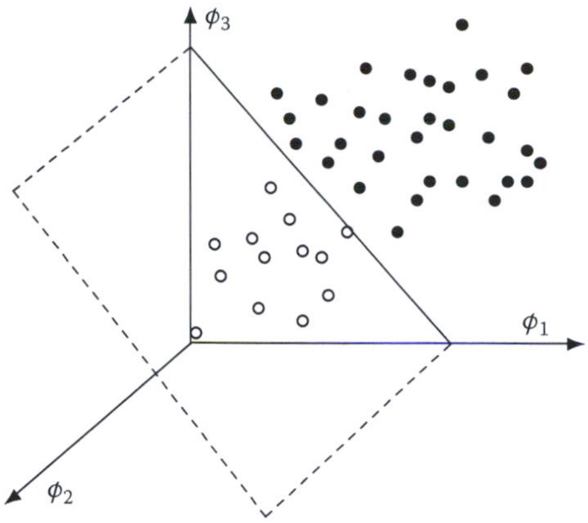

FIGURE 6.16
The classes of Figure 6.15 become linearly separable in a three-dimensional, nonlinear feature space.

In this new feature space, the observations are transformed as shown in Figure 6.16, and can be separated by a two-dimensional hyperplane. We can thus apply the support vector formalism unchanged, simply replacing $g_v^\top g_v$ by $\boldsymbol{\phi}(g(\nu))^\top \boldsymbol{\phi}(g(\nu'))$. In this case, the kernel function is given by

$$k(g(\nu), g(\nu')) = (g_1^2, \sqrt{2}g_1g_2, g_2^2)_\nu \begin{pmatrix} g_1^2 \\ \sqrt{2}g_1g_2 \\ g_2^2 \end{pmatrix}_{\nu'} = (g(\nu)^\top g(\nu'))^2,$$

called the quadratic kernel.

6.6.6 Modified SVM Classifier

The ENVI/IDL extension `SVM_RUN`, similar to `MAXLIKE_RUN` described in Section 6.3.2, provides a wrapper for ENVI's built-in SVM algorithm, separating ROI training area pixels into a training data set and a test data set in the ratio 2:1 (see Appendix C.2.11). It makes use of the ENVI batch procedure `ENVI_SVM_DOIT`.

Whereas both the maximum likelihood and Gaussian kernel training procedures have essentially no externally adjustable parameters,* this is not the case for SVMs. First, one must decide upon a kernel, and then choose associated kernel parameters as well as the soft margin penalization constant, C. In addition, the ENVI implementation provides for a multiresolution classification with a choice of different pyramid depths. This feature is mainly offered to speed up classification; however, it can influence classification accuracy

* Neural network training with backpropagation has three adjustable parameters (learn rate, momentum, and the number of hidden neurons).

as well. The kernels available in ENVI are

$$k_{\mathrm{lin}}(g_i, g_j) = g_i^\top g_j$$
$$k_{\mathrm{poly}}(g_i, g_j) = (\gamma g_i^\top g_j + r)^d$$
$$k_{\mathrm{rbf}}(g_i, g_j) = \exp(-\gamma \|g_i - g_j\|^2)$$
$$k_{\mathrm{sig}}(g_i, g_j) = \tanh(\gamma g_i^\top g_j + r).$$

The most popular kernel is the Gaussian kernel (k_{rbf} above), and this is the default for ENVI. The parameter γ essentially determines the training/generalization trade-off, with large values leading to overfitting (Shawe-Taylor and Cristianini, 2004).

`SVM_RUN` only generates classification and rule images in memory. Again, test results are saved to a file in a format consistent with that used by the other classification routines described earlier. An SVM classification is shown in Figure 6.4d.

6.7 Exercises

1. (a) Perform the integration in Equation 6.11 for the one-dimensional case:

$$p(g \mid k) = \frac{1}{\sqrt{2\pi}\sigma_k} \exp\left(-\frac{1}{2\sigma_k^2}(g - \mu_k)^2\right), \quad k = 1, 2. \tag{6.79}$$

(*Hint:* Use the definite integral $\int_{-\infty}^{\infty} \exp(ag - bg^2)dg = \sqrt{\frac{\pi}{b}} \exp(a^2/4b)$.)

(b) The Jeffries-Matusita distance between two probability densities, $p(g \mid 1)$ and $p(g \mid 2)$, is defined as

$$J = \int_{-\infty}^{\infty} \left(p(g \mid 1)^{1/2} - p(g \mid 2)^{1/2}\right)^2 dg.$$

Show that this is equivalent to the definition in Equation 6.13.

(c) A measure of the separability of $p(g \mid 1)$ and $p(g \mid 2)$ can be written in terms of the Kullback–Leibler divergence (see Equation 2.93) as follows (Richards and Jia, 2006):

$$d_{12} = \mathrm{KL}\left(p(g \mid 1), p(g \mid 2)\right) + \mathrm{KL}\left(p(g \mid 2), p(g \mid 1)\right).$$

Explain why this is a satisfactory separability measure.

(d) Show that, for the one-dimensional distributions of Equation 6.79,

$$d_{12} = \frac{1}{2}\left(\frac{1}{\sigma_1^2} - \frac{1}{\sigma_2^2}\right)(\sigma_1^2 - \sigma_2^2) + \frac{1}{2}\left(\frac{1}{\sigma_1^2} + \frac{1}{\sigma_2^2}\right)(\mu_1 - \mu_2)^2.$$

2. (Ripley, 1996) Assuming that all K land cover classes have identical covariance matrices, $\mathbf{\Sigma}_k = \mathbf{\Sigma}$,

(a) Show that the discriminant in Equation 6.17 can be replaced by the linear discriminant

$$d_k(g) = \log(\Pr(k)) - \boldsymbol{\mu}_k^\top \mathbf{\Sigma}^{-1} g + \frac{1}{2}\boldsymbol{\mu}_k^\top \mathbf{\Sigma}^{-1}\boldsymbol{\mu}_k.$$

(b) Suppose that there are just two classes, $k = 1$ and $k = 2$. Show that the maximum likelihood classifier will choose $k = 1$ if

$$h = (\boldsymbol{\mu}_1 - \boldsymbol{\mu}_2)^\top \mathbf{\Sigma}^{-1}\left(g - \frac{\boldsymbol{\mu}_1 + \boldsymbol{\mu}_2}{2}\right) > \log\left(\frac{\Pr(2)}{\Pr(1)}\right).$$

(c) The quantity $d = \sqrt{(\boldsymbol{\mu}_1 - \boldsymbol{\mu}_2)^\top \mathbf{\Sigma}^{-1}(\boldsymbol{\mu}_1 - \boldsymbol{\mu}_2)}$ is the Mahalanobis distance between the class means. Demonstrate that if g belongs to class 1, then h is the realization of a normally distributed random variable, H_1, with mean $d^2/2$ and variance d^2. What is the corresponding distribution if g belongs to class 2?

(d) Prove from the above considerations that the probability of misclassification is given by

$$\Pr(1) \cdot \Phi\left(-\frac{1}{2}d + \frac{1}{d}\log\left(\frac{\Pr(2)}{\Pr(1)}\right)\right) + \Pr(2) \cdot \Phi\left(-\frac{1}{2}d - \frac{1}{d}\log\left(\frac{\Pr(2)}{\Pr(1)}\right)\right),$$

where Φ is the standard normal distribution function.

(e) What is the minimum possible probability of misclassification?

3. (Linear separability) With reference to Figures 6.6 and 6.7,

(a) Show that the vector w is perpendicular to the hyperplane

$$I(g) = w^\top g + w_0 = 0.$$

(b) Suppose that there are just three training observations or points, g_i, $i = 1, 2, 3$, in a two-dimensional feature space, and that these do not lie on a straight line. Since there are only two possible classes, the three points can be labeled in $2^3 = 8$ possible ways. Each possibility is obviously *linearly separable*. That is to say, one can find an oriented hyperplane that will correctly classify the three points, that is, all class 1 points (if any) lie on one side and all class 2 points (if any) lie on the other side, with the vector w pointing to the class 1 side. The three points are said to be *shattered* by the set of hyperplanes. The maximum number of points that can be shattered is called the *Vapnik-Chervonenkis* (VC) dimension of the hyperplane classifier. What is the VC dimension in this case?

(c) A training set is linearly separable if a hyperplane can be found for which the smallest margin is positive. The following *perceptron algorithm* is guaranteed to find such a hyperplane, that is, to train the neuron of Figure 6.7 to classify any linearly separable training set of n observations (Cristianini and Shawe-Taylor, 2000):

i. Set $w = \mathbf{0}$, $b = 0$, and $R = \max_\nu \|g(\nu)\|$.

ii. Set $m = 0$.

iii. For $i = 1$ to n, do: if $\gamma_\nu \leq 0$ then set $w = w + \ell(\nu)g(\nu)$, $b = b + \ell(\nu)R^2$, and $m = 1$.

iv. If $m = 1$ go to ii, else stop.

The algorithm stops with a separating hyperplane, $w^\top g + b = 0$. Implement this algorithm in IDL and test it with linearly separable training data (see, e.g., Listing 3.6 and Figure 3.12).

4. (Bishop, 1995) Neural networks can also be used to approximate continuous vector functions, $\boldsymbol{h}(g)$, of their inputs, g. Suppose that, for a given observation, $g(\nu)$, the corresponding training value, $\boldsymbol{\ell}(\nu)$, is not known exactly, but that its components, $\ell_k(\nu)$, are normally and independently distributed about the (unknown) functions, $h_k(g(\nu))$, that is,

$$p(\ell_k(\nu) \mid g(\nu)) = \frac{1}{\sqrt{2\pi}\sigma} \exp\left(-\frac{(h_k(g(\nu)) - \ell_k(\nu))^2}{2\sigma^2}\right), \quad k = 1 \ldots K. \tag{6.80}$$

Show that the appropriate cost function to train the synaptic weights so as to best approximate $\boldsymbol{h}$ with the network outputs, $\boldsymbol{m}$, is the quadratic cost function (Equation 6.30). *Hint*: The probability density for a particular training pair, $(g(\nu), \boldsymbol{\ell}(\nu))$, can be written as (see Equation 2.55)

$$p(g(\nu), \boldsymbol{\ell}(\nu)) = p(\boldsymbol{\ell}(\nu) \mid g(\nu))p(g(\nu)).$$

The likelihood function for n training examples chosen independently from the same distribution is, accordingly,

$$\prod_{\nu=1}^{n} p(\boldsymbol{\ell}(\nu) \mid g(\nu))p(g(\nu)).$$

Argue that maximizing this likelihood with respect to the synaptic weights is equivalent to minimizing the cost function,

$$E = -\sum_{\nu=1}^{n} \log p(\boldsymbol{\ell}(\nu) \mid g(\nu)),$$

and then show that, with Equation 6.80, this reduces to Equation 6.30.

5. As an example of continuous function approximation, the neural network of Figure 6.10 can be used to perform a principal components analysis on sequential data. The method applied involves so-called *self-supervised* classification. The idea is illustrated in Figure 6.17 for two-dimensional observations. The training data are presented to a network with a single hidden neuron. The output is constrained to be identical to the input (self-supervision). Since the output from the hidden layer is one-dimensional, the constraint requires that as much information as possible about the input signal be coded by the hidden neuron. This is the case if the data are projected along the first principal axis. As more and more data are presented, the synaptic weight vector (w_1^h, w_2^h) will therefore point more and more in the direction of that axis. Use the object class `FFNBP` to implement the network in Figure 6.17 in IDL and test it with simulated data. (According to the previous exercise, the cross entropy cost function used in `FFNBP` is not fully appropriate, but suffices nevertheless.)
6. Demonstrate Equation 6.39.

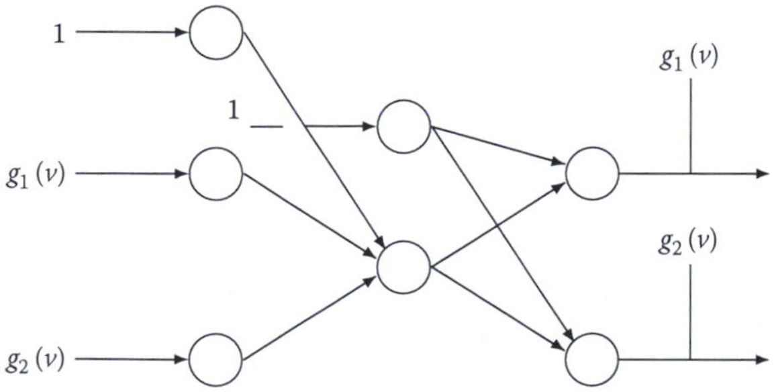

FIGURE 6.17
Principal components analysis with a neural network.

7. (Network symmetries) The hyperbolic tangent is defined as

$$\tanh(x) = \frac{e^x - e^{-x}}{e^x + e^{-x}}.$$

   (a) Show that the logistic function, $f(x)$, can be expressed in the form

$$f(x) = \frac{1}{2}\tanh\left(\frac{x}{2}\right) + \frac{1}{2}.$$

   (b) Noting that $\tanh(x)$ is an odd function, that is, $\tanh(x) = -\tanh(-x)$, argue that, for the feed-forward network of Figure 6.10, there are (at least) 2^L identical local minima in the cost function of Equation 6.33.

8. Show that Equation 6.77 follows from Equations 6.73 and 6.76.

7

Supervised Classification: Part 2

Continuing in this chapter with the subject of supervised classification, we will first of all discuss postclassification processing methods to improve classification results on the basis of contextual information. Then we turn our attention to statistical procedures for evaluating classification accuracy and for making quantitative comparisons between different classifiers. As an illustration of so-called *ensembles* or *committees* of classifiers, we next treat the *adaptive boosting* technique, applying it in particular to improve the generalization accuracy of neural network classifiers. The chapter concludes with a discussion of the problems posed by the classification of images with high spectral resolution, including an introduction to linear spectral unmixing.

7.1 Postprocessing

Intermediate resolution remote sensing satellite platforms used for land cover/land use classification, for example, LANDSAT TM, SPOT, RapidEye, ASTER, have ground sample distances (GSDs) ranging between a few to a few tens of meters, which are typically smaller than the landscape objects being classified (agricultural fields, forests, urban areas, etc.). The imagery that they generate is therefore characterized by a high degree of spatial correlation. In Chapter 8, we will see examples of how spatial or contextual information might be incorporated into unsupervised classification. The supervised classification case is somewhat different, since reference is always being made to a—generally quite small—subset of labeled training data. Two approaches can be distinguished for inclusion of contextual information into supervised classification: moving window (or filtering) methods and segmentation (or region growing) methods. Both approaches can be applied either during classification or as a postprocessing step. We will restrict ourselves here to postclassification filtering, in particular mentioning briefly the majority filtering function offered in the standard ENVI environment, and then discussing in some detail a modification of a probabilistic technique described in Richards and Jia (2006). For an overview and an example of the use of segmentation for contextual classification, see Stuckens et al. (2000) and references therein.

7.1.1 Majority Filtering

Majority postclassification filtering employs a moving window, with each central pixel assigned to the majority class of the pixels within the window. This clearly will have the effect of reducing the "salt-and-pepper" appearance typical of the thematic maps generated by pixel-oriented classifiers, and, to quote Stuckens et al. (2000), "it also results in larger classification units that might adhere more to the human perception of land cover." The ENVI implementation of the method is accessible from

```
Classification/Post Classification/Majority/Minority Analysis
```

Majority filtering merely examines the labels of neighborhood pixels. The classifiers we have discussed in the last chapter generate, in addition to class labels, class membership probability vectors for each observation. Therefore, neighboring pixels offer considerably more information than that exploited in majority filtering, information which can also be included in the relabeling process.

7.1.2 Probabilistic Label Relaxation

Recalling Figure 4.13, the 4-neighborhood $\mathcal{N}_i$ of an image pixel with an intensity vector, g_i, consists of the four pixels above, below, to the left, and to the right of the pixel. The *a posteriori* class membership probabilities of the central pixel, as approximated by one of the classifiers of Chapter 6, are given by

$$\Pr(k \mid g_i), \quad k = 1 \ldots K, \text{ where } \sum_{k=1}^{K} \Pr(k \mid g_i) = 1,$$

which, for notational convenience, we represent in the following as the K-component column vector P_i having components

$$P_i(k) = \Pr(k \mid g_i), \quad k = 1 \ldots K. \tag{7.1}$$

According to the standard decision rule (Equation 6.6), the maximum component of P_i determines the class membership of the ith pixel.

In analogy to majority filtering, we might expect that a possible misclassification of the pixel can be corrected by examining the membership probabilities in its neighborhood. If that is so, the neighboring pixels will have to in some way modify P_i such that its maximum component is more likely to correspond to the true class. We now describe a purely heuristic but nevertheless intuitively satisfying procedure to do just that, the so-called *probabilistic label relaxation* (PLR) method (Richards and Jia, 2006).

Let us postulate a multiplicative *neighborhood function* $Q_i(k)$ for the ith pixel that corrects $P_i(k)$ in the above sense, that is

$$P'_i(k) = \frac{P_i(k)Q_i(k)}{\sum_j P_i(j)Q_i(j)}, \quad k = 1 \ldots K. \tag{7.2}$$

The denominator ensures that the corrected values sum to unity and so still constitute a probability vector. In an obvious vector notation, we can write

$$\boldsymbol{P}'_i = \boldsymbol{P}_i \cdot \frac{\boldsymbol{Q}_i}{\boldsymbol{P}_i^\top \boldsymbol{Q}_i}, \tag{7.3}$$

where the dot signifies ordinary component-by-component multiplication.

The vector $\boldsymbol{Q}_i$ must somehow reflect the contextual information of the neighborhood. In order to define it, a *compatibility measure*

$$P_{ij}(k \mid m), \quad j \in \mathcal{N}_i$$

is introduced, namely, the conditional probability that the pixel i has the class label k, given that a neighboring pixel $j \in \mathcal{N}_i$ belongs to class m. A "small piece of evidence" (Richards and Jia, 2006) that i should be classified to k would then be

$$P_{ij}(k \mid m)P_j(m), \quad j \in \mathcal{N}_i.$$

This is the conditional probability that pixel i is in class k if the neighboring pixel j is in class m multiplied by the probability that pixel j actually is in class m. We obtain the component $Q_i(k)$ of the neighborhood function by summing over all pieces of evidence and then averaging over the neighborhood:

$$\begin{aligned} Q_i(k) &= \frac{1}{4} \sum_{j \in \mathcal{N}_i} \sum_{m=1}^{K} P_{ij}(k \mid m) P_j(m) \\ &= \sum_{m=1}^{K} P_{i\mathcal{N}_i}(k \mid m) P_{\mathcal{N}_i}(m). \end{aligned} \tag{7.4}$$

Here, $P_{\mathcal{N}_i}(m)$ is an average over all four neighborhood pixels:

$$P_{\mathcal{N}_i}(m) = \frac{1}{4} \sum_{j \in \mathcal{N}_i} P_j(m),$$

and $P_{i\mathcal{N}_i}(k \mid m)$ also corresponds to the *average compatibility* of pixel i with its entire neighborhood. We can write Equation 7.4 in matrix notation in the form

$$\boldsymbol{Q}_i = \boldsymbol{P}_{i\mathcal{N}_i} \boldsymbol{P}_{\mathcal{N}_i}$$

and Equation 7.3 finally as

$$P_i' = P_i \cdot \frac{P_{i\mathcal{N}_i} P_{\mathcal{N}_i}}{P_i^\top P_{i\mathcal{N}_i} P_{\mathcal{N}_i}}. \tag{7.5}$$

For supervised classification, the matrix of average compatibilities, $P_{i\mathcal{N}_i}$, is not *a priori* available. However, it may easily be estimated directly from the initially classified image by assuming that it is independent of pixel location. First, a random central pixel i is chosen and its class label $\ell_i = k$ determined. Then, again randomly, a pixel $j \in \mathcal{N}_i$ is chosen and its class label $\ell_j = m$ is also determined. Thereupon the matrix element $P_{i\mathcal{N}_i}(k \mid m)$ (which was initialized to 0) is incremented by 1. This is repeated many times and finally the rows of the matrix are normalized. All of these constitute the first step of the following algorithm:

Algorithm (Probabilistic Label Relaxation)

1. Carry out an unsupervised classification and determine the $K \times K$ compatibility matrix $P_{i\mathcal{N}_i}$.
2. For each pixel i, determine the average neighborhood vector $P_{\mathcal{N}_i}$ and replace P_i with P_i' as in Equation 7.5. Reclassify pixel i according to $\ell_i = \arg\max_k P_i'(k)$.
3. If only a few reclassifications took place, stop; otherwise go to step 2.

The stopping condition in the algorithm is obviously rather vague. Experience shows that the best results are obtained after three to four iterations (see Richards and Jia, 2006). Too many iterations lead to a widening of the effective neighborhood of a pixel to such an extent that fully irrelevant spatial information falsifies the final product.

The PLR method can be applied similarly to any unsupervised classification algorithm that generates posterior class membership probabilities.* ENVI/IDL extensions for PLR are given in Appendix C.2.12. Figure 7.1 shows a neural network classification result before and after PLR. The spatial coherence of the classes is improved.

7.2 Evaluation and Comparison of Classification Accuracy

Assuming that sufficient labeled data are available for some to be set aside for test purposes, test data can be used to make an unbiased estimate of the *misclassification rate* of a trained classifier, that is, the fraction of new data

* An example is the Gaussian mixture clustering algorithm that will be met in Chapter 8.

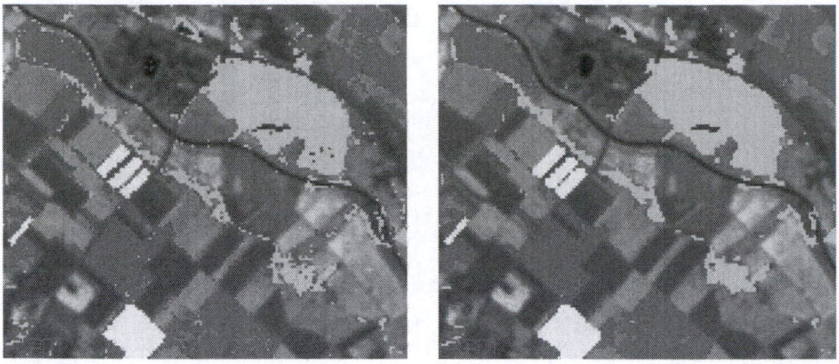

(a) (b)

FIGURE 7.1
(See color insert following page 114.) An example of postclassification processing. (a) Original classification of a portion of the Jülich ASTER scene with a neural network. The classes shown are coniferous forest (green), rapeseed (yellow), and cereal grain (red). (b) After three iterations of PLR.

that will be incorrectly classified. This quantity provides a reasonable yardstick not only for evaluating the overall accuracy of supervised classifiers but also for comparison of alternatives, for example, to compare the performance of a neural network with a maximum likelihood classifier on the same set of data.

7.2.1 Accuracy Assessment

The classification of a single test datum is a random experiment, the possible outcomes of which constitute the sample space $\{\bar{A}, A\}$, where $\bar{A}$ = *misclassified*, A = *correctly classified*. Let us define a real-valued function X on this set, that is, a random variable

$$X(\bar{A}) = 1, \quad X(A) = 0, \tag{7.6}$$

with mass function

$$\Pr(X = 1) = \theta, \quad \Pr(X = 0) = 1 - \theta.$$

The mean value of X is then

$$\langle X \rangle = 1\theta + 0(1 - \theta) = \theta, \tag{7.7}$$

and its variance is (see Equation 2.11)

$$\text{var}(X) = \langle X^2 \rangle - \langle X \rangle^2 = 1^2\theta + 0^2(1 - \theta) - \theta^2 = \theta(1 - \theta). \tag{7.8}$$

For the classification of n test data, which are represented by the i.i.d. sample $X_1 \dots X_n$, the random variable

$$Y = X_1 + X_2 + \cdots + X_n$$

corresponds to the total number of misclassifications. The random variable describing the misclassification rate is therefor Y/n having mean value

$$\left\langle \frac{1}{n} Y \right\rangle = \frac{1}{n}(\langle X_1 \rangle + \cdots + \langle X_n \rangle) = \frac{1}{n} \cdot n\theta = \theta. \tag{7.9}$$

From the independence of the X_i, $i = 1 \dots n$, the variance of Y is given by

$$\text{var}(Y) = \text{var}(X_1) + \cdots + \text{var}(X_n) = n\theta(1-\theta), \tag{7.10}$$

so the variance of the misclassification rate is

$$\sigma^2 = \text{var}\left(\frac{Y}{n}\right) = \frac{1}{n^2}\text{var}(Y) = \frac{\theta(1-\theta)}{n}. \tag{7.11}$$

For y observed misclassifications, we estimate θ as $\hat{\theta} = y/n$. Then, the estimated variance is given by

$$\hat{\sigma}^2 = \frac{\hat{\theta}(1-\hat{\theta})}{n} = \frac{\frac{y}{n}\left(1-\frac{y}{n}\right)}{n} = \frac{y(n-y)}{n^3},$$

and the estimated standard deviation by

$$\hat{\sigma} = \sqrt{\frac{y(n-y)}{n^3}}. \tag{7.12}$$

As was pointed out in Section 2.1.1, Y is (n, θ)-binomially distributed. However for a sufficiently large number n of test data, the binomial distribution is well approximated by the normal distribution. This is illustrated by the following code, which generates the mass function for the binomial distribution (Equation 2.3). The result is shown in Figure 7.2 for $n = 2000$.

```
n = 2000
theta = 0.1
x = lindgen(n)
f = fltarr(n)
FOR i=1,n-1 DO $
   f[i]=binomial(i,n,theta)-binomial(i+1,n,theta)
thisDevice =!D.Name
set_plot, 'PS'
Device, Filename='c:\temp\fig7_2.eps', $
```

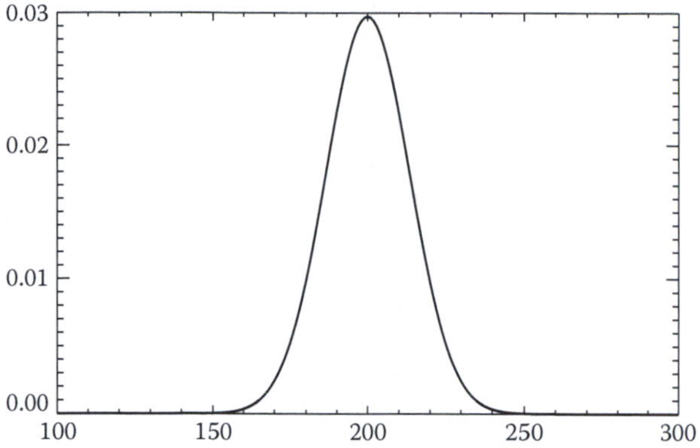

FIGURE 7.2
The binomial distribution for $n = 2000$ and $\theta = 0.1$ closely approximates a normal distribution.

```
    xsize=15,ysize=10,/Encapsulated
PLOT, x[100:300],f[100:300]
device,/close_file
set_plot,thisDevice
```

Mean and standard deviation are thus generally sufficient to characterize the distribution of misclassification rates completely. To obtain an interval estimation for θ, recalling the discussion in Section 2.3.2, we make use of the fact that the random variable $(Y/n - \theta)/\sigma$ is approximately standard normally distributed. Then

$$\Pr\left(-s < \frac{Y/n - \theta}{\sigma} \le s\right) = 2\Phi(s) - 1, \tag{7.13}$$

so that the random interval $(Y/n - s\sigma, Y/n + s\sigma)$ covers the unknown misclassification rate θ with probability $2\Phi(s) - 1$. However, note that from Equation 7.11 σ is itself a function of θ. It is easy to show (Exercise 1) that, for $1 - \alpha = 0.95 = 2\Phi(s) - 1$ (which gives $s = 1.96$ from the normal distribution table), a 95% confidence interval for θ is

$$\left(\frac{y + 1.92 - 1.96 \cdot \sqrt{0.96 + \frac{y(n-y)}{n}}}{3.84 + n}, \frac{y + 1.92 + 1.96 \cdot \sqrt{0.96 + \frac{y(n-y)}{n}}}{3.84 + n}\right). \tag{7.14}$$

This interval should be routinely stated for any supervised classification of land use/land cover.

Detailed test results for supervised classification are usually presented in the form of a *contingency table*, or *confusion matrix*, which for K classes is defined as

$$C = \begin{pmatrix} c_{11} & c_{12} & \cdots & c_{1K} \\ c_{21} & c_{22} & \cdots & c_{2K} \\ \vdots & \vdots & \ddots & \vdots \\ c_{K1} & c_{K2} & \cdots & c_{KK} \end{pmatrix}. \tag{7.15}$$

The matrix element c_{ij} is the number of test pixels with class label j which are classified as i. Note that the estimated misclassification rate is

$$\hat{\theta} = \frac{y}{n} = \frac{n - \sum_{i=1}^{K} c_{ii}}{n} = \frac{n - \text{tr}(\mathbf{C})}{n}$$

and only takes into account the diagonal elements of the confusion matrix. The so-called *Kappa-coefficient*, on the other hand, makes use of all the matrix elements. It corrects the classification rate for the possibility of chance correct classifications and is defined as follows (Cohen, 1960):

$$\kappa = \frac{\text{Pr(correct classification)} - \text{Pr(chance classification)}}{1 - \text{Pr(chance classification)}}.$$

An expression for κ can be obtained in terms of the row and column sums in the matrix $\mathbf{C}$, which we write as

$$c_{i.} = \sum_{j=1}^{K} c_{ij} \quad \text{and} \quad c_{.i} = \sum_{j=1}^{K} c_{ji},$$

respectively. For n randomly labeled test pixels, the proportion of entries in the ith row is $c_{i.}/n$ and in the ith column $c_{.i}/n$. The probability of a chance correct classification (chance coincidence of row and column index) is therefore approximately given by the sum over i of the product of these two proportions:

$$\sum_{i=1}^{K} \frac{c_{i.}\,c_{.i}}{n^2}.$$

Hence an estimate for the Kappa coefficient is

$$\hat{\kappa} = \frac{\sum_i c_{ii}/n - \sum_i c_{i.}c_{.i}/n^2}{1 - \sum_i c_{i.}c_{.i}/n^2}. \tag{7.16}$$

Again, the Kappa coefficient alone tells us little about the quality of the classifier. We require its uncertainty. This can be calculated in the large sample limit $n \to \infty$ to be (Bishop et al., 1975)

$$\hat{\sigma}_\kappa = \frac{1}{n}\left(\frac{\theta_1(1-\theta_1)}{(1-\theta_2)^2} + \frac{2(1-\theta_1)(2\theta_1\theta_2-\theta_3)}{(1-\theta_3)^3} + \frac{(1-\theta_1)^2(\theta_4-4\theta_2^2)}{(1-\theta_2)^4}\right), \quad (7.17)$$

where

$$\theta_1 = \sum_{i=1}^K c_{ii},\ \theta_2 = \sum_{i=1}^K c_{i.}c_{.i},\ \theta_3 = \sum_{i=1}^K c_{ii}(c_{i.}+c_{.i}), \quad \text{and} \quad \theta_4 = \sum_{i,j=1}^K c_{ij}(c_{j.}+c_{.i})^2.$$

The ENVI/IDL extension `CT_RUN` described in Appendix C.2.13 prints out misclassification rate $\hat{\theta}$, standard deviation $\hat{\sigma}$, the confidence interval of Equation 7.14, Kappa coefficient $\hat{\kappa}$, standard deviation $\hat{\sigma}_\kappa$, and a contingency table $\boldsymbol{C}$ from the test result files generated by the maximum likelihood classifier, the Gaussian kernel classifier, the neural network, and the support vector machine described in Sections 6.3 through 6.6. Here is a sample output for $K = 7$ classes:

```
 1 Test observations: 4505
 2 Classes: 7
 3  Misclassification rate  0.0249
 4      Standard deviation  0.0023
 5    Conf. interval (95%)  0.0207  0.0298
 6       Kappa coefficient  0.9705
 7      Standard deviation  0.0028
 8
 9 Contingency Table
10    600    12     2     4     0     0    13   631   0.951
11     14   516     0     2     0     0     4   536   0.963
12      0     0   668     0     2     0     0   670   0.997
13      0     0     0   363     0     0     0   363   1.000
14      1     0     5     0   447     0    31   484   0.924
15      0     4     0     0     0   755     0   759   0.995
16      4    11     2     0     1     0  1044  1062   0.983
17    619   543   677   369   450   755  1092  4505   0.000
18 0.969 0.950 0.987 0.984 0.993 1.000 0.956 0.000   0.000
```

The matrix $\boldsymbol{C}$ (Equation 7.15) is in the upper left 7×7 block in the table. Row 8 (line 17 in the above listing) gives the column sums $c_{.i}$ and column 8 the row sums $c_{i.}$. Row 9 (line 18) contains the ratios $c_{ii}/c_{.i}$, $i = 1\ldots 7$, which are referred to as the *producer accuracies*. Column 9 contains the *user accuracies* $c_{ii}/c_{i.}$, $i = 1\ldots 7$. The producer accuracy is the probability that an observation with label i will be classified as such. The user accuracy is the probability that the true class of an observation is i given that the classifier has labeled it as i. The rate of *correct classification*, which is more commonly quoted in the literature, is of course one minus the misclassification rate. A detailed discussion of confusion matrices for assessment of classification accuracy is provided in Congalton and Green (1999).

7.2.2 Model Comparison

A good value for a misclassification rate is $\theta \approx 0.05$. In order to claim that two rates produced by two different classifiers differ from one another significantly, a rule of thumb is that they should lie at least two standard deviations apart. A commonly used heuristic (van Niel et al., 2005) is to choose a minimum of $n \approx 30 \cdot N \cdot K$ training samples in all, where, as always, N is the data dimensionality and K is the number of classes. For the Jülich classification examples using $N = 4$ principal components and $K = 10$ classes this gives $n \approx 30 \cdot 4 \cdot 10 = 1200$. If one-third are to be reserved for testing, the number should be increased to $1200 \cdot 3/2 = 1800$. But suppose we wish to claim, for instance, that misclassification rates 0.05 and 0.06 are significantly different. Then, according to our thumb rule, their standard deviations should be no greater than 0.005. From Equation 7.11 this means $0.05(1 - 0.05)/n \approx 0.005^2$, or $n \approx 2000$ observations are needed for testing alone. Since we are dealing with pixel data, this number of test observations (assuming sufficient training areas are available) is still realistic.

In order to decide whether or not classifier A is better than classifier B in a more precise manner, a hypothesis test must be formulated. The individual misclassifications Y_A and Y_B are, as we have seen, approximately normally distributed. If they were also independent, then the test statistic

$$S = \frac{Y_A - Y_B}{\sqrt{\text{var}(Y_A) + \text{var}(Y_B)}}$$

would be standard normally distributed under the null hypothesis $\theta_A = \theta_B$. In fact, the independence of the misclassification rates is not given, since they are determined with the same set of test data.

There exist computationally expensive alternatives. The buzzwords here are *cross-validation* and *bootstrapping*, see Weiss and Kulikowski (1991, Chapter 2), for an excellent introduction. As an example, suppose that a series of p trials is carried out. In the ith trial, the training data are split randomly into training and test sets in the ratio 2:1 as before. Both classifiers are then trained and tested on these sets to give misclassifications represented by the random variables Y_A^i and Y_B^i. If the differences $Y_i = Y_A^i - Y_B^i$, $i = 1 \ldots p$, are independent and normally distributed, then the Student-t statistic (Equation 2.46) can be constructed to test the null hypothesis that the mean numbers of misclassifications are equal:

$$T = \frac{\bar{Y}}{\sqrt{S/p}}$$

$$\bar{Y} = \sum_{i=1}^{p} Y_i, \quad S = \frac{1}{p-1}\sum_{i=1}^{p}(Y_i - \bar{Y})^2.$$

T is Student-t distributed with $p - 1$ degrees of freedom.

There are some objections to this approach, the most obvious being the need to repeat the training/test cycle many times (typically $p \approx 30$). Moreover, as Dietterich (1998) points out in a comparative investigation of several such test procedures, Y_i is not normally distributed since, again, Y_A^i and Y_B^i are not independent. He recommends a *nonparametric* hypothesis test which avoids these problems and which we shall adopt here (see also Ripley, 1996).

After training of the two classifiers which are to be compared, the following events for classification of the test data can be distinguished:

$$\bar{A}B,\ A\bar{B},\ \bar{A}\bar{B},\ \text{and}\ AB.$$

The event $\bar{A}B$ is *test observation is misclassified by A and correctly classified by B*, while $A\bar{B}$ is the event *test observation is correctly classified by A and misclassified by B*, and so on. As before we define random variables:

$$X_{\bar{A}B},\ X_{A\bar{B}},\ X_{\bar{A}\bar{B}},\ \text{and}\ X_{AB}$$

where

$$X_{\bar{A}B}(\bar{A}B) = 1, \quad X_{\bar{A}B}(A\bar{B}) = X_{\bar{A}B}(\bar{A}\bar{B}) = X_{\bar{A}B}(AB) = 0,$$

with mass function

$$\Pr(X_{\bar{A}B} = 1) = \theta_{\bar{A}B}, \quad \Pr(X_{\bar{A}B} = 0) = 1 - \theta_{\bar{A}B}.$$

Corresponding definitions are made for $X_{A\bar{B}}$, $X_{\bar{A}\bar{B}}$, and X_{AB}.

Now, in comparing the two classifiers, we are interested in the events $\bar{A}B$ and $A\bar{B}$. If the number of the former is significantly smaller than the number of the latter, then A is better than B and vice versa. Events $\bar{A}\bar{B}$ in which *both* methods perform poorly are excluded.

For n test observations the random variables

$$Y_{\bar{A}B} = X_{\bar{A}B_1} + \cdots + X_{\bar{A}B_n} \quad \text{and} \quad Y_{A\bar{B}} = X_{A\bar{B}_1} + \cdots + X_{A\bar{B}_n}$$

are the frequencies of the respective events. We then have

$$\langle Y_{\bar{A}B}\rangle = n\theta_{\bar{A}B}, \quad \text{var}(Y_{\bar{A}B}) = n\theta_{\bar{A}B}(1 - \theta_{\bar{A}B})$$
$$\langle Y_{A\bar{B}}\rangle = n\theta_{A\bar{B}}, \quad \text{var}(Y_{A\bar{B}}) = n\theta_{A\bar{B}}(1 - \theta_{A\bar{B}}).$$

We expect that $\theta_{\bar{A}B} \ll 1$, that is, $\mathrm{var}(Y_{\bar{A}B}) \approx n\theta_{\bar{A}B} = \langle Y_{\bar{A}B}\rangle$. The same holds for $Y_{A\bar{B}}$. It follows that the random variables

$$\frac{Y_{\bar{A}B} - \langle Y_{\bar{A}B}\rangle}{\sqrt{\langle Y_{\bar{A}B}\rangle}} \quad \text{and} \quad \frac{Y_{A\bar{B}} - \langle Y_{A\bar{B}}\rangle}{\sqrt{\langle Y_{A\bar{B}}\rangle}}$$

are approximately standard normally distributed.

Under the null hypothesis (equivalence of the two classifiers), the expectation values of $Y_{\bar{A}B}$ and $Y_{A\bar{B}}$ satisfy

$$\langle Y_{\bar{A}B}\rangle = \langle Y_{A\bar{B}}\rangle =: \langle Y\rangle.$$

We form the *McNemar test statistic*

$$S = \frac{(Y_{\bar{A}B} - \langle Y\rangle)^2}{\langle Y\rangle} + \frac{(Y_{A\bar{B}} - \langle Y\rangle)^2}{\langle Y\rangle} \tag{7.18}$$

which is chi-square distributed with one degree of freedom (see Section 2.4.1). Let $y_{\bar{A}B}$ and $y_{A\bar{B}}$ be the number of events actually measured. Then the mean $\langle Y\rangle$ is estimated as

$$\langle \hat{Y}\rangle = \frac{y_{\bar{A}B} + y_{A\bar{B}}}{2}$$

and a realization of the test statistic S is

$$s = \frac{(y_{\bar{A}B} - \frac{y_{\bar{A}B}+y_{A\bar{B}}}{2})^2}{\frac{y_{\bar{A}B}+y_{A\bar{B}}}{2}} + \frac{(y_{A\bar{B}} - \frac{y_{\bar{A}B}+y_{A\bar{B}}}{2})^2}{\frac{y_{\bar{A}B}+y_{A\bar{B}}}{2}}.$$

With a little algebra this expression can be simplified to

$$s = \frac{(y_{\bar{A}B} - y_{A\bar{B}})^2}{y_{\bar{A}B} + y_{A\bar{B}}}. \tag{7.19}$$

A correction is usually made to Equation 7.19, writing it in the form

$$s = \frac{(|y_{\bar{A}B} - y_{A\bar{B}}| - 1)^2}{y_{\bar{A}B} + y_{A\bar{B}}}, \tag{7.20}$$

which takes into approximate account the fact that the statistic is discrete, while the chi-square distribution is continuous. From the percentiles of the chi-square distribution, the critical region for rejection of the null hypothesis of equal misclassification rates at the 5% significance level is $s \geq 3.841$. The ENVI/IDL extension `McNEMAR_RUN` (Appendix C.2.13) compares two classifiers on the basis of their test result files, printing out $y_{\bar{A}B}$, $y_{A\bar{B}}$, s, and the P-value $1 - P_{\chi^2;1}(s)$. Here is an example comparing the maximum likelihood and neural network classifiers for the Jülich ASTER scene:

```
 1 ENVI> mcnemar_run
 2 ------------------------
 3 Classification Comparison
 4 Fri Jan 09 12:53:50 2009
 5 ------------------------
 6 First classifier
 7 ; FFN test results for juelich aster\may0107_pca
 8 Second classifier
 9 ; MaxLike test results for juelich aster\may0107_pca
10 Test observations: 2391
11 Classes: 10
12      First classifier:      21
13      Second classifier:     54
14      McNemar statistic: 13.6533
15             P-value:   0.0002
```

In this case the null hypothesis can certainly be rejected in favor of the neural network. Comparing the neural network with ENVI's SVM classifier:

```
 1 ------------------------
 2 Classification Comparison
 3 Fri Jan 09 12:57:45 2009
 4 ------------------------
 5 First classifier
 6 ; FFN test results for juelich aster\may0107_pca
 7 Second classifier
 8 ; SVM test results for juelich aster\may0107_pca
 9 Test observations: 2391
10 Classes: 10
11      First classifier:      64
12      Second classifier:     88
13      McNemar statistic: 3.4803
14             P-value:   0.0621
```

where now the neural network is "better," but not at the 5% significance level.

There are still reservations to using this comparison method. The variability of the training data, which were sampled just once from their underlying distributions, has not been taken into account. If one or both of the classifiers is a neural network, we have also not considered the variability of the neural network training procedure with respect to the random initialization of the synaptic weights. Different initializations could lead to different local minima in the cost function and correspondingly different misclassification rates. Only if each of these effects are considered to be negligible can the procedure be applied. Including them properly constitutes a very computationally intensive task (Ripley, 1996).

7.3 Adaptive Boosting

Further enhancement of classifier accuracy is sometimes possible by combining several classifiers into an *ensemble* or *committee* and then applying some kind of "voting scheme" to generalize to new data. An excellent introduction to ensemble-based systems is given by Polikar (2006). The basic idea is to generate several classifiers and pool them in such a way as to improve on the performance of any single one. This implies that the pooled classifiers make errors on *different* observations, implying further that each classifier be as unique as possible, particularly with respect to misclassified instances (Polikar, 2006). One way of achieving this uniqueness is to use different training sets for each classifier, for example by resampling the training data with replacement, a procedure referred to as "bootstrap aggregation" or *bagging* (see Breiman, 1996 and Exercise 2).

Representative of ensemble methods, we consider here a powerful technique called *adaptive boosting* or *AdaBoost* for short (Freund and Shapire, 1996). It involves training a sequence of classifiers, placing increasing emphasis on hard-to-classify data, and then combining the sequence so as to reduce the overall training error. AdaBoost was originally suggested for combining binary classifiers, that is, for two-class problems. However, Freund and Shapire (1997) proposed two multiclass extensions, the more commonly used of which is *AdaBoost.M1*. In the following we shall apply AdaBoost.M1 to an ensemble of neural network classifiers. For other examples of adaptive boosting of neural networks, see Schwenk and Bengio (2000) and Murphey et al. (2001).

In order to motivate the adaptive boosting idea, consider the training of the feed-forward neural network classifier of Chapter 6 when there are just two classes to choose between. Making use as before of stochastic training, we train the network by minimizing the local cost function (Equation 6.34) on randomly selected labeled examples. To begin with, the training data are sampled uniformly, as is done, for example, in the backpropagation training algorithm of Listing 6.6. We can represent such a sampling scheme with the uniform, discrete probability distribution

$$p_1(\nu) = 1/m, \quad \nu = 1 \dots m,$$

over the m training examples. Let U_1 be the set of incorrectly classified examples after completion of the training procedure. Then the classification error is given by

$$\epsilon_1 = \sum_{\nu \in U_1} p_1(\nu).$$

Let us now find a new sampling distribution $p_2(\nu)$ such that the trained classifier would achieve an error of 50% if trained with respect to that distribution. In other words, it would perform as well as uninformed random guessing. The intention is, through the new distribution, to achieve a new classifier–training set combination which is as different as possible from the one just used. We obtain the new distribution $p_2(\nu)$ by reducing the probability for correctly classified examples by a factor $\beta_1 < 1$ so that the accuracy obtained is 1/2, that is,

$$\frac{1}{2} = \sum_{\nu \in U_1} p_2(\nu) = \sum_{\nu \notin U_1} p_2(\nu) = \frac{1}{Z} \sum_{\nu \notin U_1} \beta_1 p_1(\nu) = \frac{1}{Z}\beta_1(1 - \epsilon_1). \tag{7.21}$$

The denominator Z is a normalization which ensures that $\sum_\nu p_2(\nu) = 1$ so that $p_2(\nu)$ is indeed a probability distribution,

$$Z = \sum_{\nu \in U_1} p_1(\nu) + \beta_1 \sum_{\nu \notin U_1} p_1(\nu) = \epsilon_1 + \beta_1(1 - \epsilon_1). \tag{7.22}$$

Combining Equations 7.21 and 7.22 gives

$$\beta_1 = \frac{\epsilon_1}{1 - \epsilon_1}. \tag{7.23}$$

If we now train the network with respect to $p_2(\nu)$ we will get (it is to be hoped) a different set U_2 of incorrectly classified training examples and, correspondingly, a different classification error

$$\epsilon_2 = \sum_{\nu \in U_2} p_2(\nu).$$

This leads to a new reduction factor β_2 and the procedure is repeated. The sequence must of course terminate at i classifiers when $\epsilon_{i+1} > 1/2$, as then the incorrectly classified examples can no longer be emphasized since $\beta_{i+1} > 1$.

At the generalization phase, the "importance" of each classifier is set to some function of β_i, the smaller the β_i, the more important the classifier. As we shall see below, an appropriate weight is $\log(1/\beta_i)$. Thus if C_k is the set of networks which classify feature vector $\boldsymbol{g}$ as k, then that class receives the "vote"

$$V_k = \sum_{i \in C_k} \log(1/\beta_i), \quad k = 1, 2,$$

after which $\boldsymbol{g}$ is assigned to the class with the maximum vote.

To place things on a more precise footing, we will define the *hypothesis* generated by the neural network classifier for input observation $\boldsymbol{g}(\nu)$ as

$$h(\boldsymbol{g}(\nu)) = h(\nu) = \arg\max_k(m_k(\boldsymbol{g}(\nu)), \quad \nu = 1 \dots m. \tag{7.24}$$

This is just the index of the output neuron whose signal is largest. For a two-class problem, $h(\nu) \in \{1, 2\}$. Suppose that $k(\nu)$ is the label of observation $g(\nu)$. Following Freund and Shapire (1997), define the indicator

$$[[h(\nu) \neq k(\nu)]] = \begin{cases} 1 & \text{if the hypothesis } h(\nu) \text{ is incorrect} \\ 0 & \text{if it is correct.} \end{cases} \tag{7.25}$$

With this notation, we can give an exact formulation of the adaptive boosting algorithm for a sequence of neural networks applied to two-class problems. Then we can prove a theorem on the upper bound of the overall training error for that sequence. Here first of all is the algorithm.

Algorithm (AdaBoost)

1. Define an initial uniform probability distribution $p_1(\nu) = 1/m$, $\nu = 1 \dots m$, and the number N_c of classifiers in the sequence. Define initial weights $w_1(\nu) = p_1(\nu)$, $\nu = 1 \dots m$.
2. For $i = 1 \dots N_c$ do:
   a. Set $p_i(\nu) = w_i(\nu) / \sum_{\nu'=1}^{m} w_i(\nu')$, $\nu = 1 \dots m$.
   b. Train a network with the sampling distribution $p_i(\nu)$ to get back the hypotheses $h_i(\nu)$, $\nu = 1 \dots m$.
   c. Calculate the error $\epsilon_i = \sum_{\nu=1}^{m} p_i(\nu)[[h_i(\nu) \neq k(\nu)]]$.
   d. Set $\beta_i = \epsilon_i / (1 - \epsilon_i)$.
   e. Determine new weights w_{i+1} according to

   $$w_{i+1}(\nu) = w_i(\nu)\beta_i^{1-[[h_i(\nu) \neq k(\nu)]]}, \quad \nu = 1 \dots m.$$

3. Given an unlabeled observation g, obtain the total vote received by each class,

   $$V_k = \sum_{\{i \mid h_i(g) = k\}} \log(1/\beta_i), \quad k = 1, 2,$$

   and assign g to the class with maximum vote.

The training error in the AdaBoost algorithm is the fraction of training examples that will be incorrectly classified when put into the voting procedure in step 3 above. Before turning to the task of determining the upper bound on the training error, we need the following inequality:

$$\alpha^r \leq 1 - (1 - \alpha)r, \tag{7.26}$$

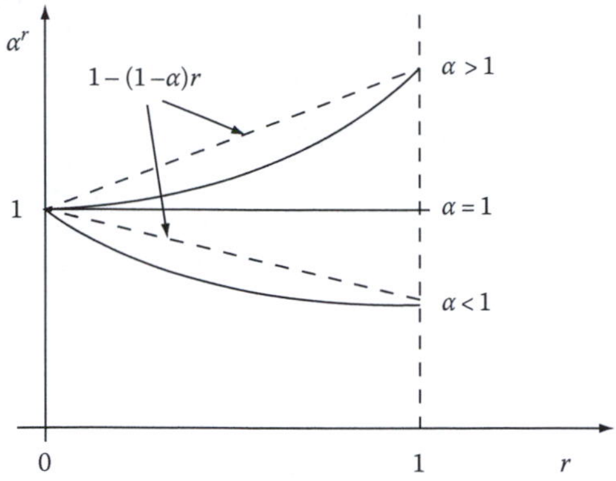

FIGURE 7.3
Illustrating Inequality 7.26.

which holds for any $\alpha \geq 0$ and $r \in [0,1]$. To see this, note that α^r is convex, that is, its second derivative is

$$\frac{d^2}{dr^2}(\alpha^r) = \alpha^r \log(\alpha)^2 \geq 0,$$

so we have the situation shown in Figure 7.3. Now we can prove the following theorem (Freund and Shapire, 1997).

THEOREM 7.1

The training error ϵ for the algorithm AdaBoost is bounded above according to

$$\epsilon \leq 2^{N_c} \prod_{i=1}^{N_c} \sqrt{\epsilon_i(1-\epsilon_i)}. \tag{7.27}$$

Proof

Applying Inequality 7.26 to step e in the AdaBoost algorithm, we have

$$\begin{aligned}
\sum_{\nu=1}^{m} w_{i+1}(\nu) &= \sum_{\nu=1}^{m} w_i(\nu)\beta_i^{1-[[h_i(\nu)\neq k(\nu)]]} \\
&\leq \sum_{\nu=1}^{m} w_i(\nu)[1-(1-\beta_i)(1-[[h_i(\nu)\neq k(\nu)]])] \\
&= \sum_{\nu=1}^{m} w_i(\nu) - (1-\beta_i)\left[\sum_{\nu=1}^{m} w_i(\nu)(1-[[h_i(\nu)\neq k(\nu)]])\right].
\end{aligned}$$

From steps a and c in the algorithm, we can write

$$\sum_{\nu=1}^{m} w_i(\nu)[[h_i(\nu) \neq k(\nu)]] = \sum_{\nu=1}^{m}\sum_{\nu'=1}^{m} w_i(\nu')p_i(\nu)[[h_i(\nu) \neq k(\nu)]] = \sum_{\nu'=1}^{m} w_i(\nu')\epsilon_i.$$

Combining the last two equations then gives

$$\sum_{\nu=1}^{m} w_{i+1}(\nu) \leq \sum_{\nu=1}^{m} w_i(\nu)[1-(1-\beta_i)(1-\epsilon_i)]. \tag{7.28}$$

If we apply this inequality successively for $i = 1 \ldots N_c$, it follows that

$$\begin{aligned}\sum_{\nu=1}^{m} w_{N_c+1}(\nu) &\leq \sum_{\nu=1}^{m} w_1(\nu)\prod_{i=1}^{m}[1-(1-\beta_i)(1-\epsilon_i)]\\ &= \prod_{i=1}^{m}[1-(1-\beta_i)(1-\epsilon_i)],\end{aligned} \tag{7.29}$$

since, in the algorithm, the initial weights sum to unity.

The final voting procedure in step 3 will make an error on training example ν if

$$\sum_{\{i\,|\,h_i(\nu)\neq k(\nu)\}} \log(1/\beta_i) \geq \sum_{\{i\,|\,h_i(\nu)=k(\nu)\}} \log(1/\beta_i).$$

Adding $\sum_{\{i\,|\,h_i(\nu)\neq k(\nu)\}} \log(1/\beta_i)$ to both sides,

$$2\left(\sum_{\{i\,|\,h_i(\nu)\neq k(\nu)\}} \log(1/\beta_i)\right) \geq \sum_{i=1}^{m} \log(1/\beta_i),$$

or equivalently

$$-2\left(\sum_{\{i\,|\,h_i(\nu)\neq k(\nu)\}} \log(\beta_i)\right) \geq -\sum_{i=1}^{m} \log(\beta_i),$$

which we can write in the form

$$-\log\left(\prod_{\{i\,|\,h_i(\nu)\neq k(\nu)\}} \beta_i\right) \geq -\frac{1}{2}\log\left(\prod_{i=1}^{N_c} \beta_i\right)$$

or, since the logarithm is a monotonic increasing function of its argument, equivalently as

$$\prod_{i=1}^{N_c} \beta_i^{[[h_i(\nu)\neq k(\nu)]]} \geq \left(\prod_{i=1}^{N_c} \beta_i\right)^{-1/2}. \tag{7.30}$$

From step e we have further

$$w_{N_c+1} = w_1(\nu) \prod_{i=1}^{N_c} \beta_i^{1-[[h_i(\nu)\neq k(\nu)]]}. \tag{7.31}$$

Now let U be the set of incorrectly classified training examples for the sequence of N_c classifiers. Then

$$\begin{aligned}
\sum_{\nu=1}^{m} w_{N_c+1}(\nu) &\geq \sum_{\nu\in U} w_{N_c+1}(\nu) \\
&= \sum_{\nu\in U} w_1(\nu) \prod_{i=1}^{N_c} \beta_i^{1-[[h_i(\nu)\neq k(\nu)]]} \quad \text{from Equation 7.31} \\
&= \sum_{\nu\in U} w_1(\nu) \prod_{i=1}^{N_c} \beta_i \prod_{i=1}^{N_c} \beta_i^{-[[h_i(\nu)\neq k(\nu)]]} \\
&\geq \sum_{\nu\in U} w_1(\nu) \left(\prod_{i=1}^{N_c} \beta_i\right)^{1/2} \quad \text{from Equation 7.30.}
\end{aligned}$$

But $\sum_{\nu\in U} w_1(\nu) = \epsilon$ and, combining this last inequality with Inequality 7.29, we have

$$\epsilon \left(\prod_{i=1}^{N_c} \beta_i\right)^{1/2} \leq \sum_{\nu=1}^{m} w_{N_c+1}(\nu) \leq \prod_{i=1}^{m} [1-(1-\beta_i)(1-\epsilon_i)]$$

or, solving for ϵ,

$$\epsilon \leq \prod_{i=1}^{N_c} \left(\frac{1-(1-\beta_i)(1-\epsilon_i)}{\sqrt{\beta_i}}\right). \tag{7.32}$$

Since each of the N_c factors is positive, the upper bound will be minimized when

$$\frac{d}{d\beta_i}\left(\frac{1-(1-\beta_i)(1-\epsilon_i)}{\sqrt{\beta_i}}\right) = 0, \quad i = 1 \ldots N_c,$$

with solution

$$\beta_i = \frac{\epsilon_i}{1-\epsilon_i}.$$

Substituting this back into Equation 7.32 gives the upper bound Equation 7.27 and the proof is complete. □

Theorem 7.1 tells us that, provided each classifier in the sequence can return an error $\epsilon_i < 1/2$, the training error will approach zero exponentially. It can be shown (Freund and Shapire, 1997) that the result is also valid for the multiclass case $K > 2$. The boosting algorithm is then referred to as AdaBoost.M1.

An ENVI/IDL extension `FFN3AB_RUN` for boosting neural networks (trained with the fast Kalman filter algorithm of Appendix B) is described in Appendix C.2.14 and in Canty (2009). A sequence of neural networks, $i = 1, 2, \ldots,$ is trained on samples chosen with respect to distributions $p_1(\nu), p_2(\nu), \ldots,$ the sequence terminating at i when $\epsilon_{i+1} \geq 1/2$ or when a maximum sequence length is reached. In order to take into account the fact that a network may become trapped in a local minimum of the cost function, training is restarted with a new random synaptic weight configuration if the current training error ϵ_i exceeds 1/2. The maximum number of restarts for a given network is five, after which the boosting terminates. The classifiers in the sequence are implemented in an object-oriented framework as instances of a neural network object class (see Chapter 6). The algorithm is as follows.

Algorithm (Adaptive Boosting of a Sequence of Neural Network Classifiers)

1. Set $p_1(\nu) = 1/m$, $\nu = 1 \ldots m$, where m is the number of observations in the set of labeled training data. Choose maximum sequence length N_{max}. Set $i = 1$.
2. Set $r = 0$.
3. Create a new neural network instance FFN(i) with random synaptic weights. Train FFN(i) with sampling distribution $p_i(\nu)$. Let U_i be the set of incorrectly classified training observations after completion of the training procedure.
4. Calculate $\epsilon_i = \sum_{\nu \in U_i} p_i(\nu)$. If $\epsilon_i < 1/2$ then continue, else if $r < 5$ then set $r = r + 1$, destroy the instance FFN(i) and go to 3, else stop.
5. Set $\beta_i = \epsilon_i/(1-\epsilon_i)$ and update the distribution:

$$p_{i+1}(\nu) = \frac{p_i(\nu)}{Z_i} \times \begin{cases} \beta_i & \text{if } \nu \notin U_i \\ 1 & \text{otherwise} \end{cases}, \quad \nu = 1 \ldots m,$$

   where $Z_i = \sum_{\nu \in U_i} p_i(\nu) + \beta_i \sum_{\nu \notin U_i} p_i(\nu)$.
6. Set $i = i + 1$. If $i > N_{\max}$ then stop, else go to 2.

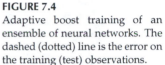

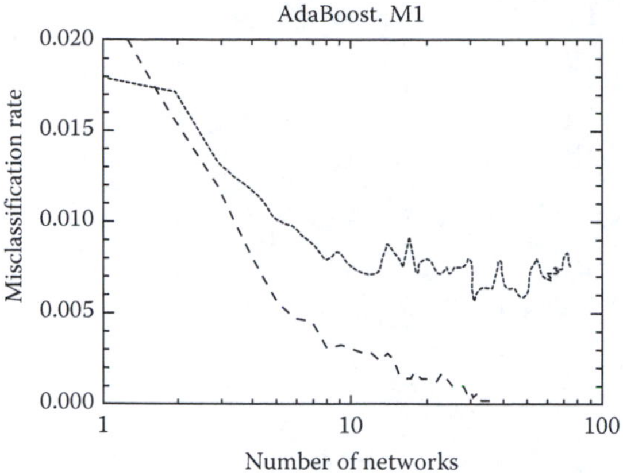

FIGURE 7.4
Adaptive boost training of an ensemble of neural networks. The dashed (dotted) line is the error on the training (test) observations.

During the training phase the program displays the cost function of the currently trained network and the training and generalization errors of the boosted sequence. Boosting can be interrupted at any time, upon which the sequence generated so far will be used to classify the input data. An example is shown in Figure 7.4 for the training/test datasets of the Jülich ASTER scene. For illustration purposes six principal components were used for training, as the boosting effect is most evident for higher dimensional input spaces. In the figure the training error is actually boosted to zero, while the generalization error is reduced by a factor of around 2.5. For a detailed comparison with other classifiers, see Canty (2009).

7.4 Hyperspectral Analysis

Hyperspectral—as opposed to multispectral—images combine both high or moderate spatial resolution with high spectral resolution. Typical remote sensing imaging spectrometers generate in excess of 200 spectral channels. Figure 7.5 shows part of a so-called *image cube* for the Airborne Visible/Infrared Imaging Spectrometer (AVIRIS) sensor taken over a region of the Californian coast. Figure 7.6 displays the spectrum of a single pixel in the image. Sensors of this kind produce much more complex data and provide correspondingly much more information about the reflecting surfaces examined than their multispectral counterparts.

The classification methods discussed in Chapter 6 and in the present chapter must in general be modified considerably in order to cope with the volume of data provided in a hyperspectral image. For example, the covariance matrix of an AVIRIS scene has dimension 224×224, so that the modeling of image or class probability distributions is much more difficult.

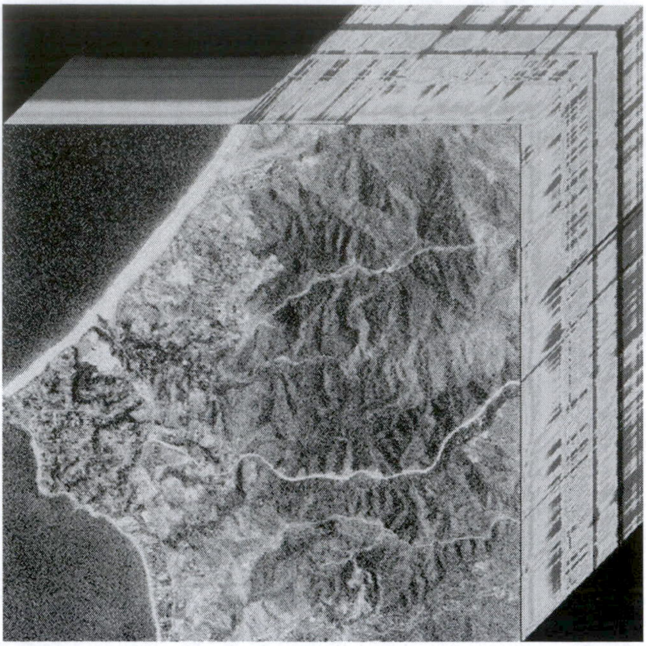

FIGURE 7.5
AVIRIS hyperspectral image cube over the Santa Monica Mountains acquired on April 7, 1997 at a GSD of 20m.

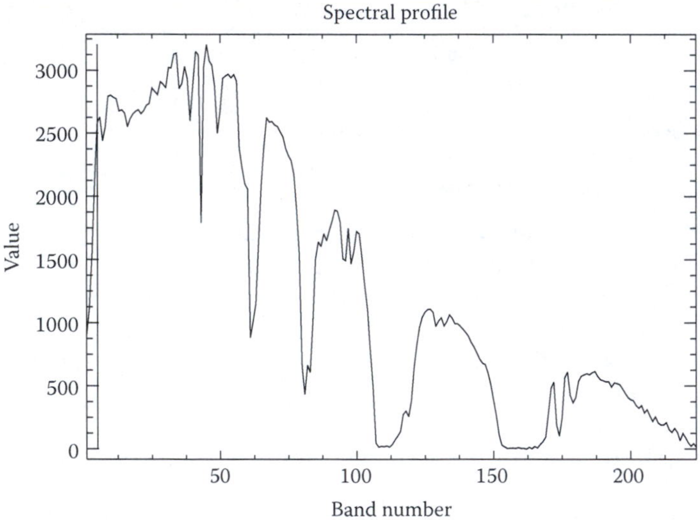

FIGURE 7.6
AVIRIS spectrum at one pixel location in Figure 7.5. There are 224 spectral bands covering the 0.4–1.5 μm wavelength interval.

Here one speaks of "ill-posed" classification problems, meaning that the available training data may be insufficient to estimate the model parameters adequately, and some sort of dimensionality reduction is needed (Richards and Jia, 2006). We will restrict discussion in the following to the concept of spectral unmixing, a method often used in lieu of "conventional" classification, and one of the most common techniques for hyperspectral image analysis.

7.4.1 Spectral Mixture Modeling

In multispectral image classification the fact that, at the scale of observation, a pixel often contains a mixture of land cover categories is generally treated as a second-order effect, and more or less ignored. When working with hyperspectral imaging spectrometers, it is possible to treat the problem of the "mixed pixel" quantitatively. While one can still speak of classification of land surfaces, the training data now consist of external spectral libraries or, in some instances, reference spectra derived from within the images themselves. Comparison with external spectral libraries requires detailed attention to atmospheric correction. The end product is not the discrete labeling of the pixels that we have become familiar with, but consists rather of image planes or maps showing, at each pixel location, the proportion of surface material contributing to the observed reflectance.

The basic premise of mixture modeling is that, within a given scene, the surface is dominated by a small number of common materials that have characteristic spectral properties. These are referred to as the *end-members* and it is assumed that the spectral variability captured by the remote sensing system can be modeled by mixtures of end-members.

Suppose that there are K end-members and N spectral bands. Denote the spectrum of the ith end-member by the column vector

$$\boldsymbol{m}^i = (m^i_1, mn_2 \dots m^i_N)^\top$$

and the matrix of end-member spectra $\boldsymbol{M}$ by

$$\boldsymbol{M} = (\boldsymbol{m}^1 \dots \boldsymbol{m}^K) = \begin{pmatrix} m^1_1 & \dots & m^K_1 \\ \vdots & \ddots & \vdots \\ m^1_N & \dots & m^K_N \end{pmatrix},$$

with one column for each end-member. For hyperspectral imagery we always have $K \ll N$, unlike the situation for multispectral data.

The measured spectrum, represented by random vector $\boldsymbol{G}$, may be modeled as a linear combination of end-members plus a residual term $\boldsymbol{R}$ which is understood to be the variation in G not explained by the mixture model:

$$G = \alpha_1 m^1 + \cdots + \alpha_K m^K + R = M\alpha + R. \tag{7.33}$$

The vector $\boldsymbol{\alpha} = (\alpha_1 \ldots \alpha_K)^\top$ contains nonnegative mixing coefficients which are to be determined. Let us assume that the residual R is normally distributed, zero-mean with covariance matrix

$$\Sigma_R = \begin{pmatrix} \sigma_1^2 & 0 & \ldots & 0 \\ 0 & \sigma_2^2 & \ldots & 0 \\ \vdots & \vdots & \ddots & \vdots \\ 0 & 0 & \ldots & \sigma_N^2 \end{pmatrix}.$$

The standardized residual is $\Sigma_R^{-1/2} R$ (why?) and the square of the standardized residual is

$$(\Sigma_R^{-1/2} R)^\top (\Sigma_R^{-1/2} R) = R^\top \Sigma_R^{-1} R. \tag{7.34}$$

The mixing coefficients $\boldsymbol{\alpha}$ may then be determined by minimizing this quantity with respect to the α_i under the condition that they sum to unity,

$$\sum_{i=1}^{K} \alpha_i = 1, \tag{7.35}$$

and are all nonnegative,

$$\alpha_i \geq 0, \quad i = 1 \ldots K. \tag{7.36}$$

Ignoring the requirement of Equation 7.36 for the time being, a Lagrange function for minimization of Equation 7.34 under the constraint of Equation 7.35 is

$$\begin{aligned} L &= R^\top \Sigma_R^{-1} R + 2\lambda \left(\sum_{i=1}^{K} \alpha_i - 1 \right) \\ &= (G - M\alpha)^\top \Sigma_R^{-1} (G - M\alpha) + 2\lambda \left(\sum_{i=1}^{K} \alpha_i - 1 \right), \end{aligned}$$

the last equality following from Equation 7.33. Solving the set of equations

$$\frac{\partial L}{\partial \boldsymbol{\alpha}} = 0, \quad \frac{\partial L}{\partial \lambda} = 0,$$

and replacing G by its realization g, we obtain the estimates for the mixing coefficients (Exercise 3)

$$\begin{aligned}\hat{\alpha} &= (M^\top \Sigma_R^{-1} M)^{-1}(M^\top \Sigma_R^{-1} g - \lambda \mathbf{1}_K) \\ \hat{\alpha}^\top \mathbf{1}_K &= 1,\end{aligned} \tag{7.37}$$

where $\mathbf{1}_K$ is a column vector of K ones. The first equation determines the mixing coefficients in terms of known quantities and λ. The second equation can be used to eliminate λ. Neglecting the constraint in Equation 7.36 is common practice. It can, however, be dealt with using appropriate numerical methods (Nielsen, 2001).

7.4.2 Unconstrained Linear Unmixing

If we work, for example, with MNF-transformed data (see Section 3.4) then we can assume that $\Sigma_R = I$.* Otherwise assume $\Sigma_R = \sigma^2 I$. If furthermore we ignore both of the constraints on α (Equations 7.35 and 7.36) which amounts to the assumption that the end-member spectra M are capable of explaining the observations completely apart from random noise, then Equation 7.37 reduces to the ordinary least squares estimate for α,

$$\hat{\alpha} = [(M^\top M)^{-1} M^\top] g. \tag{7.38}$$

The expression in square brackets is the pseudoinverse of the matrix M and the covariance matrix for α is $\sigma^2 (M^\top M)^{-1}$, as is explained in Section 2.6.2. This calculation can be invoked from the ENVI main menu under

```
Spectral/Mapping Methods/Linear Spectral Unmixing
```

which optionally can also include the constraint Equation 7.35. More sophisticated approaches, which are applicable when not all of the end-members are known, are discussed in the exercises.

7.4.3 Intrinsic End-Members and Pixel Purity

When a spectral library for all of the K end-members in M is available the mixture coefficients can be calculated directly using the above methods. The primary product of the spectral mixture analysis consists of fraction images which show the spatial distribution and abundance of the end-member components in the scene. If such external data are unavailable, there are various

* This is at least the case for the MNF transformation of Section 3.4.1. If ENVI's MNF algorithm is used, Section 3.4.2, then the components of R must first be divided by the square roots of the corresponding eigenvalues of the transformation.

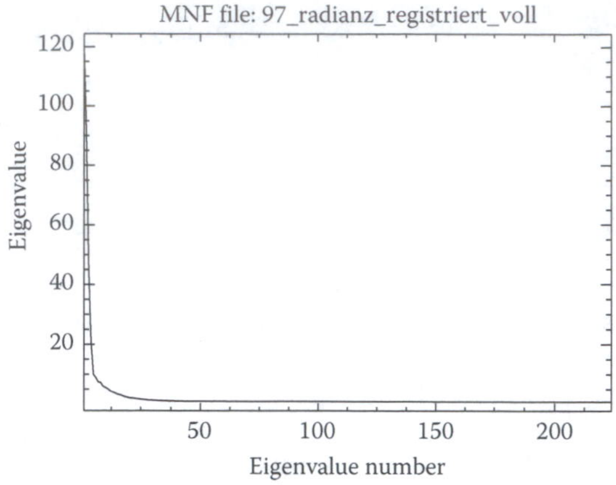

FIGURE 7.7
Eigenvalues of the MNF transformation of the image in Figure 7.5.

strategies for determining end-members from the hyperspectral image itself. We describe briefly the method recommended and implemented in ENVI.

The first step is to reduce the dimensionality of the data. This may be accomplished in ENVI with the minimum noise fraction transformation described in Section 3.4.2. By examining the eigenvalues of the transformation and retaining only the components with eigenvalues exceeding one (non-noise components), the number of dimensions can be reduced substantially (see Figure 7.7).

The so-called *pixel purity index* (PPI) is then used to find the most spectrally pure, or extreme, pixels in the reduced feature space. The most spectrally pure pixels typically correspond to end-members. These pixels must be on the corners, edges, or faces of the data cloud. The PPI is computed by repeatedly projecting n-dimensional scatter plots onto a random unit vector. The extreme pixels in each projection are noted and the number of times each pixel is marked as extreme is recorded. A threshold value is used to define how many pixels are marked as extreme at the ends of the projected vector. This value should be 2–3 times the variance in the data, which may be one when using the MNF-transformed bands. A minimum of about 5000 iterations is usually required to produce useful results.

When the iterations are completed, a PPI image is created in which the intensity of each pixel corresponds to the number of times that pixel was recorded as extreme (see Figure 7.8). Therefore bright pixels are generally end-members. (This image also hints at locations and sites that could be visited for ground truth measurements, should that be feasible.)

ENVI's n-dimensional visualizer can be used interactively to define classes of pixels corresponding to end-members and to plot their spectra. These may be saved along with their pixel locations as ROIs for later use in spectral unmixing or related procedures such as spectral angle mapping (Kruse et al., 1993). This sort of "data-driven" analysis has both advantages and disadvantages. To quote Mustard and Sunshine (1999):

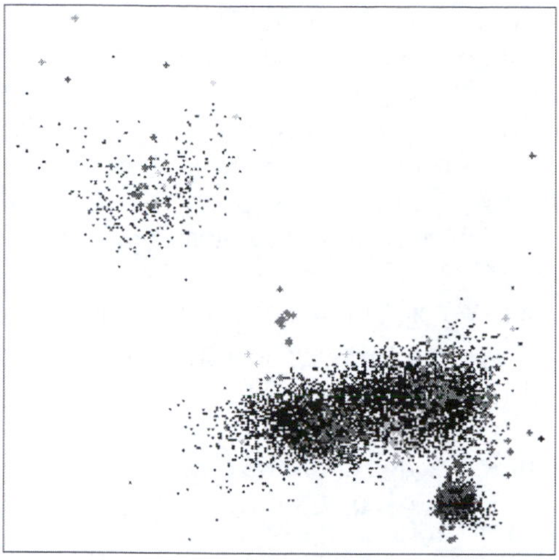

FIGURE 7.8
The *n*-D visualizer displaying PPIs.

This method is repeatable and has distinct advantages for objective analysis of a data set to assess the general dimensionality and to define end-members. The primary disadvantage of this method is that it is fundamentally a statistical approach dependent on the specific spectral variance of the scene and its components. Thus the resulting end-members are mathematical constructs and may not be physically realistic.

7.5 Exercises

1. Derive the confidence limits on the misclassification rate θ given by Equation 7.14.
2. In *bootstrap aggregation* or *bagging* (Breiman, 1996; Polikar, 2006), an ensemble of classifiers is derived from a single classifier by training it repeatedly on *bootstrapped replicas* of the training dataset, that is, on random samples drawn from the training set *with replacement*. In each replica some training examples may appear more than once, some may be missing altogether. A simple majority voting scheme is then used to classify new data with the ensemble. Breiman (1996) shows that improved classification accuracy is to be expected over that obtained by the original classifier/training dataset, especially for relatively unstable classifiers. Unstable classifiers are ones whose decision boundaries tend to be sensitive to small perturbations of the training data, and neural networks belong to this category.

(a) In the base object class `FFN` for a neural network (Listing 6.5), the function (method) `FFN::CLASSIFY(Gs,probs)` takes a row of pixels as input and returns their class labels in a byte array. For an ensemble of trained networks, one could use this function to build up a $c \times P$ array of class labels L, where c is the number of pixels in the row and P the number of classifiers. Write an IDL function `MAJORITY_VOTE(L)` which returns a $c \times 1$ array of those pixel labels which are chosen by the majority of the classifiers. You can use a loop over the class labels, but *not* over the pixels.

(b) Write an ENVI extension to implement bagging using the IDL object class `FFNKAL` described in Appendix B as the neural network classifier. Use the program `FFN_RUN.PRO` (Appendix C.2.10) as a reference. Your implementation should do the following:
   - Get the image to be classified, the ROI training data, the number P of classifiers in the ensemble, and the size $n < m$ of the bootstrapped replicas, where m is the size of the training set.
   - Create P instances of the neural network object class and train each of them on a different bootstrapped sample.
   - Classify the entire image by constructing the array L (part (a) above) for each row of the image and then passing it to the function `MAJORITY_VOTE(L)`.
   - Return the classified image to ENVI.

3. Demonstrate that the mixing coefficients of Equation 7.37 minimize the standardized residual (Equation 7.34) under constraint Equation 7.35.

4. In searching for a single spectral signature of interest $\boldsymbol{d}$, we can write Equation 7.33 in the form

$$\boldsymbol{G} = \boldsymbol{d}\alpha_d + \boldsymbol{U}\boldsymbol{\beta} + \boldsymbol{R}, \tag{7.39}$$

where
$\boldsymbol{U}$ is the $N \times (K-1)$ matrix of unwanted (or perhaps unknown) spectra
$\boldsymbol{\beta}$ is the vector of their mixing coefficients

The terms *partial unmixing* or *matched filtering* are often used to characterize methods to eliminate or reduce the effect of $\boldsymbol{U}$.

(a) *Orthogonal subspace projection* (OSP) (Harsanyi and Chang, 1994). Show that the matrix

$$\boldsymbol{P} = \boldsymbol{I} - \boldsymbol{U}(\boldsymbol{U}^\top\boldsymbol{U})^{-1}\boldsymbol{U}^\top, \tag{7.40}$$

where $\boldsymbol{I}$ is the $N \times N$ identity matrix, "projects out" the unwanted components. (Hilger and Nielsen (2000) give an example of the use of this transformation for the suppression of cloud cover from a multispectral image.)

(b) Write an IDL function `OSP(G,U)` which takes a data matrix $\mathcal{G}$ and the matrix of undesired spectra $\boldsymbol{U}$ as input and returns the OSP projection $\boldsymbol{P}\mathcal{G}^\top$.

(c) *Constrained energy minimization* (CEM) (Harsanyi, 1993; Nielsen, 2001). OSP still requires knowledge of all of the end-member spectra. In fact, Settle (1996) shows that it is fully equivalent to linear unmixing. If the spectra $\boldsymbol{U}$ are unknown, CEM reduces the influence of the undesired spectra by finding a projection direction $\boldsymbol{w}$ for which

$$\boldsymbol{w}^\top \boldsymbol{d} = 1$$

while at the same time minimizing the "mean of the output energy" $\langle(\boldsymbol{w}^\top G)^2\rangle$. Show that

$$\boldsymbol{w} = \frac{\boldsymbol{\Sigma}_R^{-1/2}\boldsymbol{d}}{\boldsymbol{d}^\top \boldsymbol{\Sigma}_R^{-1/2}\boldsymbol{d}}$$

satisfies these conditions.

(d) *Spectral angle mapping* (SAM) (Kruse et al., 1993). In order to establish a measure of closeness to a desired spectrum, one can compare the angle θ between the desired spectrum $\boldsymbol{d}$ and each pixel vector $\boldsymbol{g}$. Show that this is equivalent to CEM with a diagonal covariance matrix $\boldsymbol{\Sigma}_R = \sigma^2 \boldsymbol{I}$, that is, to CEM without any allowance for covariance between the spectral bands.

8

Unsupervised Classification

Supervised classification of multispectral remote sensing imagery, the subject of the previous two chapters, involves the use of a training dataset consisting of labeled pixels representative of each land cover category of interest in an image. We saw how to use these data to generalize to a complete labeling, or thematic map, for an entire scene. The choice of training areas that adequately represent the spectral characteristics of each category is very important for supervised classification, as the quality of the training set has a profound effect on the validity of the result. Finding and verifying training areas can be laborious, since the analyst must select representative pixels for each of the classes by visual examination of the image and by information extraction from additional sources such as ground reference data (ground truth), aerial photos, or existing maps.

Unlike supervised classification, unsupervised classification, or *clustering* as it is often called, requires no reference information at all. Instead, the attempt is made to find an underlying class structure automatically by organizing the data into groups sharing similar (e.g., spectrally homogeneous) characteristics. Often one only needs to specify beforehand the number, K, of classes present. Unsupervised classification plays an especially important role when very little *a priori* information about the data is available. A primary objective of using clustering algorithms for multispectral remote sensing data is to obtain useful information for the selection of training regions in a subsequent supervised classification.

We can view the basic problem of unsupervised classification at the individual pixel level as the partitioning of a set of samples (the image pixel intensity vectors $g(\nu)$, $\nu = 1 \ldots m$) into K disjoint subsets, called classes or clusters. The members of each class are to be in some sense more similar to one another than to the members of the other classes. If one wishes to take the spatial context of the image into account, then the problem also becomes one of labeling a regular lattice. Here the random field concept introduced in Chapter 4 can be used to advantage.

Clearly, a criterion is needed that will determine the quality of any given partitioning, along with means to determine the partitioning that optimizes it. The *sum of squares* cost function will provide us with a sufficient basis to justify some of the most popular algorithms for clustering of multispectral imagery, so we begin with its derivation. For a broad overview of clustering techniques for multispectral images, see Tran et al. (2005).

8.1 Simple Cost Functions

The N-dimensional observations that are to be partitioned comprise the set

$$\{g(\nu) \mid \nu = 1 \dots m\},$$

which we can conveniently represent as the $m \times N$ data matrix, $\mathcal{G}$. A given partitioning may be written in the form

$$C = [C_1, \dots C_k, \dots C_K],$$

where C_k is the set of indices $\{\ \nu \mid \nu = 1 \dots m,\ g(\nu)$ is in class $k\}$. Our strategy will be to maximize the posterior probability, $\Pr(C \mid \mathcal{G})$, for observing the partitioning, C, given the data, $\mathcal{G}$. From Bayes' theorem, we can write

$$\Pr(C \mid \mathcal{G}) = \frac{p(\mathcal{G} \mid C)\Pr(C)}{p(\mathcal{G})}. \tag{8.1}$$

$\Pr(C)$ is the prior probability for C. The quantity $p(\mathcal{G} \mid C)$ is the probability density function for the observations when the partitioning is C, also referred to as the *likelihood* of partitioning C given the data $\mathcal{G}$, while $p(\mathcal{G})$ is a normalization independent of C.

Following Fraley (1996), we first of all make the strong assumption that the observations are chosen independently from K multivariate normally distributed populations corresponding the K land cover categories present in the image. Under this assumption, the $g(\nu)$ are realizations of random vectors

$$G_k \sim \mathcal{N}(\mu_k, \Sigma_k), \quad k = 1 \dots K,$$

with multivariate normal probability densities, which we denote $p(g \mid k)$. The likelihood is the product of the individual probability densities given the partitioning, that is,

$$\begin{aligned} p(\mathcal{G} \mid C) &= \prod_{k=1}^{K} \prod_{\nu \in C_k} p(g(\nu) \mid k) \\ &= \prod_{k=1}^{K} \prod_{\nu \in C_k} (2\pi)^{-N/2} |\Sigma_k|^{-1/2} \exp\left(-\frac{1}{2}(g(\nu) - \mu_k)^\top \Sigma_k^{-1} (g(\nu) - \mu_k)\right). \end{aligned}$$

Forming the product in this way is justified by the independence of the observations. Taking the logarithm gives the log-likelihood:

$$\mathcal{L}(C) = \sum_{k=1}^{K} \sum_{\nu \in C_k} \Big(-\frac{N}{2}\log(2\pi) - \frac{1}{2}\log|\boldsymbol{\Sigma}_k| - \frac{1}{2}(\boldsymbol{g}(\nu) - \boldsymbol{\mu}_k)^\top \boldsymbol{\Sigma}_k^{-1}(\boldsymbol{g}(\nu) - \boldsymbol{\mu}_k) \Big). \tag{8.2}$$

Maximizing the log-likelihood is obviously equivalent to maximizing the likelihood. From Equation (8.1), we can then write

$$\log \Pr(C \mid \mathcal{G}) = \mathcal{L}(C) + \log \Pr(C) - \log p(\mathcal{G}). \tag{8.3}$$

Since the last term is independent of C, maximizing $\Pr(C \mid \mathcal{G})$ with respect to C is equivalent to maximizing $\mathcal{L}(C) + \log \Pr(C)$. If the even stronger assumption is now made that all K classes exhibit identical covariance matrices given by

$$\boldsymbol{\Sigma}_k = \sigma^2 \boldsymbol{I}, \quad k = 1 \ldots K, \tag{8.4}$$

where $\boldsymbol{I}$ is the identity matrix, then $\mathcal{L}(C)$ is maximized when the expression

$$\sum_{k=1}^{K} \sum_{\nu \in C_k} (\boldsymbol{g}(\nu) - \boldsymbol{\mu}_k)^\top \left(\frac{1}{2\sigma^2}\boldsymbol{I} \right) (\boldsymbol{g}(\nu) - \boldsymbol{\mu}_k) = \sum_{k=1}^{K} \sum_{\nu \in C_k} \frac{\|\boldsymbol{g}(\nu) - \boldsymbol{\mu}_k\|^2}{2\sigma^2}$$

is minimized. Finally, with Equation (8.3), $\Pr(C \mid \mathcal{G})$ itself is maximized by minimizing the *cost function*:

$$E(C) = \sum_{k=1}^{K} \sum_{\nu \in C_k} \frac{\|\boldsymbol{g}(\nu) - \boldsymbol{\mu}_k\|^2}{2\sigma^2} - \log \Pr(C). \tag{8.5}$$

Now let us introduce a "hard" class dependency in the form of a matrix $\boldsymbol{U}$ with elements

$$u_{k\nu} = \begin{cases} 1 & \text{if } \nu \in C_k \\ 0 & \text{otherwise.} \end{cases} \tag{8.6}$$

These matrix elements are required to satisfy the conditions

$$\sum_{k=1}^{K} u_{k\nu} = 1, \quad \nu = 1 \ldots m, \tag{8.7}$$

meaning that each pixel $\boldsymbol{g}(\nu)$, $\nu = 1 \ldots m$, belongs to precisely one class, and

$$\sum_{\nu=1}^{m} u_{k\nu} = m_k > 0, \quad k = 1 \ldots K, \tag{8.8}$$

meaning that no class C_k is empty. The sum in Equation 8.8 is the number m_k of pixels in the kth class. Maximum likelihood estimates (Section 2.5) for the mean of the kth cluster can be then written in the form

$$\boldsymbol{m}_k = \frac{1}{m_k} \sum_{\nu \in C_k} \boldsymbol{g}(\nu) = \frac{\sum_{\nu=1}^{m} u_{k\nu} \boldsymbol{g}(\nu)}{\sum_{\nu=1}^{m} u_{k\nu}}, \quad k = 1 \ldots K, \tag{8.9}$$

and for the covariance matrix as

$$\boldsymbol{C}_k = \frac{\sum_{\nu=1}^{m} u_{k\nu} (\boldsymbol{g}(\nu) - \boldsymbol{m}_k)(\boldsymbol{g}(\nu) - \boldsymbol{m}_k)^\top}{\sum_{\nu=1}^{m} u_{k\nu}}, \quad k = 1 \ldots K. \tag{8.10}$$

The cost function (Equation 8.5) can also be expressed in terms of the class dependencies $u_{k\nu}$ as

$$E(C) = \sum_{k=1}^{K} \sum_{\nu=1}^{m} u_{k\nu} \frac{\|\boldsymbol{g}(\nu) - \boldsymbol{m}_k\|^2}{2\sigma^2} - \log \Pr(C). \tag{8.11}$$

The parameter σ^2 can be thought of as the average within-cluster or image noise variance. If we have no prior information on the class structure, we can simply say that all partitionings C are *a priori* equally likely. Then the last term in Equation 8.11 is independent of C and, dropping it and the multiplicative constant $1/2\sigma^2$, we get the *sum of squares* cost function

$$E(C) = \sum_{k=1}^{K} \sum_{\nu=1}^{m} u_{k\nu} \|\boldsymbol{g}(\nu) - \boldsymbol{m}_k\|^2. \tag{8.12}$$

8.2 Algorithms That Minimize the Simple Cost Functions

The problem to find the partitioning which minimizes the cost functions (Equation 8.11 or 8.12) is unfortunately impossible to solve. The number of conceivable partitions, while obviously finite, is in any real situation astronomical. For example, for $m = 1000$ pixels and just $K = 2$ possible classes, there are $2^{1000-1} - 1 \approx 10^{300}$ possibilities (Duda and Hart, 1973). Direct enumeration is therefore not feasible. The line of attack most frequently taken is to start with some initial clustering and its associated cost function and then attempt to minimize the latter iteratively. This will always find a local minimum, but there is no guarantee that a global minimum for the cost function will be reached. Therefore one can never know if the best solution has been found. Nevertheless the approach is used because the computational burden is acceptable.

We shall follow the iterative approach in this section, beginning with the well-known K-means algorithm (including a kernelized version), followed by consideration of a variant due to Palubinskas (1998) which uses the cost function of Equation 8.11 and for which the number of clusters is determined automatically. Then we discuss a common example of bottom-up or agglomerative hierarchical clustering and conclude with a "fuzzy" version of the K-means algorithm.

8.2.1 K-Means Clustering

The K-means (KM) clustering algorithm (sometimes referred to as *basic ISODATA* (Duda and Hart, 1973) or *migrating means* (Richards and Jia, 2006)) is based on the sum of squares cost function (Equation 8.12). After some random initialization of the cluster centers $\boldsymbol{m}_k$ and setting $\boldsymbol{U} = \boldsymbol{0}$, the distance measure corresponding to a minimization of Equation 8.12, namely

$$d(\boldsymbol{g}(\nu), k) = \|\boldsymbol{g}(\nu) - \boldsymbol{m}_k\|^2, \tag{8.13}$$

is used to cluster the pixel vectors. Specifically, set $u_{k\nu} = 1$, where

$$k = \arg\min_k d(\boldsymbol{g}(\nu), k) \tag{8.14}$$

for $\nu = 1 \dots m$. Then Equation 8.9 is invoked to recalculate the cluster centers. This procedure is iterated until the class labels cease to change. A popular extension, referred to as ISODATA (Iterative Self-Organizing Data Analysis), involves splitting and merging of the clusters in repeated passes, albeit at the cost of setting additional parameters. K-means (and ISODATA) clustering may be performed within the standard ENVI environment from the main menu

```
Classification/Unsupervised/K-Means(ISODATA)
```

8.2.2 Kernel K-Means Clustering

To kernelize the K-means algorithm (Shawe-Taylor and Cristianini, 2004) we require the dual formulation for Equation 8.13. With Equation 8.8, the matrix

$$\boldsymbol{M} = [\mathrm{Diag}(\boldsymbol{U}\boldsymbol{1}_m)]^{-1},$$

where $\boldsymbol{1}_m$ is an m-component vector of ones, can be seen to be a $K \times K$ diagonal matrix with inverse class populations along the diagonal,

$$\boldsymbol{M} = \begin{pmatrix} 1/m_1 & 0 & \dots & 0 \\ 0 & 1/m_2 & \dots & 0 \\ \vdots & \vdots & \ddots & \vdots \\ 0 & 0 & \dots & 1/m_K \end{pmatrix}.$$

Therefore, from the rule for matrix multiplication (Equation 1.13) we can express the class means given by Equation 8.9 as the columns of the $N \times K$ matrix

$$(m_1, m_2 \dots m_K) = \mathcal{G}^\top \mathcal{U}^\top M.$$

We then obtain the dual formulation of Equation 8.13 as

$$\begin{aligned} d(g(\nu), k) &= \|g(\nu)\|^2 - 2g(\nu)^\top m_k + \|m_k\|^2 \\ &= \|g(\nu)\|^2 - 2g(\nu)^\top [\mathcal{G}^\top \mathcal{U}^\top M]_{\cdot k} + [M\mathcal{U}\mathcal{G}\mathcal{G}^\top \mathcal{U}^\top M]_{kk} \\ &= \|g(\nu)\|^2 - 2[(g(\nu)^\top \mathcal{G}^\top)\mathcal{U}^\top M]_k + [M\mathcal{U}\mathcal{G}\mathcal{G}^\top \mathcal{U}^\top M]_{kk}\,. \end{aligned} \tag{8.15}$$

In the second line above, $[\ ]_{\cdot k}$ denotes the kth column. In Equation 8.15 the observations appear only as inner products: $g(\nu)^\top g(\nu)$ (first term), in the Gram matrix $\mathcal{G}\mathcal{G}^\top$ (last term), and in $g(\nu)^\top \mathcal{G}^\top$ (second term). This latter expression is just the νth row of the Gram matrix, that is,

$$g(\nu)^\top \mathcal{G}^\top = [\mathcal{G}\mathcal{G}^\top]_{\nu\cdot}.$$

For kernel K-means, where we work in an (implicit) nonlinear feature space $\boldsymbol{\phi}(g)$, we substitute $\mathcal{G}\mathcal{G}^\top \to \mathcal{K}$ to get

$$d(\boldsymbol{\phi}(g(\nu)), k) = [\mathcal{K}]_{\nu\nu} - 2[\mathcal{K}_{\nu\cdot}\mathcal{U}^\top M]_k + [M\mathcal{U}\mathcal{K}\mathcal{U}^\top M]_{kk}, \tag{8.16}$$

where

$\mathcal{K}$ is the kernel matrix

$\mathcal{K}_{\nu\cdot}$ is its νth row

Since the first term in Equation 8.16 does not depend on the class index k, the clustering rule for kernel K-means is to assign observation $g(\nu)$ to class k, where

$$k = \arg\min_k \left([M\mathcal{U}\mathcal{K}\mathcal{U}^\top M]_{kk} - 2[\mathcal{K}_{\nu\cdot}\mathcal{U}^\top M]_k \right). \tag{8.17}$$

An ENVI extension `GPUKKMEANS` for kernel K-means clustering is described in Appendix C.2.15. As in the case of kernel PCA (Chapter 4), memory restrictions require that one work with only a relatively small training sample of m pixel vectors. A portion of the code is shown in Listing 8.1. The variable `Z` in line 161 is a $K \times m$ array of current values of the expression on the right-hand side of Equation 8.17, with one row for each of the training observations and one column for each class label. In line 163 the class labels corresponding to the minimum distance to the mean are extracted into the variable `labels1` (see the definition of the IDL function `MIN()`) and converted to k-values in line 164. This avoids an expensive `FOR`-loop over the training data. The iteration terminates when the class labels cease to change.

Listing 8.1
Excerpt from the program GPUKKMEANS_RUN.PRO.

```
labels = ceil(randomu(seed,m)*K)-1
; iteration
change = 1
iter = 0
ones = fltarr(1,m)+1
WHILE change AND (iter LT 100) DO BEGIN
   progressbar->Update,iter, $
      text='Training:␣iter='+strtrim(iter,2)
   change = 0
   U = fltarr(m,K)
   FOR i=0,m-1 DO U[i,labels[i]] = 1
   MU = diag_matrix(1/(total(U,1)+1))##U
; Z is a K by m array
   Z = ones##diag_matrix(MU##KK##transpose(MU)) $
                    - 2*KK##transpose(MU)
   _ = min(Z,labels1,dimension=1)
   labels1 = labels1 MOD K
   IF total(labels1 NE labels) THEN change=1
   labels=labels1
   IF progressbar->CheckCancel() THEN BEGIN
      PRINT,'interrupted...'
      iter=99
   ENDIF
   iter++
ENDWHILE
```

After convergence, unsupervised classification of the remaining pixels can again be achieved with Equation 8.17, merely replacing the row vector $\mathcal{K}_{\nu.}$ by

$$\big(\kappa(g,g(1)),\kappa(g,g(2))\dots\kappa(g,g(m))\big),$$

where g is a pixel to be classified. The processing of the entire image thus requires evaluation of the kernel for every image pixel with every training pixel. Therefore, it is best to classify the image row-by-row, as was the case for kernel PCA. The ENVI extension makes use of GPULib (Appendix C and Section 4.4.2) if a CUDA-enabled GPU is available. This accelerates the classification step substantially. Figure 8.1a shows an unsupervised classification of the Jülich image of Figure 6.1 with kernel K-means clustering.

8.2.3 Extended K-Means Clustering

Starting this time from the cost function Equation 8.11, denote by $p_k = \Pr(C_k)$ the prior probability for cluster C_k. The entropy H associated with this prior

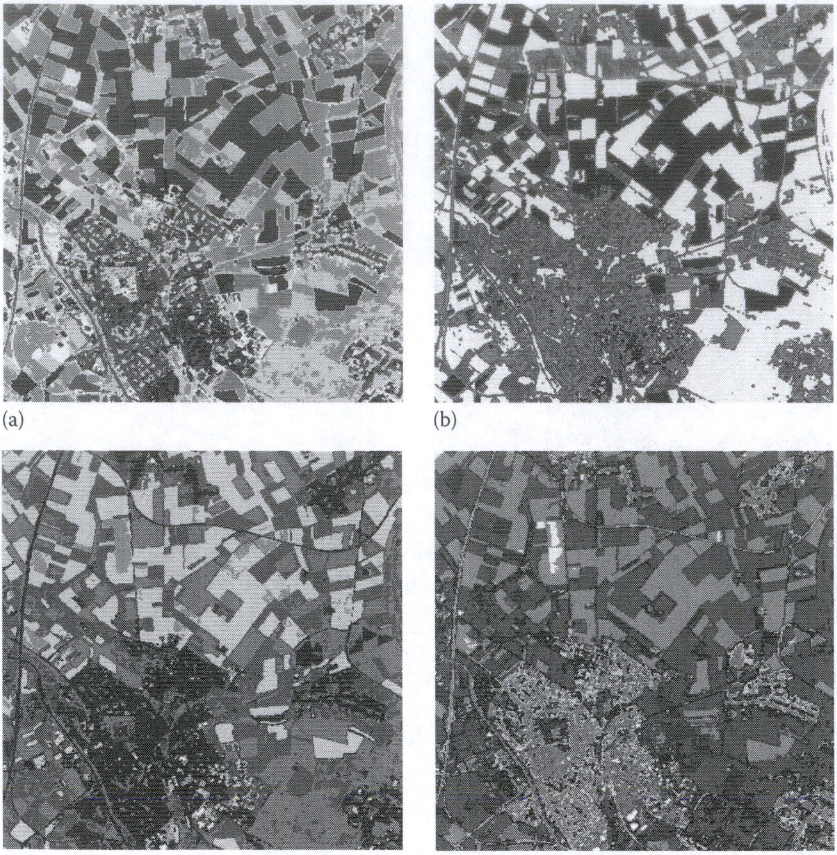

(a) (b) (c) (d)

FIGURE 8.1
(See color insert following page 114.) Unsupervised classification of a spatial subset of the principal components of the fused May 1, 2007 ASTER image over Jülich (Figure 6.1) using eight classes. (a) Kernel K-means clustering on the first four components; (b) extended K-means clustering on the first component with $\tilde{K} = 8$ (see Equation 8.22); (c) agglomerative hierarchical clustering on the first four components; and (d) fuzzy K-means clustering on the first four components.

distribution is given by

$$H = -\sum_{k=1}^{K} p_k \log p_k \tag{8.18}$$

(see Equation 2.89). It was shown in Section 2.7 that distributions with high entropy are those for which the p_k are all similar. So in this case high entropy corresponds to the pixels being distributed evenly over all available clusters, whereas low entropy means that most of the data are concentrated in very few clusters. Following Palubinskas (1998) we choose a prior distribution Pr(C) in the cost function of Equation 8.11 for which few clusters (low entropy) are more probable than many clusters (high entropy), namely

$$\Pr(C) = A\exp(-\alpha_E H) = A\exp\left(\alpha_E \sum_{k=1}^{K} p_k \log p_k\right),$$

where A and α_E are parameters. The cost function can then be written in the form

$$E(C) = \sum_{k=1}^{K}\sum_{\nu=1}^{m} u_{k\nu}\frac{\|\boldsymbol{g}(\nu) - \boldsymbol{m}_k\|^2}{2\sigma^2} - \alpha_E \sum_{k=1}^{K} p_k \log p_k, \tag{8.19}$$

dropping the $\log A$ term, which is independent of the clustering. With

$$p_k = \frac{m_k}{m} = \frac{1}{m}\sum_{\nu=1}^{m} u_{k\nu}, \tag{8.20}$$

Equation 8.19 is equivalent to the cost function

$$E(C) = \sum_{k=1}^{K}\sum_{\nu=1}^{m} u_{k\nu}\left[\frac{\|\boldsymbol{g}(\nu) - \boldsymbol{m}_k\|^2}{2\sigma^2} - \frac{\alpha_E}{m}\log p_k\right]. \tag{8.21}$$

An estimate for the parameter α_E in Equation 8.21 may be obtained as follows (Palubinskas, 1998). With the approximation

$$\sum_{\nu=1}^{m} u_{k\nu}\|\boldsymbol{g}(\nu) - \boldsymbol{m}_k\|^2 \approx m_k\sigma^2 = mp_k\sigma^2,$$

we can write Equation 8.21 as

$$E(C) \approx \sum_{k=1}^{K}\left[\frac{mp_k}{2} - \alpha_E p_k \log p_k\right].$$

Equating the likelihood and prior terms in this expression to give $E(C) = 0$ and taking $p_k \approx 1/\tilde{K}$, where $\tilde{K}$ is some *a priori* expected number of clusters, then gives

$$\alpha_E \approx -\frac{m}{2\log(1/\tilde{K})}. \tag{8.22}$$

The parameter σ^2 in Equation 8.21 (the within-cluster variance) can be estimated from the data; see below.

The *extended K-means* (EKM) algorithm is then as follows. First an initial configuration $\boldsymbol{U}$ with a very large number of clusters K is chosen (for single-band data this might conveniently be the $K \le 256$ gray values that an image with 8-bit quantization can possibly have) and initial values

$$\boldsymbol{m}_k = \frac{1}{m_k}\sum_{\nu=1}^{m} u_{k\nu}\boldsymbol{g}(\nu),\ p_k = \frac{m_k}{m}, \quad k = 1\ldots K, \tag{8.23}$$

are determined. Then the data are reclustered according to the distance measure which minimizes Equation 8.21:

$$k = \arg\min_k \left(\frac{\|\boldsymbol{g}(\nu) - \boldsymbol{m}_k\|^2}{2\sigma^2} - \frac{\alpha_E}{m}\log p_k \right), \quad \nu = 1\ldots m. \tag{8.24}$$

The prior term tends to put more observations into fewer clusters. Any cluster for which, in the course of the iteration, m_k equals zero (or is less than some threshold) is simply dropped from the calculation so that the final number of clusters is determined by the data. The condition that no class be empty (Equation 8.8) is thus relaxed. The explicit choice of the number of clusters K is replaced by the necessity of choosing a value for the "meta-parameter" α_E (or $\tilde{K}$ if Equation 8.22 is used). This has the advantage that one can use a single parameter for a wide variety of images and let the algorithm itself decide on the actual value of K in any given instance. Iteration of Equations 8.23 and 8.24 continues until the cluster labels cease to change.

An implementation of the EKM algorithm in IDL for grayscale, or one-band, images is shown in Listing 8.2. The variance σ^2 is estimated (naively, see Exercise 5) by calculating the variance of the difference of the image with a copy of itself shifted by one pixel, program line 23. An *a priori* number of classes $\tilde{K} = 8$ determines the meta-parameter α_E, line 24. The initial number K of clusters is chosen as the number of nonempty bins in the 256-bin histogram of the linearly stretched image, while the initial cluster means m_k and prior probabilities p_k are the corresponding bin numbers respectively the bin contents divided by the number of pixels, lines 27–30. The iteration of Equations 8.23 and 8.24 is carried out in the `WHILE` loop, lines 37–70. Clusters with prior probabilities less than 0.01 are discarded in line 44. The `REPLICATE` command in lines 54 and 55 avoids the use of a `FOR` loop over the number of pixels. Instead, the array of distances $d(i,k)$ is manipulated directly in the variable `Ds` with IDL's much more efficient array functions. The variable `INDICES` in line 60 locates the pixels, if any, which belong to cluster j for $j = 1\ldots K$. For convenience, the termination condition is that there be no significant change in the cluster means, line 68. A classification result is shown in Figure 8.1b.

Listing 8.2
EKM clustering.

```
PRO ex8_1
; extended K-means

envi_select, title='Choose␣multispectral␣band', $
             fid=fid, dims=dims,pos=pos, /band_only
IF (fid EQ -1) THEN RETURN
num_cols = dims[2]-dims[1]+1
num_rows = dims[4]-dims[3]+1
num_pixels = float(num_cols*num_rows)

; map tie point
map_info = envi_get_map_info(fid=fid)
envi_convert_file_coordinates, fid, $
   dims[1], dims[3], e, n, /to_map
map_info.mc[2:3]= [e,n]

; image band
G = bytscl(envi_get_data(fid=fid,dims=dims,pos=0))
; classification image
labels = G*0

; parameters
sigma2 = stddev(float(G)-shift(G,[1,0]))^2
alphaE = -1/(2*alog(1.0/8))

; initial K, means and priors
hist = histogram(G,nbins=256)
indices = where(hist,K)
means = (findgen(256))[indices]
priors = hist[indices]/num_pixels

; iteration
progressbar = Obj_New('progressbar', Color='blue', $
  title='EKM␣clustering:␣delta...',xsize=250,ysize=20)
progressbar->start
delta = 100.0 & iter = 0
WHILE (delta GT 1.0) AND (iter LE 100) DO BEGIN
   IF progressbar->CheckCancel() THEN BEGIN
      PRINT,'clustering␣aborted'
      progressbar->Destroy & RETURN
   ENDIF
   progressbar->Update,iter,text=strtrim(delta,2)
; drop (almost) empty clusters
   indices = where(priors GT 0.01, K)
; distance array
   ds = fltarr(num_pixels,K)
```

Listing 8.2
EKM clustering (continued).

```
   means = means[indices]
   priors = priors[indices]
   means1 = means
   priors1 = priors
   means = means*0.0
   priors = priors*0.0
   FOR j=0,K-1 DO BEGIN
      ms = replicate(means1[j],num_pixels)
      logps = replicate(alog(priors1[j]),num_pixels)
      ds[*,j] = (G-ms)^2/(2*sigma2) - alphaE*logps
   ENDFOR
   min_ds = min(ds,dimension=2)
   FOR j=0,K-1 DO BEGIN
      indices = where(ds[*,j] EQ min_ds,count)
      IF count GT 0 THEN BEGIN
         nj = n_elements(indices)
         priors[j] = nj/num_pixels
         means[j] = total(G[indices])/nj
         labels[indices] = j+1
      ENDIF
   ENDFOR
   delta = max(abs(means - means1))
   iter++
ENDWHILE
progressbar->Destroy

envi_enter_data, labels, file_type=3, $
  num_classes=K+1, $
  class_names='class␣'+strtrim(indgen(K+1),2), $
  lookup=class_lookup_table(indgen(K+1)), $
  map_info=map_info

END
```

8.2.4 Agglomerative Hierarchical Clustering

The unsupervised classification algorithm that we consider next is, as for ordinary K-means clustering, based on the cost function of Equation 8.12. Let us first write it in a more convenient form:

$$E(C) = \sum_{k=1}^{K} E_k, \tag{8.25}$$

where E_k is given by

$$E_k = \sum_{\nu \in C_k} \|\boldsymbol{g}(\nu) - \boldsymbol{m}_k\|^2. \tag{8.26}$$

The algorithm is initialized by assigning each observation to its own class. At this stage, the cost function is zero, since $C_\nu = \{\nu\}$ and $\boldsymbol{g}(\nu) = \boldsymbol{m}(\nu)$. Every agglomeration of clusters to form a smaller number will increase $E(C)$. We therefore seek a prescription for choosing two clusters for combination that increases $E(C)$ *by the smallest amount possible* (Duda and Hart, 1973).

Suppose that, at some stage of the algorithm, clusters k with m_k members and j with m_j members are merged, where $k < j$, and the new cluster is labeled k. Then the mean of the merged cluster is the weighted average of the original cluster means:

$$\boldsymbol{m}_k \rightarrow \frac{m_k \boldsymbol{m}_k + m_j \boldsymbol{m}_j}{m_k + m_j} = \bar{\boldsymbol{m}}.$$

Thus, after the agglomeration, E_k increases to

$$E_k = \sum_{\nu \in C_k \cup C_j} \|\boldsymbol{g}(\nu) - \bar{\boldsymbol{m}}\|^2$$

and E_j disappears. The net change in $E(C)$ is therefore, after some algebra (Exercise 7),

$$\begin{aligned}\Delta(k,j) &= \sum_{\nu \in C_k \cup C_j} \|\boldsymbol{g}(\nu) - \bar{\boldsymbol{m}}\|^2 - \sum_{\nu \in C_k} \|\boldsymbol{g}(\nu) - \boldsymbol{m}_k\|^2 - \sum_{\nu \in C_j} \|\boldsymbol{g}(\nu) - \boldsymbol{m}_j\|^2 \\ &= \frac{m_k m_j}{m_k + m_j} \cdot \|\boldsymbol{m}_k - \boldsymbol{m}_j\|^2. \end{aligned} \tag{8.27}$$

The minimum increase in $E(C)$ is achieved by combining those two clusters k and j which minimize the above expression. Given two alternative candidate cluster pairs with similar combined memberships $m_k + m_j$ and whose means have similar separations $\|\boldsymbol{m}_k - \boldsymbol{m}_j\|$, this prescription obviously favors combining that pair having the larger discrepancy between m_k and m_j. Thus similar-sized clusters are preserved and smaller clusters are absorbed by larger ones.

Let $\langle k, j\rangle$ represent the cluster formed by combination of the clusters k and j. Then the increase in cost incurred by combining this cluster with some other cluster r can be determined from Equation 8.27 as (Fraley, 1996)

$$\Delta(\langle k,j\rangle, r) = \frac{(m_k + m_r)\Delta(k,r) + (m_j + m_r)\Delta(j,r) - m_r \Delta(k,j)}{m_k + m_j + m_r}. \tag{8.28}$$

To see this, note that with Equation 8.27 the right-hand side of Equation 8.28 is, apart from the factor $1/(m_k + m_j + m_r)$, equivalent to

$$m_k m_r \|\boldsymbol{m}_k - \boldsymbol{m}_r\|^2 + m_j m_r \|\boldsymbol{m}_j - \boldsymbol{m}_r\|^2 - \frac{m_r m_k m_j}{m_k + m_j} \|\boldsymbol{m}_k - \boldsymbol{m}_j\|^2$$

and the left-hand side similarly is equivalent to

$$m_r(m_k + m_j) \left\| \frac{m_k \boldsymbol{m}_k + m_j \boldsymbol{m}_j}{m_k + m_j} - \boldsymbol{m}_r \right\|^2.$$

The identity can be established by comparing coefficients of the vector products. For example, the coefficient of $\|\boldsymbol{m}_k\|^2$ is $m_r m_k^2/(m_k + m_j)$ on both sides.

Once the quantities

$$\Delta(k, j) = \frac{1}{2} \|g(k) - g(j)\|^2$$

have been initialized from Equation 8.27 for all possible combinations of observations, $k = 2 \ldots m$, $j = 1 \ldots k - 1$, the recursive Equation 8.28 can be used to calculate the cost function efficiently for any further merging without reference to the original data. The algorithm terminates when the desired number of clusters has been reached or continues until a single cluster has been formed. Assuming that the data consist of $\tilde{K}$ compact and well-separated clusters, the slope of $E(C)$ vs. the number of clusters K should decrease (become more negative) for $K \leq \tilde{K}$. This provides, at least in principle, a means to decide on the optimal number of clusters.* An ENVI/IDL extension for agglomerative hierarchic clustering is given in Appendix C.2.16. Figure 8.1c shows an example of a classified image.

8.2.5 Fuzzy K-Means Clustering

Introducing the parameter $q > 1$, we can write the cluster means and the sum of squares cost function equivalently as (Dunn, 1973)

$$\boldsymbol{m}_k = \frac{\sum_{i=\nu}^{m} u_{k\nu}^q g(\nu)}{\sum_{\nu=1}^{m} u_{k\nu}^q}, \quad k = 1 \ldots K, \tag{8.29}$$

$$E(C) = \sum_{\nu=1}^{m} \sum_{k=1}^{K} u_{k\nu}^q \|g(\nu) - \boldsymbol{m}_k\|^2. \tag{8.30}$$

* For multispectral imagery, however, well-separated clusters seldom occur.

Since $u_{k\nu} \in \{0, 1\}$, these equations are identical to Equations 8.9 and 8.12. The transition from "hard" to "fuzzy" clustering is effected by label relaxation, that is, by replacing the class memberships by continuous variables

$$0 \le u_{k\nu} \le 1, \quad k = 1 \ldots K, \ \nu = 1 \ldots m, \tag{8.31}$$

but retaining the requirements in Equations 8.7 and 8.8. The parameter q now has an effect and determines the "degree of fuzziness." It is often chosen as $q = 2$. The matrix $\boldsymbol{U}$ is referred to as a *fuzzy class membership* matrix. The classification problem is to find that value of $\boldsymbol{U}$ which minimizes Equation 8.30. Minimization can be achieved by finding values for the $u_{k\nu}$ which solve the minimization problems

$$E_\nu = \sum_{k=1}^{K} u_{k\nu}^q \|\boldsymbol{g}(\nu) - \boldsymbol{m}_k\|^2 \rightarrow \min, \quad \nu = 1 \ldots m,$$

under the constraints imposed by Equation 8.7. Accordingly, we define the Lagrange function which takes the constraints into account,

$$L_\nu = E_\nu - \lambda \left(\sum_{k=1}^{K} u_{k\nu} - 1 \right),$$

and solve the unconstrained problem $L_\nu \rightarrow \min$ by setting the derivative with respect to u_{ki} equal to zero:

$$\frac{\partial L_\nu}{\partial u_{k\nu}} = q(u_{k\nu})^{q-1} \|\boldsymbol{g}_\nu - \boldsymbol{m}_k\|^2 - \lambda = 0, \quad k = 1 \ldots K.$$

This gives an expression for fuzzy memberships in terms of λ,

$$u_{k\nu} = \left(\frac{\lambda}{q} \right)^{1/(q-1)} \left(\frac{1}{\|\boldsymbol{g}_\nu - \boldsymbol{m}_k\|^2} \right)^{1/(q-1)}. \tag{8.32}$$

The Lagrange multiplier λ, in turn, is determined by the constraint

$$1 = \sum_{k=1}^{K} u_{k\nu} = \left(\frac{\lambda}{q} \right)^{1/(q-1)} \sum_{k=1}^{K} \left(\frac{1}{\|\boldsymbol{g}(\nu) - \boldsymbol{m}_k\|^2} \right)^{1/(q-1)}.$$

If we solve for λ and substitute it into Equation 8.32, we obtain

$$u_{k\nu} = \frac{\left(\frac{1}{\|\boldsymbol{g}(\nu) - \boldsymbol{m}_k\|^2} \right)^{1/(q-1)}}{\sum_{k'=1}^{K} \left(\frac{1}{\|\boldsymbol{g}(\nu) - \boldsymbol{m}_{k'}\|^2} \right)^{1/(q-1)}}, \quad k = 1 \ldots K, \ \nu = 1 \ldots m. \tag{8.33}$$

The *fuzzy K-means* (FKM) algorithm consists of a simple iteration of Equations 8.29 and 8.33. The iteration terminates when the cluster centers $\boldsymbol{m}_k$—or alternatively the matrix elements u_{ki}—cease to change significantly. After convergence each pixel vector is labeled with the index of the class with the highest membership probability,

$$\ell_\nu = \arg\max_k u_{k\nu}, \quad \nu = 1 \dots m.$$

This algorithm should give similar results to the K-means algorithm, Section 8.2.1. However, because every pixel "belongs" to some extent to all of the classes, one expects the algorithm to be less likely to become trapped in a local minimum of the cost function. An ENVI/IDL extension for fuzzy K-means clustering is given in Appendix C.2.17, and an example is shown in Figure 8.1d.

We might interpret the fuzzy memberships as "reflecting the relative proportions of each category within the spatially and spectrally integrated multispectral vector of the pixel" (Schowengerdt, 1997). This interpretation will be more plausible when, in the next section, we replace the fuzzy memberships $u_{k\nu}$ by posterior class membership probabilities.

8.3 Gaussian Mixture Clustering

The unsupervised classification algorithms of the preceding section are based on the cost functions of Equation 8.11 or 8.12. These favor the formation of hyperspherical clusters having similar radii, due to the assumption made in Equation 8.4. The use of multivariate Gaussian probability densities to model the classes allows for ellipsoidal clusters of arbitrary extent and is considerably more flexible. Rather than deriving an algorithm from first principles (see Exercise 9), we can obtain one directly from the FKM algorithm. We merely replace Equation 8.33 for the class memberships $u_{k\nu}$ by the posterior probability $\Pr(k \mid \boldsymbol{g}(\nu))$ for class k given the observation $\boldsymbol{g}(\nu)$ (Gath and Geva, 1989). That is, using Bayes' theorem,

$$u_{k\nu} \to \Pr(k \mid \boldsymbol{g}(\nu)) \propto p(\boldsymbol{g}(\nu) \mid k)\Pr(k).$$

Then $p(\boldsymbol{g}(\nu) \mid k)$ is chosen to be a multivariate normal density function. Its estimated mean $\boldsymbol{m}_k$ and covariance matrix $\boldsymbol{C}_k$ are given by Equations 8.9 and 8.10, respectively. Thus, apart from a normalization factor, we have

$$\begin{aligned} u_{k\nu} &= p(\boldsymbol{g}(\nu) \mid k)\Pr(k) \\ &= \frac{1}{\sqrt{|\boldsymbol{C}_k|}} \exp\left[-\frac{1}{2}(\boldsymbol{g}(\nu) - \boldsymbol{m}_k)^\top \boldsymbol{C}_k^{-1}(\boldsymbol{g}(\nu) - \boldsymbol{m}_k)\right] p_k. \end{aligned} \tag{8.34}$$

In the above expression the prior probability Pr(k) has been replaced by

$$p_k = \frac{m_k}{m} = \frac{1}{m}\sum_{\nu=1}^{m} u_{k\nu}. \tag{8.35}$$

The algorithm consists of an iteration of Equations 8.9, 8.10, 8.34, and 8.35 with the same termination condition as for the FKM algorithm. After each iteration the columns of $\boldsymbol{U}$ are normalized according to Equation 8.7. Unlike the FKM classifier, the memberships $u_{k\nu}$ are now functions of the directionally sensitive Mahalanobis distance

$$d = \sqrt{(\boldsymbol{g}(\nu) - \boldsymbol{m}_k)^\top \boldsymbol{C}_k^{-1}(\boldsymbol{g}(\nu) - \boldsymbol{m}_k)}.$$

Because of the exponential dependence of the memberships on d^2 in Equation 8.34, the computation is very sensitive to initialization conditions and can even become unstable. To avoid this problem, one can first obtain initial values for $\boldsymbol{U}$ by preceding the calculation with the FKM algorithm (see Gath and Geva, 1989, who referred to this algorithm as *fuzzy maximum likelihood expectation* [FMLE] clustering). Explicitly, then, the algorithm is as follows:

Algorithm (FMLE Clustering)

1. Determine starting values for the initial memberships $u_{k\nu}$, for example, from the FKM algorithm.
2. Determine the cluster centers $\boldsymbol{m}_k$ with Equation 8.9, the covariance matrices $\boldsymbol{C}_k$ with Equation 8.10, and the priors p_k with Equation 8.35.
3. Calculate with Equation 8.34 the new class membership probabilities $u_{k\nu}$. Normalize the columns of $\boldsymbol{U}$.
4. If $\boldsymbol{U}$ has not changed significantly, stop, else go to 2.

8.3.1 Expectation Maximization

Since the above algorithm has nothing to do with the simple cost functions of Section 8.1, we might ask why it should converge at all. The FMLE algorithm is in fact exactly equivalent to the application of the *expectation maximization* (EM) algorithm (Redner and Walker, 1984) to a Gaussian mixture model of the clustering problem. In that formulation, the probability density $p(\boldsymbol{g})$ for observing a value $\boldsymbol{g}$ in a given clustering configuration is modeled as a superposition of class-specific, multivariate normal probability density functions $p(\boldsymbol{g} \mid k)$ with mixing coefficients p_k:

$$p(\boldsymbol{g}) = \sum_{k=1}^{K} p(\boldsymbol{g} \mid k) p_k. \tag{8.36}$$

This expression has the same form as Equation 2.59, with the mixing coefficients p_k playing the role of the prior probabilities. To obtain the parameters of the model, the likelihood for the data,

$$\prod_{\nu=1}^{m} p(\mathbf{g}(\nu)) = \prod_{\nu=1}^{m}\left[\sum_{k=1}^{K} p(\mathbf{g}(\nu) \mid k)p_k\right], \tag{8.37}$$

is maximized. The maximization takes place in alternating *expectation* and *maximization* steps. In step 3 of the FMLE algorithm, the class membership probabilities $u_{k\nu} = p(k \mid \mathbf{g}(\nu))$ are computed. This corresponds to the expectation step of the EM algorithm in which the terms $p(\mathbf{g} \mid k)p_k$ of the mixture model are recalculated. In step 2 of the FMLE algorithm, the parameters for $p(\mathbf{g} \mid k)$, namely $\boldsymbol{m}_k$ and $\boldsymbol{C}_k$, and the mixture coefficients p_k are estimated for $k = 1 \ldots K$. This corresponds to the EM maximization step (see also Exercise 9).

The likelihood increases on each iteration and therefore the algorithm will indeed converge to a (local) maximum in the likelihood. To demonstrate this, we follow Bishop (2006) and write Equation 8.37 in the more general form

$$p(\mathcal{G} \mid \theta) = \prod_{\nu=1}^{m}\left[\sum_{k=1}^{K} p(\mathbf{g}(\nu) \mid \theta_k)p_k\right], \tag{8.38}$$

where

$\mathcal{G}$ is the observed dataset
θ_k represents the parameters $\mu_k, \boldsymbol{\Sigma}_k$ of the kth Gaussian distribution
θ is the set of all parameters of the model, $\theta = \{\mu_k, \boldsymbol{\Sigma}_k, p_k \mid k = 1 \ldots K\}$

Now suppose that the class labels were known. That is, in the notation of Chapter 6, suppose we were given

$$\ell = \{\boldsymbol{\ell}_1 \ldots \boldsymbol{\ell}_m\}, \quad \boldsymbol{\ell}_\nu = (0, \ldots 1, \ldots 0)^\top,$$

where, if observation ν is in class k, $(\boldsymbol{\ell}_\nu)_k = \ell_{\nu k} = 1$ and the other components are 0. Then it would only be necessary to maximize the likelihood function

$$p(\mathcal{G}, \ell \mid \theta) = \prod_{\nu=1}^{m}\left[\sum_{k=1}^{K} \ell_{\nu k} p(\mathbf{g}(\nu) \mid \theta_k)p_k\right] = \prod_{\nu=1}^{m}\prod_{k=1}^{K} p(\mathbf{g}(\nu) \mid \theta_k)^{\ell_{\nu k}} p_k^{\ell_{\nu k}}, \tag{8.39}$$

which is straightforward, since it can be done separately for each class. That is, taking the logarithm of Equation 8.39, we get the log-likelihood

$$\log p(\mathcal{G}, \ell \mid \theta) = \sum_{\nu}\sum_{k} \log\left[p(\mathbf{g}(\nu) \mid \theta_k)p_k\right]\ell_{\nu k} = \sum_{k}\sum_{\nu \in C_k} \log\left[p(\mathbf{g}(\nu) \mid \theta_k)p_k\right].$$

However, the variables ℓ are unknown; they are referred to as *latent* variables (Bishop, 2006). Let us postulate an unknown mass function $q(\ell)$ for the latent variables. Then the likelihood function $p(\mathcal{G} \mid \theta)$ (Equation 8.38) can be expressed as an average of $p(\mathcal{G}, \ell \mid \theta)$ over $q(\ell)$:

$$p(\mathcal{G} \mid \theta) = \sum_{\ell} p(\mathcal{G} \mid \ell, \theta) q(\ell) = \sum_{\ell} p(\mathcal{G}, \ell \mid \theta). \tag{8.40}$$

The second equality above follows from the definition of conditional probability (Equation 2.53). The log-likelihood may now be separated into two terms,

$$\log p(\mathcal{G} \mid \theta) = \mathcal{L}(q, \theta) + \mathrm{KL}(q, p), \tag{8.41}$$

where

$$\mathcal{L}(q, \theta) = \sum_{\ell} q(\ell) \log \left(\frac{p(\mathcal{G}, \ell \mid \theta)}{q(\ell)} \right) \tag{8.42}$$

and

$$\mathrm{KL}(q, p) = -\sum_{\ell} q(\ell) \log \left(\frac{p(\ell \mid \mathcal{G}, \theta)}{q(\ell)} \right). \tag{8.43}$$

This decomposition can be seen immediately by writing

$$p(\mathcal{G}, \ell \mid \theta) = p(\ell \mid \mathcal{G}, \theta) p(\mathcal{G} \mid \theta),$$

which again follows from the definition of conditional probability, and expanding Equation 8.42:

$$\begin{aligned}\mathcal{L}(q, \theta) &= \sum_{\ell} q(\ell)[\log p(\ell \mid \mathcal{G}, \theta) + \log p(\mathcal{G} \mid \theta) - \log q(\ell)] \\ &= -\mathrm{KL}(q, p) + \sum_{\ell} q(\ell) \log p(\mathcal{G} \mid \theta) = -\mathrm{KL}(q, p) + \log p(\mathcal{G} \mid \theta).\end{aligned}$$

This is just Equation 8.41.

From Section 2.7 we recognize $\mathrm{KL}(q, p)$ as the Kullback–Leibler divergence between $q(\ell)$ and the posterior density $p(\ell \mid \mathcal{G}, \theta)$. Since $\mathrm{KL}(q, p) \geq 0$, with equality only when the two densities are equal, it follows that

$$\log p(\mathcal{G} \mid \theta) \geq \mathcal{L}(q, \theta),$$

that is, $\mathcal{L}(q, \theta)$ is a lower bound for $\log p(\mathcal{G} \mid \theta)$. The EM procedure is then as follows:

- **E-step:** Let the current best set of parameters be θ'. The lower bound on the log-likelihood is maximized by estimating $q(\ell)$ as $p(\ell \mid \mathcal{G}, \theta')$, causing the $\mathrm{KL}(q, p)$ term to vanish and not affecting the log-likelihood since it does not depend on $q(\ell)$ (see Equation 8.40). At this stage $\log p(\mathcal{G} \mid \theta') = \mathcal{L}(q, \theta')$.

- **M-step:** Now $q(\ell)$ is held fixed and $p(\mathcal{G}, \ell \mid \theta)$ is maximized wrt θ (which, as we saw above, is straightforward), giving a new set of parameters θ'', and causing the lower bound $\mathcal{L}(q, \theta'')$ again to increase (unless it is already at a maximum). This will necessarily cause the log-likelihood to increase, since $q(\ell)$ is not the same as $p(\ell \mid \mathcal{G}, \theta'')$, and therefore the Kullback–Leibler divergence will be positive. Now the E-step can be repeated. Iteration of the two steps will therefore always cause the log-likelihood to increase until a maximum is reached.

At any stage of the iteration of the FMLE algorithm, the current estimate for the log-likelihood for the data set can be shown to be given by (Bishop, 1995)

$$\sum_{\nu=1}^{m} \log p(g(\nu)) = \sum_{k=1}^{K} \sum_{\nu=1}^{n} [u_{k\nu}]^o \log u_{k\nu}. \tag{8.44}$$

Here $[u_{k\nu}]^o$ is the "old" value of the posterior class membership probability, that is, the value determined on the previous step.

8.3.2 Simulated Annealing

Notwithstanding the initialization procedure, the FMLE (or EM) algorithm, like all iterative methods, may be trapped in a local optimum. A remedial scheme is to apply a technique called *simulated annealing*. The membership probabilities in the early iterations are given a strong random component and only gradually are the calculated class values allowed to influence the estimation of the class means and covariance matrices. The rate of reduction of randomness may be determined by a temperature parameter T, for example,

$$u_{k\nu} \rightarrow u_{k\nu}(1 - r^{1/T}) \tag{8.45}$$

on each iteration, where $r \in [0, 1)$ is a uniformly distributed random number and where T is initialized to some maximum value T_0 (Hilger, 2001). The temperature T is reduced at each iteration by a factor $c < 1$ according to $T \rightarrow cT$. As T approaches zero, $u - k\nu$ will be determined more and more by the probability distribution parameters in Equation 8.34 alone.

8.3.3 Partition Density

Since the sum of squares cost function $E(C)$ in Equation 8.12 is no longer relevant, we choose with Gath and Geva (1989) the *partition density* as a possible criterion for selecting the best number of clusters. To obtain it, first of all define the *fuzzy hypervolume* as

$$FHV = \sum_{k=1}^{K} \sqrt{|\mathcal{C}_k|}. \tag{8.46}$$

This expression is proportional to the volume in feature space occupied by the ellipsoidal clusters generated by the algorithm. For instance, for a two-dimensional cluster with a normal (elliptical) probability density we have, in its principal axis coordinate system,

$$\sqrt{|\mathcal{C}|} = \sqrt{\begin{vmatrix} \sigma_1^2 & 0 \\ 0 & \sigma_2^2 \end{vmatrix}} = \sigma_1\sigma_2 \approx \text{ Area (volume) of the ellipse.}$$

Next, let s be the sum of all membership probabilities of observations that lie within unit Mahalanobis distance of a cluster center:

$$s = \sum_{\nu\in\mathcal{N}} \sum_{k=1}^{K} u_{\nu k}, \quad \mathcal{N} = \{\nu \mid (\boldsymbol{g}(\nu) - \boldsymbol{m}_k)^\top \mathcal{C}_k^{-1}(\boldsymbol{g}(\nu) - \boldsymbol{m}_k) < 1\}.$$

Then the partition density is defined as

$$P_D = s/FHV. \tag{8.47}$$

Assuming that the data consist of $\tilde{K}$ well-separated clusters of approximately multivariate normally distributed pixels, the partition density should exhibit a maximum at $K = \tilde{K}$.

8.3.4 Implementation Notes

An ENVI/IDL extension for FMLE (or EM) clustering, called `EM_RUN`, is described in Appendix C.2.18. The need for continual reestimation of $\mathcal{C}_k$ is computationally expensive, so the algorithm is not suited for clustering large and/or high-dimensional datasets. The necessity to invert the covariance matrices can also occasionally lead to instability. Listing 8.3 shows an excerpt from the module `EM.PRO` which implements the EM algorithm. In the listing, `NN` is the dimension of the observations and `K` is the number of classes. `MASK` indexes the portion of the image which is to be classified (usually all of it). By means of the keyword `UNFROZEN`, the observations partaking in the clustering process can be additionally specified. In program line 186, only the membership probabilities $u_{j\nu}$ for "unfrozen" observations are recalculated, the remaining observations are "frozen" to their current values (see, e.g., Bruzzone and Prieto, 2000). This will be made use of in Section 8.4.1. The code snippet also reveals methods used to avoid `FOR` loops over

Listing 8.3
Excerpt from the program module EM.PRO.

```
162 ; loop over the cluster index for the case NN>1
163   IF NN GT 1 THEN FOR j=0,K-1 DO BEGIN
164     Ms[j,*] = temporary(Ms[j,*])/ns[j]
165     W = (fltarr(n)+1)##transpose(Ms[j,*])
166     Ds[*,mask] = G[*,mask] - W[*,mask]
167
168 ;   covariance matrix
169     FOR i=0,NN-1 DO W[i,mask] = $
170          sqrt(U[mask,j])*Ds[i,mask]
171     C = matrix_multiply(W[*,mask],W[*,mask], $
172                                   /btranspose)/ns[j]
173     Cs[*,*,j] = C
174     sqrtdetC = sqrt(determ(C,/double))
175     Cinv = invert(C,/double)
176     qf[mask] = total(Ds[*,mask]*(Ds[*,mask]##Cinv),1)
177
178 ;   class hypervolume and partition density
179     fhv[j] = sqrtdetC
180     indices = where(qf LT 1.0, count)
181     IF (count GT 0) THEN BEGIN
182        pdens[j] = total(U[indices,j])/fhv[j]
183     ENDIF
184
185 ;   new memberships
186     U[unfrozen,j] = $
187           exp(-qf[unfrozen]/2.0)*(Ps[j]/sqrtdetC)
188
189 ;   random membership for annealing
190     IF T GT 0.0 THEN BEGIN
191       Ur=1.0-randomu(seed,n_elements(unfrozen))^(1.0/T)
192       U[unfrozen,j] = temporary(U[unfrozen,j])*Ur
193     ENDIF
194   ENDFOR $
```

the pixel index i. For instance, after line 166 the variable `Ds` is an array of distance vectors $(\boldsymbol{g}(\nu) - \boldsymbol{m}_j)^\top$, $\nu = 1 \ldots m$, from the mean of cluster j. In lines 169–173 the covariance matrix $\boldsymbol{C}_j$ (the IDL array `Cs[*,*,j]`) is then calculated using array operations involving `Ds` without a loop over the pixel index ν. A similar technique is used in line 176 to calculate the quadratic forms $(\boldsymbol{g}(\nu) - \boldsymbol{m}_j)^\top \boldsymbol{C}_j^{-1}(\boldsymbol{g}(\nu) - \boldsymbol{m}_j)$, $\nu = 1 \ldots m$.

Since the cluster memberships are multivariate probability densities, the program optionally generates a rule image (see Chapters 6 and 7). This can be postprocessed, for example, with the probabilistic label relaxation filter discussed in Section 7.1.2.

8.4 Including Spatial Information

All of the clustering algorithms described so far make use exclusively of the spectral properties of the individual observations (pixel vectors). Spatial relationships within an image such as scale, coherent regions, or textures are not taken into account. In the following we look at two refinements which deal with spatial context. They will be illustrated in connection with FMLE classification, but could equally well be applied to other clustering approaches.

8.4.1 Multiresolution Clustering

Prior to performing FMLE clustering, an image pyramid for the chosen scene may be created using, for example, the discrete wavelet transform (DWT) filter bank described in Chapter 4. Clustering will then begin at the coarsest resolution and proceed to finer resolutions, with the membership probabilities passed on to each successive scale (Hilger, 2001). In our implementation, the membership probabilities from preceding scales which, after upsampling to the next scale exceed some threshold (e.g., 0.9), are frozen in the manner discussed in Section 8.3.4. The depth of the pyramid can be chosen by the user.

The relevant program code is shown in Listing 8.4. The initial clustering carried out at the lowest resolution (the call to the procedure `EM` in lines 179–183) is gradually refined, whereby pixel vectors determined to have sufficiently high membership probabilities (>0.9, line 213) at the lower resolutions are not reclassified. In line 187 the clusters are sorted according to decreasing partition density.* Apart from improving the rate of convergence of the calculation, scaling has the intended effect that the "spatial awareness" extracted at coarse resolution is passed up to the finer scales.

8.4.2 Spatial Clustering

As described in Chapter 4, class labels for multispectral images can be represented by realizations of a Markov random field, for which the label of a given pixel may be influenced only by the class labels of other pixels in its immediate neighborhood. According to Gibbs–Markov equivalence, Theorem 4.3, the probability density for any complete labeling, ℓ, of the image is given by

$$p(\ell) = \frac{1}{Z}\exp(-\beta U(\ell)), \tag{8.48}$$

* This will be particularly useful in Chapter 9 when the algorithm is used to cluster multispectral change images, since the no-change pixels usually form the most dense cluster.

Listing 8.4
Excerpt from the program module EM_RUN.PRO.

```
; initial clustering of maximally compressed image
window,11,xsize=600,ysize=400,title='Log␣Likelihood'
EM,  s_image, U, Ms, Ps, Cs, T0=T0, wnd=11, $
     pb_msg='EM:␣Pass␣1␣of␣'+ $
     strtrim(max_scale-min_scale+1,2), $
     num_cols=s_num_cols, num_rows=s_num_rows, $
     beta=beta, mask=mask, pdens=pdens,status=status
IF status THEN RETURN

; sort clusters wrt partition density
U = U[*,reverse(sort(pdens))]

; clustering at increasing scales
FOR scale=1,max_scale-min_scale DO BEGIN
; upsample class memberships (bilinear interpolation)
   U = reform(U,s_num_cols,s_num_rows,K,/overwrite)
   s_num_cols = s_num_cols*2
   s_num_rows = s_num_rows*2
   s_num_pixels = s_num_cols*s_num_rows
   U = shift(rebin(U,s_num_cols,s_num_rows,K),1,1,0)
   U = reform(U,s_num_pixels,K,/overwrite)

; expand the image and mask
   s_image = fltarr(num_bands,s_num_pixels)
   FOR i=0,num_bands-1 DO BEGIN
      DWTBands[i]->expand
      s_image[i,*] = DWTbands[i]-> get_quadrant(0)
   ENDFOR
   IF (m_fid NE -1) THEN BEGIN
      DWTmask -> expand
      mask = where(floor(DWTmask -> get_quadrant(0)> 0) $
            GT 0,count)
      IF count EQ 0 THEN mask=lindgen(s_num_pixels)
   END ELSE mask=lindgen(s_num_pixels)

; cluster
   unfrozen = where(max(U,dimension=2) LE 0.90, count)
   IF count GT 0 THEN $
   EM, s_image, U, Ms, Ps, Cs, miniter=5, wnd=11, $
    unfrozen=unfrozen, T0=0.0, pb_msg= $
    'EM:␣Pass␣'+strtrim(scale+1,2)+'␣of␣'+ $
    strtrim(max_scale-min_scale+1,2),$
    num_cols=s_num_cols, num_rows=s_num_rows, $
    beta=beta, mask=mask, status=status
   IF status THEN RETURN
ENDFOR
```

where Z is a normalization and the energy function, $U(\ell)$, is given by a sum over clique potentials,

$$U(\ell) = \sum_{c \in \mathcal{C}} V_c(\ell), \tag{8.49}$$

relative to a neighborhood system, $\mathcal{N}$. If we restrict discussion to 4-neighborhoods, the only possible cliques are singletons and pairs of vertically or horizontally adjacent pixels (see Figure 4.14a). If, furthermore, the potential for singleton cliques is set to zero and the random field is isotropic (independent of clique orientation), then we can write Equation 8.49 in the form

$$U(\ell) = \sum_{\nu \in \mathcal{I}} \sum_{\nu' \in \mathcal{N}_\nu} V_2(\ell_\nu, \ell_{\nu'}), \tag{8.50}$$

where

$\mathcal{I}$ is the complete image lattice

$\mathcal{N}_\nu$ is the neighborhood of pixel ν

$V_2(\ell_\nu, \ell_{\nu'})$ is the clique potential for two neighboring sites

Let us now choose

$$V_2(\ell_\nu, \ell_{\nu'}) = \frac{1}{4}(1 - u_{\ell_\nu \nu'}), \tag{8.51}$$

where $u_{\ell_\nu \nu'}$ is an element of the cluster membership probability matrix $\boldsymbol{U}$. This says that, when the probability $u_{\ell_\nu \nu'}$ is large that neighboring site ν' has the same label ℓ_ν as site ν, then the clique potential is small. Combining Equations 8.50 and 8.51,

$$U(\ell) = \sum_{\nu \in \mathcal{I}} \frac{1}{4}\left(4 - \sum_{\nu' \in \mathcal{N}_\nu} u_{\ell_\nu \nu'}\right) = \sum_{\nu \in \mathcal{I}} (1 - u_{\ell_\nu \mathcal{N}_\nu}), \tag{8.52}$$

where $u_{\ell_\nu \mathcal{N}_\nu}$ is the averaged membership probability for ℓ_ν within the neighborhood,

$$u_{\ell_\nu \mathcal{N}_\nu} = \frac{1}{4} \sum_{\nu' \in \mathcal{N}_\nu} u_{\ell_\nu \nu'}\ . \tag{8.53}$$

Substituting Equation 8.52 into Equation 8.48, we obtain

$$p(\ell) = \frac{1}{Z} \prod_{\nu \in \mathcal{I}} \exp(-\beta(1 - u_{\ell_\nu \mathcal{N}_\nu})). \tag{8.54}$$

Equation 8.54 is reminiscent of the likelihood functions that we have been using for pixel-based clustering and suggests the following heuristic ansatz

(Hilger, 2001): Along with the *spectral* class membership probabilities $u_{k\nu}$ that characterize the FMLE algorithm and which are given by Equation 8.34, introduce a *spatial* class membership probability $v_{k\nu}$,

$$v_{k\nu} \propto \exp(-\beta(1 - u_{k\mathcal{N}_\nu})). \tag{8.55}$$

A combined *spectral–spatial* class membership probability for the νth observation is then determined by replacing $u_{k\nu}$ with

$$\frac{u_{k\nu} v_{k\nu}}{\sum_{k'=1}^{K} u_{k'\nu} v_{k'\nu}}, \tag{8.56}$$

apart from which the algorithm proceeds as before. Note that, from Equations 8.34 and 8.55, the combined membership probability above is now proportional to

$$\frac{1}{\sqrt{|\mathcal{C}_k|}} \cdot \frac{m_k}{m} \cdot \exp\left(-\frac{1}{2}(g(\nu) - \boldsymbol{m}_k)^\top \mathcal{C}_k^{-1}(g(\nu) - \boldsymbol{m}_k) - \beta(1 - u_{k\mathcal{N}_\nu})\right).$$

This way of folding spatial information with the spectral similarity measure (here the Mahalanobis distance) is referred to in Tran et al. (2005) as the *addition form*.

The quantity $u_{k\mathcal{N}_\nu}$ appearing in Equation 8.55 can be determined for an entire image simply by convolving the two-dimensional kernel

$$\begin{matrix} 0 & \frac{1}{4} & 0 \\ \frac{1}{4} & 0 & \frac{1}{4} \\ 0 & \frac{1}{4} & 0 \end{matrix}$$

with the array $\boldsymbol{U}$, after reforming the latter to the correct image dimensions (see Listing 8.5, lines 220–224). (The kernel is stored in the variable `Nb`.)

Figure 8.2 shows an example, again using the Jülich ASTER image. The classification was determined with an image pyramid of depth 1 (i.e., two levels) and with $\beta = 1.0$.

8.5 Benchmark

Together with the built-in ENVI algorithms K-means and ISODATA, we now have no less than seven clustering methods at our disposal. So which one should we use? The degree of success of unsupervised image classification is notoriously difficult to quantify (see Duda and Canty, 2002). This is because there is, by definition, no prior information with regard to what one might expect to be a "reasonable" result. Ultimately, judgement is qualitative, almost a question of esthetics.

Listing 8.5
Excerpt from the program module EM.PRO.

```
; determine spatial membership matrix
  IF beta GT 0 THEN BEGIN
    FOR j=0,K-1 DO BEGIN
      U_N=1.0-convol(reform(U[*,j],num_cols, $
                              num_rows),Nb)
      V[*,j] = exp(-beta*U_N)
    ENDFOR
;  combine spectral/spatial for unfrozen
    U[unfrozen,*]=temporary(U[unfrozen,*])* $
                              V[unfrozen,*]
  ENDIF

; normalize all
  a = 1/total(U,2)
  void=where(finite(a),complement=complement, $
    ncomplement=ncomplement)
  IF ncomplement GT 0 THEN a[complement]=0.0
  FOR j=0,K-1 DO U[*,j]=temporary(U[*,j])*a

; log likelihood
  indices=where(U,count)
  IF count GT 0 THEN LL[iter] =  $
       total(Uold[indices]*alog(U[indices]))
```

Nevertheless, in order to compare the algorithms we have looked at so far more objectively, we will generate a "toy" benchmark image with the code in Listing 8.6. The image is shown in the upper left-hand corner of Figure 8.3. It consists of three "spectral bands" and three classes, namely the dark background, the four points of the cross, and the cross center. The cluster sizes differ considerably (approximately in the ratio 12:4:1) with a large overlap, and the bands are strongly correlated. The program in Listing 3.5 estimates the image noise covariance matrix as

```
ENVI> EX3_4
Noise covariance matrix, file [Memory1] (800x800x3)
       1.0041422       1.0038484        0.5020555
       1.0038484       2.0039095        1.0036373
       0.5020555       1.0036373        1.5035802
```

Some results are shown in Figure 8.3. Neither the K-means variants nor the Gaussian mixture algorithm without scaling identify all three classes. Agglomerative hierarchical clustering succeeds reasonably well. The "best" classification is obtained with the Gaussian mixture model when both multiresolution and spatial clustering are employed. The success of Gaussian

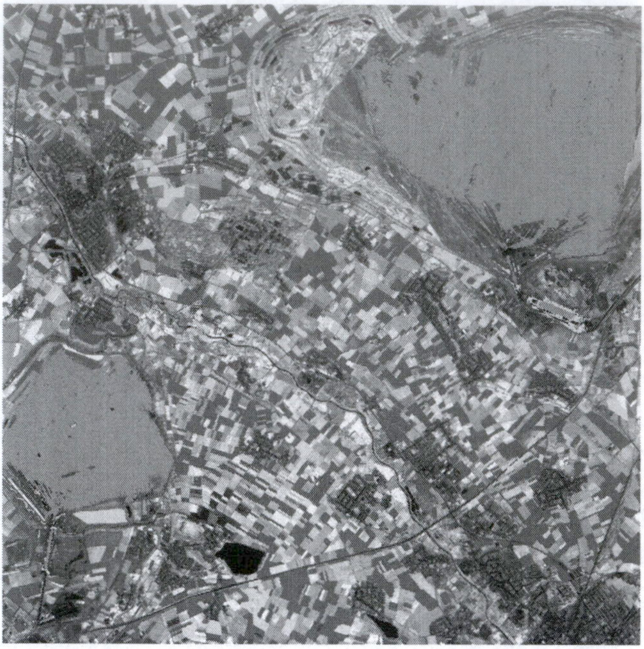

FIGURE 8.2
(See color insert following page 114.) Gaussian mixture clustering of the first four principal components of the Jülich ASTER scene, eight clusters. Three clusters associated with new sugarbeet plantation (maroon), rapeseed (yellow), and open cast coal mining (coral) are shown superimposed on VNIR spectral band 3N (gray).

Listing 8.6

Generating a toy image.

```
PRO toy_image
 image = fltarr (800 ,800 ,3)
 b = 2.0
 image [99:699 ,299:499 ,*] = b
 image [299:499 ,99:699 ,*] = b
 image [299:499 ,299:499 ,*] = 2*b
 image = smooth (image ,[13 ,13 ,1] ,/ edge_truncate )
 n1 = randomu (seed ,800 ,800 ,/ normal )
 n2 = randomu (seed ,800 ,800 ,/ normal )
 n3 = randomu (seed ,800 ,800 ,/ normal )
 image [* ,* ,0] = image [* ,* ,0] + n1
 image [* ,* ,1] = image [* ,* ,1] + n2 + n1
 image [* ,* ,2] = image [* ,* ,2] + n3 + n1 /2 + n2 /2
 envi_enter_data , image
END
```

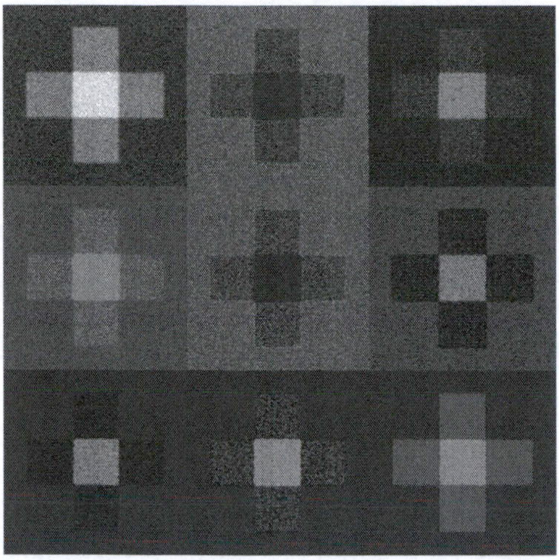

FIGURE 8.3
(See color insert following page 114.) Unsupervised classification of a toy image. Row-wise, left to right, top to bottom: The toy image, K-means, kernel K-means, EKM (on first principal component), FKM, agglomerative hierarchical, Gaussian mixture with depth 0 and $\beta = 1.0$, Gaussian mixture with depth 2 and $\beta = 0.0$, and Gaussian mixture with depth 2 and $\beta = 1.0$.

mixture classification is not particulary surprising since the toy image classes are multivariate normally distributed. However, if the other methods exhibit inferior performance on normal distributions, then they might be expected to be similarly inferior when presented with real, non-Gaussian data.

8.6 Kohonen Self-Organizing Map

The *Kohonen self-organizing map* (SOM), a simple example of which is sketched in Figure 8.4 , belongs to a class of neural networks that are trained by *competitive learning* (Kohonen, 1989; Hertz et al., 1991). It is very useful as a visualization tool for exploring the class structure of multispectral imagery. The layer of neurons shown in the figure can have any geometry, but usually a one-, two-, or three-dimensional array is chosen. The input signal is the observation vector $\boldsymbol{g} = (g_1, g_2 \dots g_N)^\top$, where, in the figure, $N = 2$. With each input to a neuron is associated a *synaptic weight* so that, for K neurons, the synaptic weights can be represented as an $(N \times K)$ matrix

$$w = \begin{pmatrix} w_{11} & w_{12} & \cdots & w_{1K} \\ w_{21} & w_{22} & \cdots & w_{2K} \\ \vdots & \vdots & \vdots & \vdots \\ w_{N1} & w_{N2} & \cdots & w_{NK} \end{pmatrix}. \tag{8.57}$$

The components of the vector $\boldsymbol{w}_k = (w_{1k}, w_{2k} \dots w_{Nk})^\top$ are the synaptic weights of the kth neuron. The set of observations $\{\boldsymbol{g}(\nu) \mid \nu = 1 \dots m\}$, where $m > K$, comprise the training data for the network. The synaptic weight vectors are to be adjusted so as to reflect in some way the class structure of the training data in an N-dimensional feature space.

The training procedure is as follows. First of all the neurons' weights are initialized from a random subset of K training observations, setting

$$w_k = g(k), \quad k = 1 \ldots K.$$

Then, when a new training vector, $g(\nu)$, is presented to the input of the network, the neuron whose weight vector, w_k, lies nearest to $g(\nu)$ is designated to be the "winner." Distances are given by $\|g(\nu) - w_k\|$. Suppose that the winner's index is k^*. Its weight vector is then shifted a small amount in the direction of the training vector:

$$w_{k^*}(\nu + 1) = w_{k^*}(\nu) + \eta(g(\nu) - w_{k^*}(\nu)), \quad (8.58)$$

where $w_{k^*}(\nu + 1)$ is the weight vector after presentation of the νth training vector (see Figure 8.5). The parameter η is called the *learning rate* of the network. The intention is to repeat this learning procedure until the synaptic weight vectors reflect the class structure of the training data, thereby achieving what is often referred to as a *vector quantization* of the feature space (Hertz et al., 1991). In order for this method to converge, it is necessary to allow the learning rate to decrease gradually during the training process. A convenient function for this is

$$\eta(\nu) = \eta_{max} \left(\frac{\eta_{min}}{\eta_{max}} \right)^{\nu/m}.$$

However, the SOM algorithm goes a step further and attempts to map the *topology* of the feature space onto the network as well. This is achieved by defining a neighborhood function for the winner neuron on the network of neurons. Usually a Gaussian of the form

$$\mathcal{N}(k^*, k) = \exp(-d^2(k^*, k)/2\sigma^2)$$

is chosen, where $d^2(k^*, k)$ is the square of the distance between neurons k^* and k in the network array. For example, for a two-dimensional array of

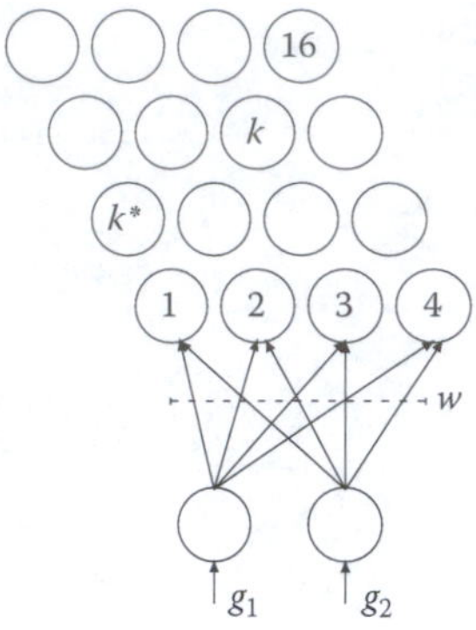

FIGURE 8.4
The Kohonen SOM in two dimensions with a two-dimensional input. The inputs are connected to all 16 neurons (only the first four connections are shown).

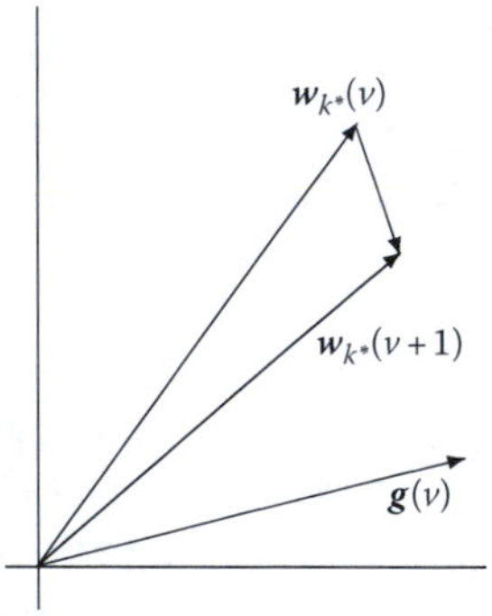

FIGURE 8.5
Movement of a synaptic weight vector in the direction of a training vector.

$n \times n$ neurons $d^2(k^*, k)$ can by calculated in integer arithmetic as

$$d^2(k^*, k) = [(k^* - 1) \bmod n - (k - 1) \bmod n]^2 + [(k^* - 1)/n - (k - 1)/n]^2. \quad (8.59)$$

During the learning phase not only the winner neuron, but also the neurons in its neighborhood are moved in the direction of the training vectors by an amount proportional to the value of the neighborhood function:

$$w_k(\nu + 1) = w_k(\nu) + \eta(\nu)\mathcal{N}(k^*, k)(g(\nu) - w_k(\nu)), \quad k = 1 \ldots K. \quad (8.60)$$

Finally, the extent of the neighborhood is allowed to shrink steadily as well:

$$\sigma(\nu) = \sigma_{max}\left(\frac{\sigma_{min}}{\sigma_{max}}\right)^{\nu/m}.$$

Typically, $\sigma_{max} \approx n/2$ and $\sigma_{min} \approx 1/2$. The neighborhood is initially the entire network, but toward the end of training it becomes very localized.

For clustering of multispectral satellite imagery a cubic network geometry is useful (Groß and Seibert, 1993). After training on some representative sample of pixel vectors, the entire image is classified by associating each pixel vector with the neuron having the closest synaptic weight vector. Then the pixel is colored by mapping the position of that neuron in the cube to coordinates in RGB color space. Thus, the N-dimensional feature space is "projected" onto the three-dimensional RGB color cube and pixels that are close together in feature space are given similar colors. An ENVI/IDL extension for the Kohonen SOM is given in Appendix C.2.19 and an example is shown in Figure 8.6. The positions and colors of the neurons in feature space after completion of training are shown in Figure 8.7.

8.7 Image Segmentation

The term *image segmentation* refers, in its broadest sense, to the process of partitioning a digital image into multiple regions. Certainly all of the clustering algorithms that we have considered till now fall within this category. However, in the present section, we will use the term in a more restrictive way, namely as referring to the partitioning of pixels not only by spectral similarity (essentially what has been discussed so far) but also by spatial proximity. Segmentation plays a major role in low-level computer vision, and in autonomous image processing in general. Popular methods include the use of edge detectors, watershed segmentation, region growing algorithms, and morphology, see Gonzalez and Woods (2002, Chapter 10), for a good overview.

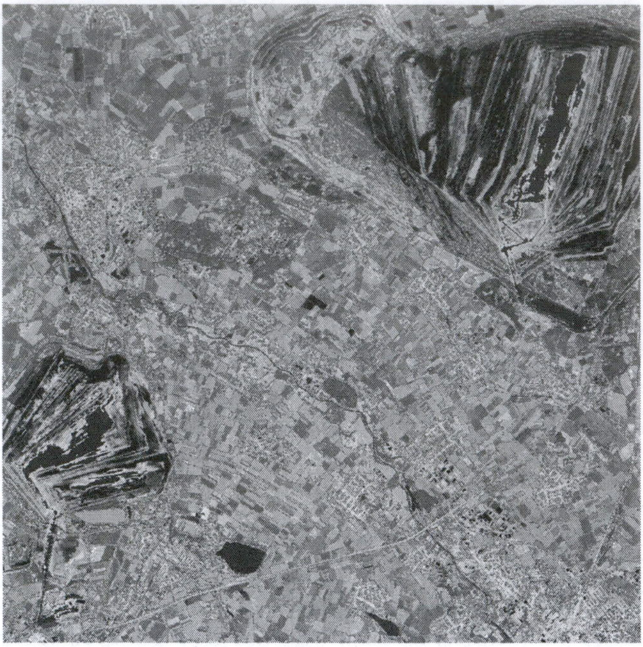

FIGURE 8.6
(See color insert following page 114.) Kohonen SOM of the nine VNIR and sharpened SWIR spectral bands of the Jülich ASTER scene. The network is a cube having dimensions 6 × 6 × 6.

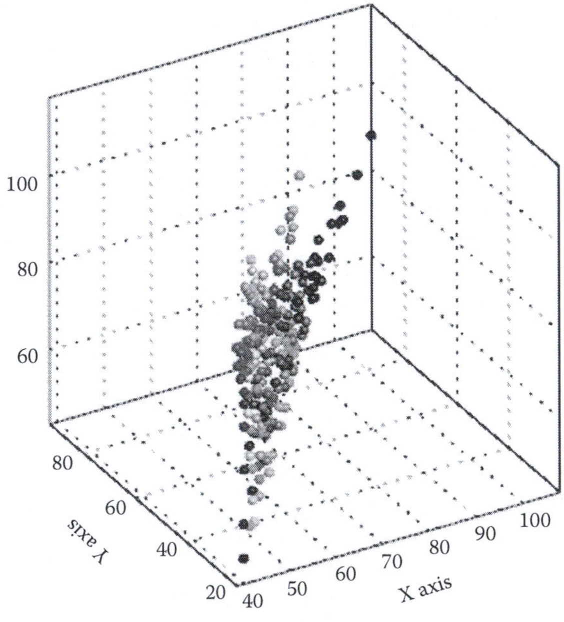

FIGURE 8.7
(See color insert following page 114.) The $6^3 = 216$ SOM neurons in the feature space of the three VNIR spectral bands of the Jülich scene (see Figure 8.6).

Listing 8.7
Excerpt from the program SEGMENT_CLASS_RUN.PRO.

```
115 ; segment first class
116 limg = label_region(climg EQ cl[0],/ULONG, $
117         all_neighbors=all_neighbors)
118 ; segment others, if any
119 FOR i = 1, K - 1 DO BEGIN
120   mxlb   = max(limg)
121   climgi = climg EQ cl[i]
122   limg   = temporary(limg) + $
123         label_region(climgi, /ULONG, $
124          all_neighbors=all_beighbors) $
125         + climgi*mxlb
126 ENDFOR
127 ; re-label segments larger than min_size
128 indices = where(histogram(limg,reverse_indices=r) $
129                            GE min_size,count)
130 limg=limg*0
131 IF count GT 0 THEN FOR i=1L,count-1 DO BEGIN
132    j = indices[i]         ;label of segment
133    p = r[r[j]:r[j+1]-1] ;pixels in segment
134    limg[p]=i             ;new label
135 ENDFOR
```

8.7.1 Segmenting a Classified Image

The ENVI menu command

`Classification/Post Classification/Segmentation Image`

provides a simple mechanism to convert a classification image into a segmented image by means of *blobbing*, that is, finding connected regions. Unfortunately this command does not respect class labels. The program SEGMENT_CLASS_RUN, that is described in Appendix C.2.20 and an excerpt of which is shown in Listing 8.7, may be used as a replacement. In lines 116–126 of the listing the IDL function LABEL_REGION() is used to segment each selected class successively. Connected regions smaller than the user-defined value MIN_SIZE are then eliminated in lines 128–135. Here use is made of the powerful, if somewhat intimidating,* HISTOGRAM() function.

Figure 8.8 shows an example. A single cluster associated with a mixture of green vegetation (mixed forest, winter wheat, and grassland) has been segmented into about 350 connected regions. Note that the final segmentation has been achieved in two distinct steps: spectral clustering (in this case

* See the highly recommended and, for insiders, notorious *HISTOGRAM: The Breathless Horror and Disgust* by J. D. Smith, http://www.dfanning.com/tips/histogram_tutorial.html

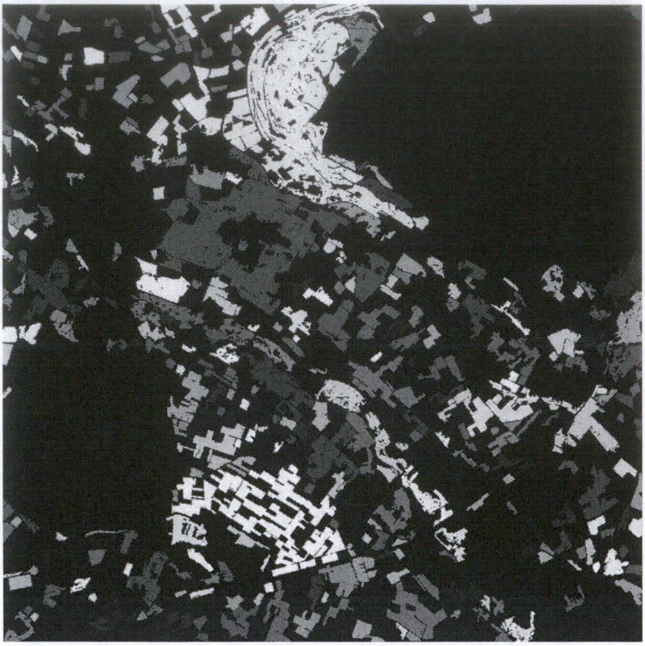

FIGURE 8.8
Postclassification segmentation of the Gaussian mixture cluster associated with green vegetation in the Jülich ASTER scene. Each connected region has been assigned a random label for display purposes.

with the FMLE or EM algorithm) followed by identification of connected regions. In Section 8.7.3, these steps are merged into a single procedure.

8.7.2 Object-Based Classification

Segmentation of a classification image, or segmentation by any other means, offers the possibility of a refinement of classification on the basis of characteristics of the individual segments not necessarily related to pixel intensities, such as shape, compactness, proximity to other segments, etc. One speaks generally of *object-based* classification. Probably the best-known commercial implementation of object-based classification is the Definiens AG software suite (see Benz et al., 2004). In the simple example shown in Listing 8.8, the segments in Figure 8.8 are clustered into three classes on the basis of their Hu invariant moments (see Section 5.2.1). After reading in a segmented image (line 10) a table of the logarithms of the Hu moments is constructed (lines 14–26). Then the agglomerative hierarchical clustering algorithm of Section 8.2.4 is called on the standardized (zero mean, unit variance) logarithms (line 28). The result is shown in Figure 8.9. The three clusters are seen to vary from

Listing 8.8
Clustering segments on the basis of their invariant moments.

```
PRO Ex8_2
; number of clusters
   K = 3
; select segment file
   envi_select,fid=fid,pos=pos,dims=dims, $
     /no_dims,/no_spec,/band_only, $
     title='Select␣segment␣file'
   IF fid EQ -1 THEN RETURN
   num_cols = dims[2]-dims[1]+1
   seg_im = envi_get_data(fid=fid,dims=dims,pos=pos)
   num_segs = max(seg_im)
; calculate table of invariant moments
   moments = fltarr(7,num_segs)
   FOR i=0L, num_segs-1 DO BEGIN
      indices = where(seg_im EQ i+1,count)
      IF count GT 0 THEN BEGIN
         X = indices MOD num_cols
         Y = indices/num_cols
         A = seg_im[min(X):max(X)+1,min(Y):max(Y)+1]
         indices1=where(A EQ i+1, complement=comp,$
                                  ncomplement=ncomp)
         A[indices1]=1.0
         IF ncomp GT 0 THEN A[comp]=0.0
         moments[*,i] = hu_moments(A,/log)
      ENDIF
   ENDFOR
; agglomerative hierarichic clustering
   HCL,standardize(moments),K,Ls
; re-label segments with cluster label
   _ = histogram(seg_im,reverse_indices=r)
   FOR i=1L,num_segs DO BEGIN
       p = r[r[i]:r[i+1]-1] ;pixels in segment
       seg_im[p]=Ls[i-1]+1  ;new label
   ENDFOR
; output to memory
   envi_enter_data, seg_im, $
      file_type=3,num_classes=K+1, $
      class_names=string(indgen(K+1)), $
      lookup=class_lookup_table(indgen(K+1))
END
```

compact and nearly square-shaped (green) to extended and multiply connected (red) with the white segment cluster intermediate between the two. The segments are plotted in the feature space of the standardized logarithms of the first three invariant moments in Figure 8.10.

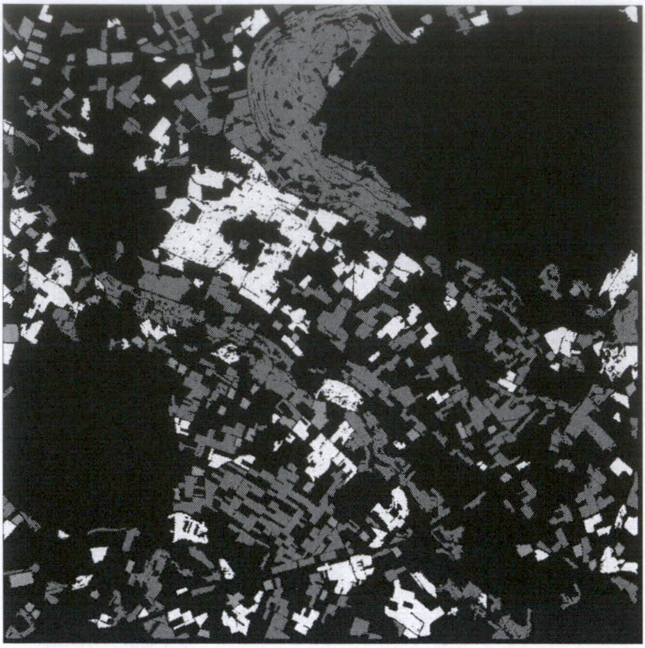

FIGURE 8.9
(See color insert following page 114.) Classification of the segments in Figure 8.8 on the basis of their invariant moments.

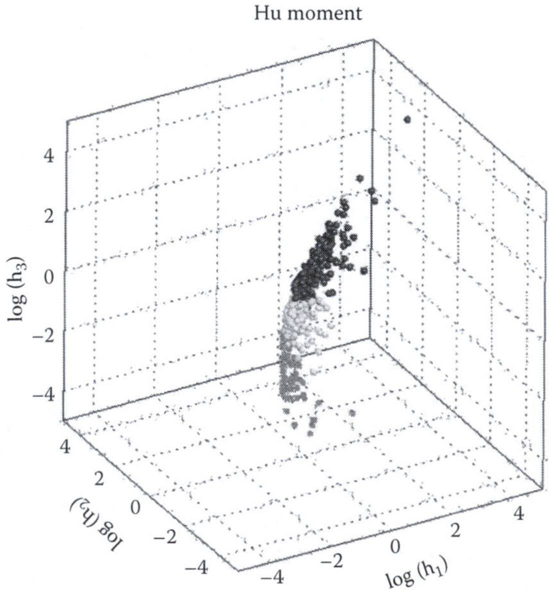

FIGURE 8.10
(See color insert following page 114.) Standardized logarithms of the first three Hu moments of the classified segments of Figure 8.9.

8.7.3 Mean Shift

We conclude the present chapter with the description of an especially popular, nonparametric segmentation algorithm called the *mean shift* (Fukunaga and Hostetler, 1975; Comaniciu and Meer, 2002). The algorithm is nonparametric in the sense that no assumptions are made regarding the probability density of the observations being clustered. The mean shift partitions pixels in an N-dimensional feature space by associating each pixel with a local maximum in the estimated probability density, called a *mode*. For each pixel, the associated mode is determined by defining a (hyper-)sphere of radius $r_{spectral}$ centered at the pixel and calculating the mean of the pixels that lie within the sphere. Then the sphere's center is shifted to that mean. This continues until convergence, that is, until the mean shift is less than some threshold. At each iteration, the sphere moves to a region of higher probability density until a mode is reached. The pixel is assigned that mode.

It will be apparent from the above that mean shift clustering *per se* will not lead to image segmentation in the restricted sense that we are discussing here. However, simply by extending the feature space to include the spatial position of the pixels, an elegant segmentation algorithm emerges. We only need distinguish, additionally to $r_{spectral}$, a spatial radius $r_{spatial}$ for determining the mean shift. After an appropriate normalization of the spectral and spatial distances, for example by rescaling the pixel intensities according to

$$g(\nu) \rightarrow g(\nu)\frac{r_{spatial}}{r_{spectral}}, \quad \nu = 1 \dots m,$$

the mean shift procedure is carried out in the concatenated, $(N+2)$-dimensional, spectral–spatial feature space using a hyper-sphere of radius $r = r_{spatial}$.

Specifically, let $\boldsymbol{y}_\nu = \begin{pmatrix} g(\nu) \\ \boldsymbol{x}_\nu \end{pmatrix}$, $\nu = 1 \dots m$, represent the image pixels in the combined feature space, let $\boldsymbol{z}_\nu$, $\nu = 1 \dots m$, denote the mean shift filtered pixels after the segmentation has concluded, and define $S(\boldsymbol{w})$ as the set of pixels within a radius r of a point $\boldsymbol{w}$ (cardinality $|S(\boldsymbol{w})|$). Then the algorithm is as follows:

Algorithm (Mean Shift Segmentation)

For each $\nu = 1 \dots m$

1. Set $k = 1$ and $\boldsymbol{w}_k = \boldsymbol{y}_\nu$.
2. Repeat:
   $\boldsymbol{w}_{k+1} = \frac{1}{|S(\boldsymbol{w}_k)|}\sum_{\boldsymbol{y}_\nu \in S(\boldsymbol{w}_k)} \boldsymbol{y}_\nu$
   $k = k + 1$
   until convergence to $\boldsymbol{w} = \begin{pmatrix} \boldsymbol{f} \\ \boldsymbol{x} \end{pmatrix}$.
3. Assign $\boldsymbol{z}_\nu = \begin{pmatrix} \boldsymbol{f} \\ \boldsymbol{x}_\nu \end{pmatrix}$.

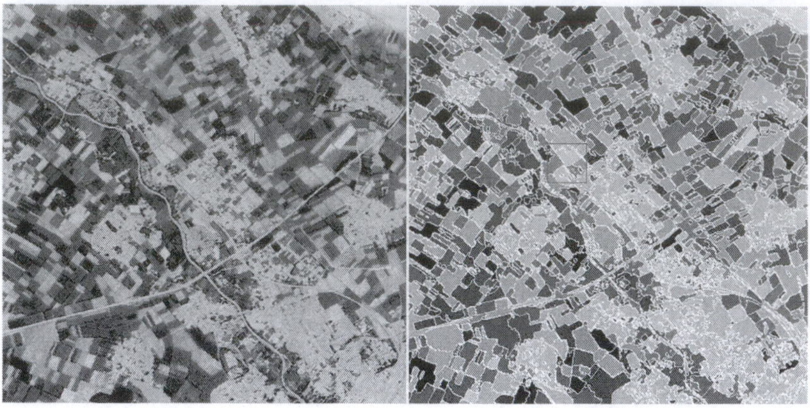

FIGURE 8.11
(See color insert following page 114.) Left, a spatial subset of the Jülich principal component image of Figure 6.1. Right, the mean shift filtered image and the segment boundaries.

The last step assigns the filtered spectral component f (the spectral mode) to the original pixel location $\boldsymbol{x}_\nu$. Segmentation simply involves identifying all pixels with the same mode as a segment.

A direct implementation of this algorithm would be prohibitively slow, since mean shifts are to be computed for all of the pixels. In the ENVI/IDL extension `MEAN_SHIFT_RUN` described in Appendix C.2.21, two approximations are employed to speed things up. First, all pixels within radius r of convergence point w are assumed also to converge to that mode and are excluded from further processing. Second, all pixels lying sufficiently close (within a distance $r/2$) to the path traversed from the initial vector $\boldsymbol{y}_\nu$ to the mode w are similarly assigned to that mode and excluded from the calculation. Additionally, a user-defined minimum segment size can be specified. Segments with smaller extent than this minimum are assigned to the nearest segment in the combined spatial–spectral feature space. An example of the mean shift applied to the first three principal components of the Jülich ASTER scene is shown in Figure 8.11.

8.8 Exercises

1. (a) Show that the sum of squares cost function, Equation (8.12), can be expressed equivalently as

$$E(C) = \frac{1}{2}\sum_{k=1}^{K} \bar{s}_k \tag{8.61}$$

Listing 8.9
Generating another toy image.

```
PRO toy_image1
 image = fltarr (400 ,400 ,2)
 n = randomu(seed ,400 ,400,/normal )
 image[*,*,0] = n+8
 image[*,*,1]=n^2+randomu(seed,400,400,/normal )+8
 image[0:199,*,0]=randomu(seed, 200,400,/normal )/2+8
 image[0:199,*,1]=randomu(seed, 200,400,/normal )+14
 envi_enter_data,image
END
```

where $\bar{s}_k$ is the average squared distance between points in the kth cluster,

$$\bar{s}_k = \frac{1}{m_k}\sum_{i=1}^{m}\sum_{i'=1}^{m} u_{ki}u_{ki'}\|\boldsymbol{g}_i - \boldsymbol{g}_{i'}\|^2. \tag{8.62}$$

(b) Recall that, in the notation of Section 8.2.2,

$$\mathcal{G}^\top \boldsymbol{U}^\top \boldsymbol{M} = (\boldsymbol{m}_1 \ldots \boldsymbol{m}_k).$$

It follows that $(\mathcal{G}^\top \boldsymbol{U}^\top \boldsymbol{M})\boldsymbol{U}$ is an $N \times m$ matrix whose νth column is the mean vector associated with the νth observation. Use this to demonstrate that the sum of squares cost function can also be written in the form

$$E(C) = \mathrm{tr}\big(\mathcal{G}\mathcal{G}^\top\big) - \mathrm{tr}\big(\boldsymbol{U}^\top \boldsymbol{M}\boldsymbol{U}\mathcal{G}\mathcal{G}^\top\big). \tag{8.63}$$

2. (a) The ENVI/IDL extension GPUKKMEANS_RUN makes use of the Gaussian kernel only. Modify it to include the option of using polynomial kernels (see Chapter 4, Exercise 8).

   (b) The program in Listing 8.9 generates a two-band toy image with the two nonlinearly separable clusters shown in Figure 8.12. Experiment with GPUKKMEANS_RUN to achieve a good clustering of the image. Compare your result with ENVI's K-means tool.

3. Kernel K-means clustering is closely related to *spectral clustering* (see Dhillon et al. (2005)). Let $\mathcal{K}$ be an $m \times m$ Gaussian kernel matrix. Define the diagonal *degree* matrix

$$\boldsymbol{D} = \mathrm{Diag}(d_1 \ldots d_m),$$

where $d_i = \sum_{j=1}^{m}(\mathcal{K})_{ij}$, and the symmetric *Laplacian* matrix

$$\boldsymbol{L} = \boldsymbol{D} - \mathcal{K}. \tag{8.64}$$

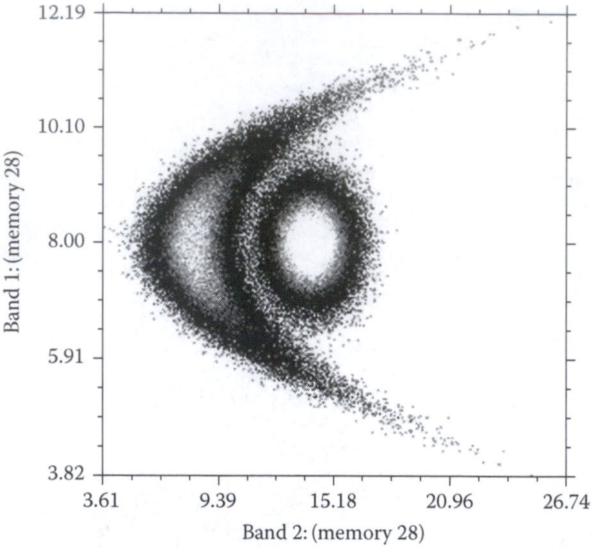

FIGURE 8.12
Two clusters which are not linearly separable.

Minimization of the kernelized version of the sum of squares cost function, namely

$$E(C) = \mathrm{tr}(\mathcal{K}) - \mathrm{tr}(M^{1/2}U\mathcal{K}U^\top M^{1/2})$$

(see Equation 8.63), can be shown to be equivalent to ordinary (linear) clustering of the m training observations in the space of the first K eigenvectors of $\boldsymbol{L}$ (Shawe-Taylor and Cristianini, 2004; von Luxburg, 2006).

(a) Show that $\boldsymbol{L}$ is positive semidefinite and that its smallest eigenvalue is zero with corresponding eigenvector $\mathbf{1}_m$.

(b) Write an IDL function SPECTRAL_CLUSTER(G,K) to implement the following algorithm. (The input observations $g(\nu)$, $\nu = 1 \ldots m$, are passed to the function in the form of a data matrix $\mathcal{G}$ along with K, the number of clusters.)

Algorithm (Unnormalized Spectral Clustering)

1. Determine the Laplacian matrix $\boldsymbol{L}$ as given by Equation 8.64.
2. Compute the $m \times K$ matrix $V = (v_1 \ldots v_K)$, the columns of which are the eigenvectors of $\boldsymbol{L}$ corresponding to the K smallest eigenvalues. Let $y(\nu)$ be the K-component vector consisting of the νth row of V, $\nu = 1 \ldots m$.
3. Partition the vectors $y(\nu)$ with the K-means algorithm into clusters $C_1 \ldots C_K$. (*Hint:* see IDL functions CLT_WTS() AND CLUSTER().)
4. Output the cluster labels for $g(\nu)$ as those for $y(\nu)$.

(c) Thinking of $\mathcal{G}$ in the algorithm as a training dataset sampled from a multispectral image, suggest how the clustering result might be generalized to the whole image.

4. Write an IDL routine to give a first order estimate of the entropy of an image spectral band.* Use the definition in Equation 8.18 and interpret p_i as the probability that a pixel in the image has intensity i, $i = 0 \ldots 255$. (*Hint:* Some intensities may have zero probabilities. The logarithm of zero is undefined, but the product $p \cdot \log p \to 0$ as $p \to 0$.)
5. Modify the program in Listing 8.2 to give a more robust estimation of the variance, σ^2, by masking out edge pixels with a thresholded Sobel filter (see Section 5.2).
6. Using the modified program from Exercise 5 as a starting point, write an IDL extension to the ENVI environment to perform EKM clustering on image data with more than one spectral band.
7. Derive the expression in Equation 8.27 for the net change in the sum of squares cost function when clusters k and j are merged.
8. Recall the definition of fuzzy hypervolume in Section 8.3.3, namely

$$FHV = \sum_{k=1}^{K} \sqrt{|\mathcal{C}_k|},$$

where $\mathcal{C}_k$ is the covariance matrix for the kth cluster (Equation 8.10). This quantity might itself be used as a cost function, since it measures the compactness of the clusters corresponding to some partitioning C. Suppose that the observations, $\boldsymbol{g}(\nu)$, are rescaled by some arbitrary nonsingular transformation matrix $\boldsymbol{T}$, that is, $\boldsymbol{g}'(\nu) = \boldsymbol{T}\boldsymbol{g}(\nu)$. Show that any algorithm which achieves partitioning, C, on the basis of the original observations, $\boldsymbol{g}(\nu)$, by minimizing FHV will produce the same partitioning with the transformed observations, $\boldsymbol{g}'(\nu)$.

9. (EM algorithm) Consider a set of one-dimensional observations $\{g(\nu) \mid \nu = 1 \ldots m\}$ that are to be clustered into K classes according to a Gaussian mixture model having the form

$$p(g) = \sum_{k=1}^{K} p(g \mid k) \Pr(k),$$

* *First order* means that one assumes that all pixel intensities are statistically independent.

where the $\Pr(k)$ are weighting factors, $\sum_{k=1}^{K} \Pr(k) = 1$, and

$$p(g \mid k) = \frac{1}{\sqrt{2\pi}\,\sigma_k} \exp\left(-\frac{(g-\mu_k)^2}{2\sigma_k^2}\right).$$

The log-likelihood for the observations is

$$\mathcal{L} = \log \prod_{\nu=1}^{m} p(g(\nu)) = \sum_{\nu=1}^{m} \log\left(\sum_{k=1}^{K} p(g(\nu) \mid k)\Pr(k)\right).$$

(a) Show, with the help of Bayes' theorem, that this expression is maximized by the following values for the model parameters μ_k, σ_k, and $\Pr(k)$:

$$\mu_k = \frac{\sum_\nu \Pr(k \mid g(\nu))g(\nu)}{\sum_\nu \Pr(k \mid g(\nu))}$$

$$\sigma_k^2 = \frac{\sum_\nu \Pr(k \mid g(\nu))(g(\nu)-\mu_k)^2}{\sum_\nu \Pr(k \mid g(\nu))}$$

$$\Pr(k) = \frac{1}{m}\sum_\nu \Pr(k \mid g(\nu)).$$

(b) The EM algorithm consists of iterating these three equations (in the order given), together with

$$\Pr(k \mid g(\nu)) \propto p(g(\nu) \mid k)\,\Pr(k), \quad \sum_k \Pr(k \mid g(\nu)) = 1,$$

until convergence. Explain why this is identical to the FMLE algorithm for one-dimensional data.

10. Give an integer arithmetic expression for $d^2(k^*, k)$ for a cubic geometry SOM.
11. The traveling salesperson problem is to find the shortest closed route between n points on a map (cities) which visits each point exactly once. Program an approximate solution to the TSP using a one-dimensional self-organizing map in the form of a closed loop.
12. (a) The mean shift segmentation of Section 8.7.3 algorithm is said to be "edge-preserving." Explain why this is so.

(b) A uniform kernel was used for simplicity to determine the mean shift, the hypersphere, S, in the algorithm, which we can represent as

$$S(\boldsymbol{w} - \boldsymbol{y}_\nu) = \begin{cases} 1 & \text{for } \|\boldsymbol{w} - \boldsymbol{y}_\nu\| \le r \\ 0 & \text{otherwise.} \end{cases}$$

We could alternatively have used the radial basis (Gaussian) kernel

$$k(w - y_\nu) = \exp\left(-\frac{\|w - y_\nu\|^2}{2r^2}\right).$$

The density estimate at the point w is (see Section 6.4)

$$p(w) = \frac{1}{m(2\pi r^2)^{(N+2)/2}} \sum_{\nu=1}^{m} \exp\left(-\frac{\|w - y_\nu\|^2}{2r^2}\right)$$

and the estimated gradient at w, that is, the direction of maximum change in $p(w)$, is

$$\frac{\partial p(w)}{\partial w}.$$

Show that the mean shift always is in the direction of the gradient. (This demonstrates that the mean shift algorithm will find the modes of the distribution without actually estimating its density, and can be shown to be a general result (Comaniciu and Meer, 2002).)

9

Change Detection

To quote a well-known, if now rather dated, review article on change detection (Singh, 1989),

> The basic premise in using remote sensing data for change detection is that changes in land cover must result in changes in radiance values [...] [which] must be large with respect to radiance changes from other factors.

When comparing multispectral images of a given scene taken at different times, it is therefore desirable to correct the pixel intensities as much as possible for uninteresting differences such as those due to illumination, atmospheric conditions, or sensor calibration. If comparison is on a pixel-by-pixel basis, then the images must also be coregistered to high accuracy in order to avoid spurious signals resulting from misaligned pixels. Some of the required preprocessing steps were discussed in Chapter 5. Two coregistered satellite images are shown in Figure 9.1.

After having performed the necessary preprocessing, it is common to examine various functions of the spectral bands involved (differences, ratios, or linear combinations), which, in some way, bring the change information contained within them to the fore. The shallow flooding at the western edge of the reservoir in Figure 9.1 is evident at a glance. However other changes have occurred between the two acquisition times that require more image processing to be made visible. In this chapter, we will mention some commonly used digital techniques for enhancing "change signals" in bitemporal satellite images, and then focus our particular attention on the *multivariate alteration detection* (MAD) algorithm of Nielsen et al. (1998). The chapter concludes with an "inverse" application of change detection, in which *unchanged* pixels are used for automatic relative radiometric normalization of multitemporal imagery. For recent reviews of change detection in a more general context, see Radke et al. (2005) or Coppin et al. (2004).

9.1 Algebraic Methods

A simple way to detect changes in two corrected and coregistered multispectral images, represented by N-dimensional random vectors $\boldsymbol{F}$ and $\boldsymbol{G}$, is to

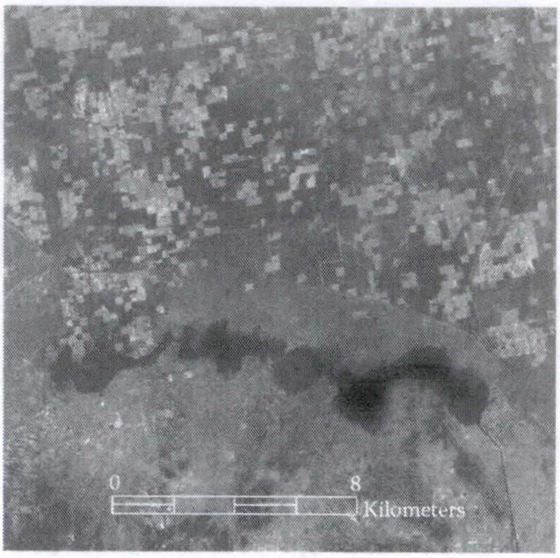

(a)

(b)

FIGURE 9.1
LANDSAT 5 TM images (band 4) over a water reservoir in India, (a) acquired on March 29, 1998, (b) on May 16, 1998.

subtract them from each other component by component, and then examine the N difference images

$$F - G = (F_1 - G_1, F_2 - G_2 \ldots F_N - G_N)^\top. \tag{9.1}$$

Small intensity differences indicate no change, large positive or negative values indicate change, and decision thresholds can be set to define significance. The thresholds are usually expressed in terms of standard deviations from the mean difference value, which is taken to correspond to no change. If the significant difference signatures in the spectral channels are combined so as to try to characterize the kinds of changes that have taken place, one speaks of *spectral change vector analysis* (Jensen, 2005). Alternatively, ratios of intensities

$$\frac{F_k}{G_k}, \quad k = 1 \ldots N \tag{9.2}$$

are sometimes formed between successive images. Ratios near unity correspond to no change, while small and large values indicate change. A disadvantage of this method is that ratios of random variables are not normally distributed even if the random variables themselves are, so that symmetric threshold values defined in terms of standard deviations are not valid.*

* More formally, for data with additive errors like multispectral images, the best statistic for rejecting a no-change hypothesis is a linear combination of observations, not a ratio.

More complicated algebraic combinations, such as differences in vegetation indices or tasseled cap transforms, are also in use. All of these band manipulations can of course be performed conveniently within the ENVI environment using the `BAND MATH` facility, as well as with the main menu command:

```
Basic Tools/Change Detection/Compute Difference Map
```

9.2 Postclassification Comparison

If two coregistered satellite images have been classified to yield thematic maps, then the class labels can be compared to determine land cover changes. If classification is carried out at the pixel level (as opposed to using segments or objects), then classification errors (typically >5%) may corrupt or even dominate the true change signal, depending on the strength of the latter. ENVI offers the function

```
Basic Tools/Change Detection/Change Detection Statistics
```

for the statistical analysis of postclassification change detection.

9.3 Principal Components Analysis

In a scatterplot of pixel intensities in band i for two different times, each point is a realization of the random vector $(F_i, G_i)^\top$. Since unchanged pixels will be highly correlated over time, they will lie in a narrow, elongated cluster along the principal axis, whereas changed pixels will be scattered some distance away from it (see Figure 9.2). The second principal component, which measures intensities at right angles to the first principal axis, will therefore quantify the degree of change associated with a given pixel and may serve as a change image. A change detector based on principal components analysis (PCA) is implemented in the ENVI SPEAR Tools menu:

```
Spectral/SPEAR Tools/Change Detection
```

9.3.1 Iterated PCA

Since the principal axes are determined by diagonalization of the covariance matrix for all of the pixels, the no-change axis may be poorly defined. To overcome this problem, the principal components can be calculated iteratively using weights for each pixel determined by the probability of no change obtained from the preceding iteration.

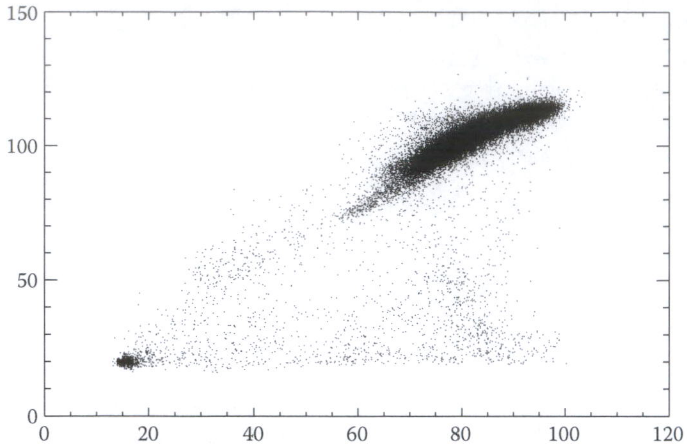

FIGURE 9.2
Scatterplot of the two spectral bands of Figure 9.1.

Listing 9.1 give an IDL routine for performing change detection with iterated PCA. After an initial principal components transformation, lines 25–32, the FMLE routine described in Section 8.3 is used to cluster the second principal component into two classes, lines 48 and 49. The probability array `U[*,0]` of membership of the pixels to the central cluster is then roughly their probability of no change. The complement array `U[*,1]` contains the change probabilities. The probabilities of no change for observations having a second principal component within one standard deviation of the first principal axis are "frozen" to the value 1 (lines 43–46) in order to accelerate convergence. Using the no-change probabilities as weights, the covariance matrix is recalculated (lines 51 and 52) and the PCA repeated. This procedure is iterated five times in the example. Results are shown in Figures 9.3 and 9.4. In Figure 9.4, the quantity `U[*,1]` is displayed rather than the second principal component, as it better highlights the changes. The method can be generalized to treat all multispectral bands together (Wiemker, 1997).

9.3.2 Kernel PCA

For highly nonlinear data, change detection based on linear transformations such as PCA will be expected to give inferior results. As an illustration (Nielsen and Canty, 2008), consider the bitemporal scene in Figure 9.5. These images were recorded with the airborne DLR 3K-camera system from the German Aerospace Center (DLR) (Kurz et al., 2007), a system consisting of three off-the-shelf cameras arranged on a mount with one camera looking in the nadir direction and two cameras tilted approximately 35° across track. The 600-by-600 pixel sub-images shown in the figure were acquired 0.7 s

Listing 9.1
Iterated principal components.

```
PRO ex9_1

; get an image band for T1
envi_select, title='Choose␣multispectral␣band␣for␣T1',$
             fid=fid1, dims=dims1,pos=pos1, /band_only
IF (fid1 EQ -1) THEN RETURN

; get an image band for T2
envi_select, title='Choose␣multispectral␣band␣for␣T2',$
             fid=fid2, dims=dims2,pos=pos2, /band_only
IF (fid2 EQ -1) THEN RETURN

IF (dims1[2]-dims1[1] NE dims2[2]-dims2[1]) OR $
   (dims1[4]-dims1[3] NE dims2[4]-dims2[3]) THEN RETURN
num_cols = dims1[2]-dims1[1]+1
num_rows = dims1[4]-dims1[3]+1
num_pixels = num_cols*num_rows
bitemp = fltarr(2,num_pixels)
bitemp[0,*] = envi_get_data(fid=fid1,dims=dims1,pos=0)
bitemp[1,*] = envi_get_data(fid=fid2,dims=dims2,pos=0)
bitemp[0,*] = bitemp[0,*] - mean(bitemp[0,*])
bitemp[1,*] = bitemp[1,*] - mean(bitemp[1,*])

; initial principal components transformation
C = correlate(bitemp,/covariance)
eivs = eigenql(C, eigenvectors=V, /double)
PCs = bitemp##transpose(V)
window,11,xsize=600,ysize=400, $
   title='Iterated␣principal␣components'
PLOT, [-1,1],[-abs(V[0,1]/V[0,0]),abs(V[0,1]/V[0,0])],$
  xrange=[-1,1],yrange=[-1,1],color=0, $
         background='FFFFFF'XL

; iterated principal components
covpm = Obj_New('COVPM',2)
iter = 0L
niter = 5
; begin iteration
REPEAT BEGIN
; determine weights via clustering
   sigma1 = sqrt(eivs[1])
   U = randomu(seed,num_pixels,2)
   unfrozen = where( abs(PCs[1,*]) GE sigma1, $
      complement = frozen)
```

Listing 9.1
Iterated principal components (continued).

```
45     U[frozen,0] = 1.0
46     U[frozen,1] = 0.0
47     FOR j=0,1 DO U[*,j]=U[*,j]/total(U,2)
48     EM, PCs[1,*], U, Ms, Ps, Fs, T0=0, $
49        unfrozen=unfrozen
50 ; weighted covariance matrix
51     covpm->update, bitemp, weights=U[*,0]
52     C = covpm->covariance()
53     eivs = eigenql(C, eigenvectors=V, /double)
54     OPLOT, [-1,1],[-abs(V[0,1]/V[0,0]), $
55        abs(V[0,1]/V[0,0])],linestyle=2, color=0
56     PCs = bitemp##transpose(V)
57     iter++
58 END UNTIL iter GE niter
59
60 ; return result to ENVI
61 out_array = fltarr(num_cols,num_rows,2)
62 FOR i=0,1 DO out_array[*,*,i] = $
63     reform(U[*,i],num_cols,num_rows)
64 envi_enter_data, out_array
65
66 END
```


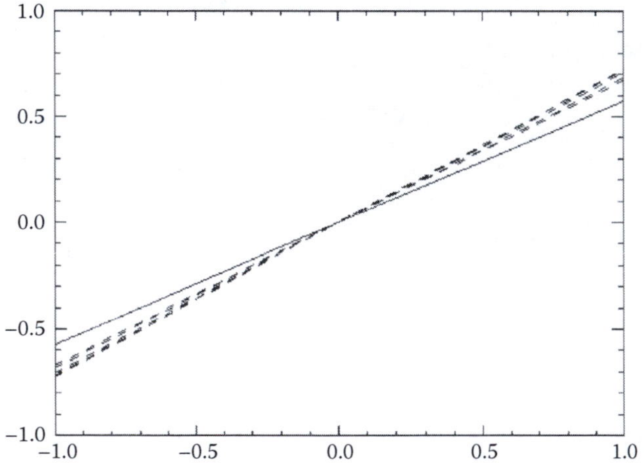

FIGURE 9.3
Iterated PCA. The solid line is the initial principal axis; the dashed lines correspond to five subsequent iterations.

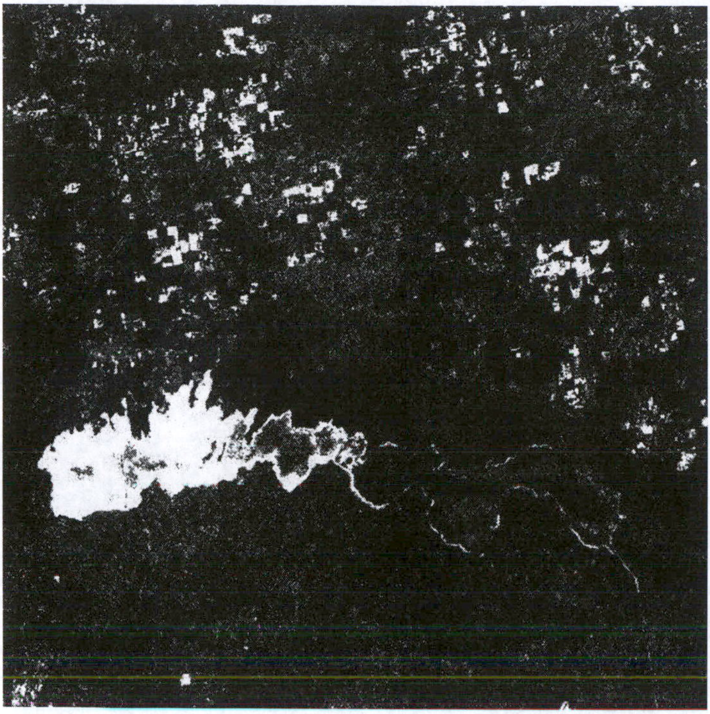

FIGURE 9.4
The change probability for the bitemporal image in Figure 9.1 after five iterations of PCA.

FIGURE 9.5
Two traffic scenes taken 0.7 s apart.

apart, and cover a busy motorway near Munich, Germany. The images were registered to one another with subpixel accuracy. The only physical changes on the ground are due to the motion of the automobiles.

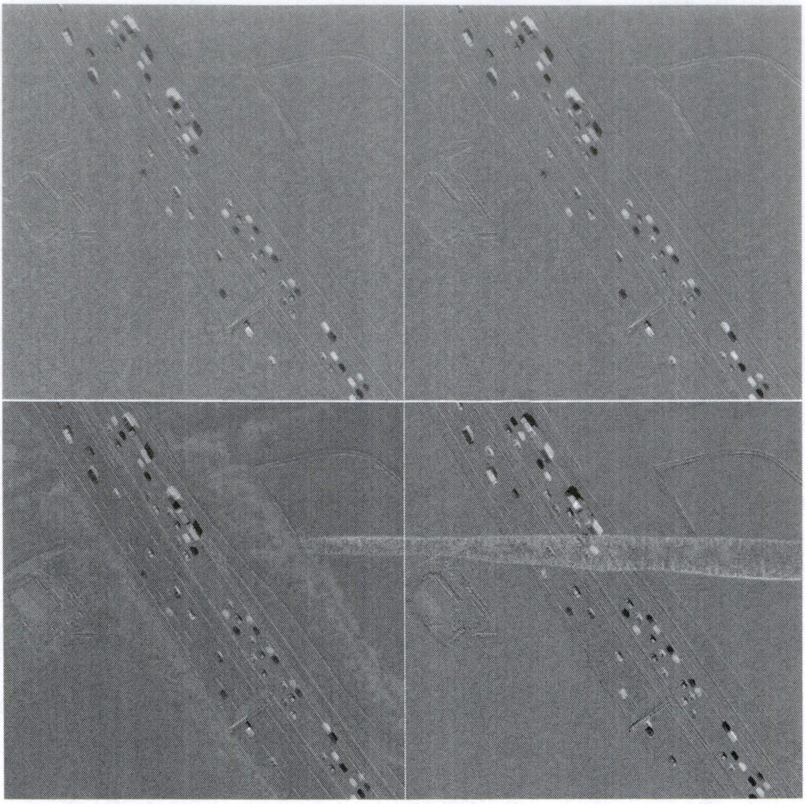

FIGURE 9.6
Change detection with nonlinear PCA. Top row: linear data, bottom row: artificial nonlinear data, left column: second principal component for linear PCA, and right column: fifth principal component for kernel PCA.

The upper row in Figure 9.6 shows an ordinary PCA (left) and kernel PCA (right). Kernel PCA was discussed in Chapter 4. In the upper left-hand image, the second principal component is displayed, signaling changes as bright and dark pixels, with no-change indicated as middle gray. In the upper right-hand image, the fifth principal component in the kernel-induced nonlinear feature space is displayed and is seen to return the same change information. In the lower row of the figure, the data were artificially "nonlinearized" by raising the intensities in the second image to the third power, while leaving those of the first image unchanged. On the lower left (linear PCA), there are now bright and dark pixels falsely signaling change, whereas on the lower right (kernel PCA), the fifth kernel principal component still indicates changes correctly. Figure 9.7 is a contour plot of the fifth kernel principal component with the sampled training pixels superimposed and marking the no-change axis.

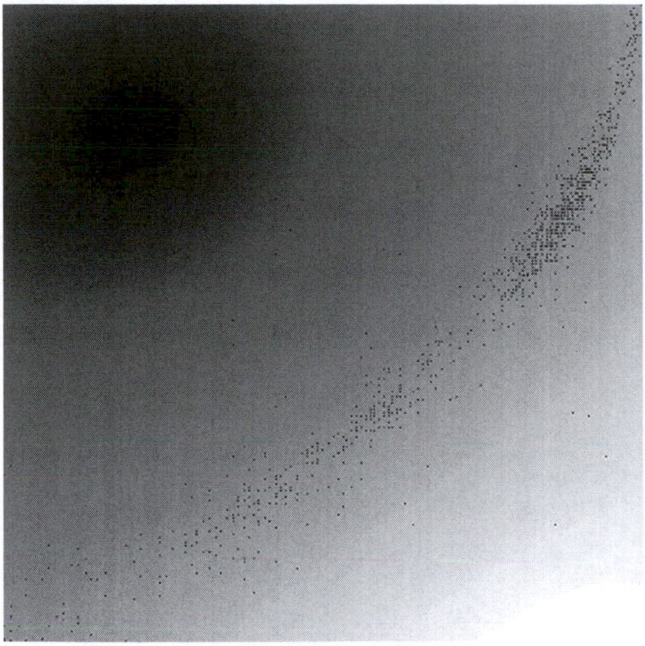

FIGURE 9.7
The fifth kernel principal component for the nonlinearized bitemporal traffic scene. The brightness (darkness) indicates the value of the projection onto the fifth principal axis. The dots are the sampled pixels.

9.4 Multivariate Alteration Detection

Let us consider two N-band multi- or hyperspectral images of the same scene acquired at different times, between which ground reflectance changes have occurred at some locations but not everywhere. We make a linear combination of the intensities for all N bands in the first image, represented by the random vector $\boldsymbol{G}_1$, thus creating a scalar image characterized by the random variable

$$U = \boldsymbol{a}^\top \boldsymbol{G}_1.$$

The vector of coefficients, $\boldsymbol{a}$, is as yet unspecified. We do the same for the second image, represented by $\boldsymbol{G}_2$, forming the linear combination

$$V = \boldsymbol{b}^\top \boldsymbol{G}_2,$$

and then look at the scalar difference image, $U - V$. The change information is now contained in a single image. One has, of course, to choose the coefficients $\boldsymbol{a}$ and $\boldsymbol{b}$ in some suitable way. In Nielsen et al. (1998), it is suggested that they be determined by applying standard canonical correlation analysis (CCA), first described by Hotelling (1936), to the two sets of random variables represented by vectors $\boldsymbol{G}_1$ and $\boldsymbol{G}_2$. The resulting linear combinations are then, as we shall see, ordered by similarity (correlation) rather than, as in the original images, by wavelength. This provides a more natural framework in which to look for change. Forming linear combinations has an additional advantage that, due to the central limit theorem (Theorem 2.2), the quantities involved are increasingly well described by the normal distribution.

To anticipate somewhat, in performing CCA on a bitemporal image one maximizes the correlation, ρ, between the random variables U and V. The correlation is given by (see Chapter 2)

$$\rho = \frac{\text{cov}(U, V)}{\sqrt{\text{var}(U)}\sqrt{\text{var}(V)}}. \tag{9.3}$$

Arbitrary multiples of U and V would clearly have the same correlation, so a constraint must be chosen. A convenient one is

$$\text{var}(U) = \text{var}(V) = 1. \tag{9.4}$$

Note that, under this constraint, the variance of the difference image is

$$\text{var}(U - V) = \text{var}(U) + \text{var}(V) - 2\text{cov}(U, V) = 2(1 - \rho). \tag{9.5}$$

Therefore, vectors $\boldsymbol{a}$ and $\boldsymbol{b}$ which maximize the correlation (Equation 9.3) under the constraints of Equation 9.4 will in fact minimize the variance of the difference image.

9.4.1 Canonical Correlation Analysis

CCA entails a linear transformation of each set of image bands $(G_{1_1} \dots G_{1_N})$ and $(G_{2_1} \dots G_{2_N})$ such that, rather than being ordered according to wavelength, the transformed components are ordered according to their mutual correlation (see especially Anderson, 2003).

The bitemporal, multispectral image may be represented by the combined random vector, $\begin{pmatrix} \boldsymbol{G}_1 \\ \boldsymbol{G}_2 \end{pmatrix}$. This random vector has the covariance matrix

$$\boldsymbol{\Sigma} = \begin{pmatrix} \boldsymbol{\Sigma}_{11} & \boldsymbol{\Sigma}_{12} \\ \boldsymbol{\Sigma}_{12}^\top & \boldsymbol{\Sigma}_{22} \end{pmatrix},$$

where, assuming that the means have been subtracted from the image data, $\boldsymbol{\Sigma}_{11} = \langle \boldsymbol{G}_1 \boldsymbol{G}_1^\top \rangle$ is the covariance matrix of the first image, $\boldsymbol{\Sigma}_{22} = \langle \boldsymbol{G}_2 \boldsymbol{G}_2^\top \rangle$ that of the second image, and $\boldsymbol{\Sigma}_{12} = \langle \boldsymbol{G}_1 \boldsymbol{G}_2^\top \rangle$ is the matrix of covariances between the two. We then have, for the transformed variables U and V,

$$\text{var}(U) = \boldsymbol{a}^\top \boldsymbol{\Sigma}_{11} \boldsymbol{a}, \quad \text{var}(V) = \boldsymbol{b}^\top \boldsymbol{\Sigma}_{22} \boldsymbol{b}, \quad \text{cov}(U, V) = \rho = \boldsymbol{a}^\top \boldsymbol{\Sigma}_{12} \boldsymbol{b}.$$

CCA now consists of maximizing the covariance $\boldsymbol{a}^\top \boldsymbol{\Sigma}_{12}\boldsymbol{b}$ under constraints $\boldsymbol{a}^\top \boldsymbol{\Sigma}_{11}\boldsymbol{a} = 1$ and $\boldsymbol{b}^\top \boldsymbol{\Sigma}_{22}\boldsymbol{b} = 1$. If, following the usual procedure, we introduce the Lagrange multipliers, $\nu/2$ and $\mu/2$, for each of the two constraints, then the problem becomes one of maximizing the unconstrained Lagrange function

$$L = \boldsymbol{a}^\top \boldsymbol{\Sigma}_{12}\boldsymbol{b} - \frac{\nu}{2}(\boldsymbol{a}^\top \boldsymbol{\Sigma}_{11}\boldsymbol{a} - 1) - \frac{\mu}{2}(\boldsymbol{b}^\top \boldsymbol{\Sigma}_{22}\boldsymbol{b} - 1).$$

Setting derivatives with respect to $\boldsymbol{a}$ and $\boldsymbol{b}$ equal to zero gives

$$\begin{aligned} \frac{\partial L}{\partial \boldsymbol{a}} &= \boldsymbol{\Sigma}_{12}\boldsymbol{b} - \nu\boldsymbol{\Sigma}_{11}\boldsymbol{a} = \boldsymbol{0} \\ \frac{\partial L}{\partial \boldsymbol{b}} &= \boldsymbol{\Sigma}_{12}^\top\boldsymbol{a} - \mu\boldsymbol{\Sigma}_{22}\boldsymbol{b} = \boldsymbol{0}. \end{aligned} \tag{9.6}$$

Multiplying the first of the above equations from the left with the vector $\boldsymbol{a}^\top$, the second with $\boldsymbol{b}^\top$, and using the constraints leads immediately to

$$\nu = \mu = \boldsymbol{a}^\top \boldsymbol{\Sigma}_{12}\boldsymbol{b} = \rho.$$

Therefore, we can write Equations 9.6 in the form

$$\begin{aligned} \boldsymbol{\Sigma}_{12}\boldsymbol{b} - \rho\boldsymbol{\Sigma}_{11}\boldsymbol{a} &= \boldsymbol{0} \\ \boldsymbol{\Sigma}_{12}^\top\boldsymbol{a} - \rho\boldsymbol{\Sigma}_{22}\boldsymbol{b} &= \boldsymbol{0}. \end{aligned} \tag{9.7}$$

Next, multiply the first of Equations 9.7 by ρ and the second from the left by $\boldsymbol{\Sigma}_{22}^{-1}$. This gives

$$\rho\boldsymbol{\Sigma}_{12}\boldsymbol{b} = \rho^2\boldsymbol{\Sigma}_{11}\boldsymbol{a} \tag{9.8}$$

and

$$\boldsymbol{\Sigma}_{22}^{-1}\boldsymbol{\Sigma}_{12}^\top\boldsymbol{a} = \rho\boldsymbol{b}. \tag{9.9}$$

Finally, combining Equations 9.8 and 9.9, we obtain the following equation for the transformation coefficient, $\boldsymbol{a}$:

$$\boldsymbol{\Sigma}_{12}\boldsymbol{\Sigma}_{22}^{-1}\boldsymbol{\Sigma}_{12}^\top\boldsymbol{a} = \rho^2\boldsymbol{\Sigma}_{11}\boldsymbol{a}. \tag{9.10}$$

A similar argument (Exercise 3) leads to the corresponding equation for $\boldsymbol{b}$, namely

$$\boldsymbol{\Sigma}_{12}^\top\boldsymbol{\Sigma}_{11}^{-1}\boldsymbol{\Sigma}_{12}\boldsymbol{b} = \rho^2\boldsymbol{\Sigma}_{22}\boldsymbol{b}. \tag{9.11}$$

Solution of these equations will indeed lead to a maximum in the correlation ρ since, from Equation 9.6, we have

$$\frac{\partial^2 L}{\partial \boldsymbol{a}^\top \partial \boldsymbol{a}} = -\rho\boldsymbol{\Sigma}_{11}, \quad \frac{\partial^2 L}{\partial \boldsymbol{b}^\top \partial \boldsymbol{b}} = -\rho\boldsymbol{\Sigma}_{22}$$

so that the Hessian of L is negative definite.

Equations 9.10 and 9.11 are generalized eigenvalue problems similar to those that we already met for the maximum noise fraction (MNF) and maximum autocorrelation factor (MAF) transformations of Chapter 3 (see Equations 3.53 and 3.70). Note, however, that they are coupled via the eigenvalue ρ^2. The desired projections $U = \boldsymbol{a}^\top \boldsymbol{G}_1$ are given by the eigenvectors $\boldsymbol{a}_1 \dots \boldsymbol{a}_N$ of Equation 9.10 corresponding to eigenvalues $\rho_1^2 \geq \rho_2^2 \geq \dots \geq \rho_N^2$. Similarly the desired projections $V = \boldsymbol{b}^\top \boldsymbol{G}_2$ are given by the eigenvectors $\boldsymbol{b}_1 \dots \boldsymbol{b}_N$ of Equation 9.11 corresponding to the *same* eigenvalues.

Solution of the eigenvalue problems generates new multispectral images $\boldsymbol{U} = (U_1 \dots U_N)^\top$ and $\boldsymbol{V} = (V_1 \dots V_N)^\top$, the components of which are called the *canonical variates* (CVs). The CVs are ordered by similarity (correlation) rather than, as in the original images, by wavelength. The canonical correlations $\rho_i = \text{corr}(U_i, V_i)$, $i = 1 \dots N$, are the square roots of the eigenvalues of the coupled eigenvalue problem. The pair (U_1, V_1) is maximally correlated, the pair (U_2, V_2) is maximally correlated subject to being orthogonal to (uncorrelated with) both U_1 and V_1, and so on. Taking paired differences (in reverse order) then generates a sequence of transformed difference images

$$M_i = U_{N-i+1} - V_{N-i+1}, \quad i = 1 \dots N, \tag{9.12}$$

referred to as the MAD *variates* (Nielsen et al., 1998). Since we are dealing with change detection, we want the pairs of canonical variates U_i and V_i to be positively correlated, just like the original image bands. This is easily achieved by appropriate choice of the relative signs of the eigenvector pairs $\boldsymbol{a}_i$ and $\boldsymbol{b}_i$ so as to ensure that $\boldsymbol{a}_i^\top \boldsymbol{\Sigma}_{12} \boldsymbol{b}_i > 0$, $i = 1 \dots N$.

An ENVI/IDL extension `MAD_RUN` for MAD is described in Appendix C.2.22. Figure 9.8 gives an example of its application to the bitemporal data of Figure 9.1. Only the last (least correlated) and first (most correlated) CVs are shown. Corresponding scatterplots are displayed in Figure 9.9.

9.4.2 Orthogonality Properties

Equations 9.10 and 9.11 are of the form

$$\boldsymbol{\Sigma}_1 \boldsymbol{a} = \rho^2 \boldsymbol{\Sigma} \boldsymbol{a}, \tag{9.13}$$

where both $\boldsymbol{\Sigma}_1$ and $\boldsymbol{\Sigma}$ are symmetric and $\boldsymbol{\Sigma}$ is positive definite. Equation 9.13 can be solved in the same way as for the MNF transformation in Chapter 3. We repeat the procedure here for convenience. First, write Equation 9.13 in the form

$$\boldsymbol{\Sigma}_1 \boldsymbol{a} = \rho^2 \boldsymbol{L} \boldsymbol{L}^\top \boldsymbol{a},$$

where $\boldsymbol{\Sigma}$ has been replaced by its Cholesky decomposition $\boldsymbol{L}\boldsymbol{L}^\top$. The matrix $\boldsymbol{L}$ is positive definite, lower triangular. Equivalently,

$$\boldsymbol{L}^{-1} \boldsymbol{\Sigma}_1 (\boldsymbol{L}^\top)^{-1} \boldsymbol{L}^\top \boldsymbol{a} = \rho^2 \boldsymbol{L}^\top \boldsymbol{a}$$

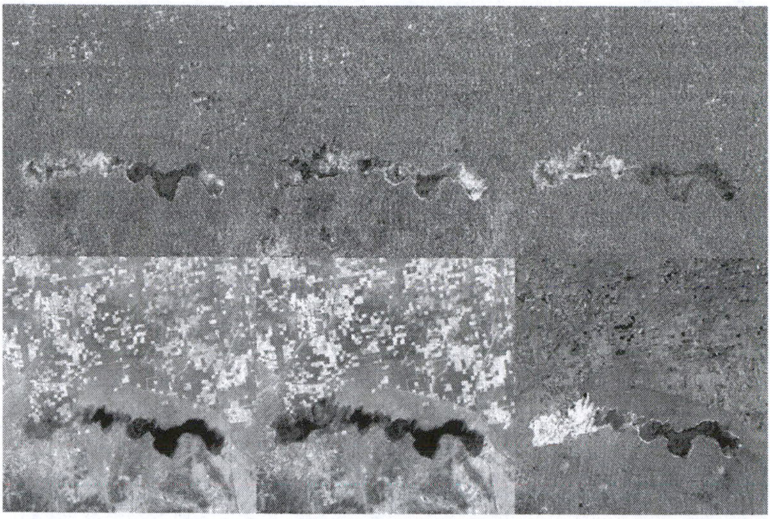

FIGURE 9.8
MAD change detection using the six nonthermal bands of the bitemporal scene of Figure 9.1. The top row shows, left to right, the canonical variates U_6 and V_6 and their difference M_1. The bottom row shows U_1, V_1, and M_6. All images are displayed with a 2% saturated linear histogram stretch.

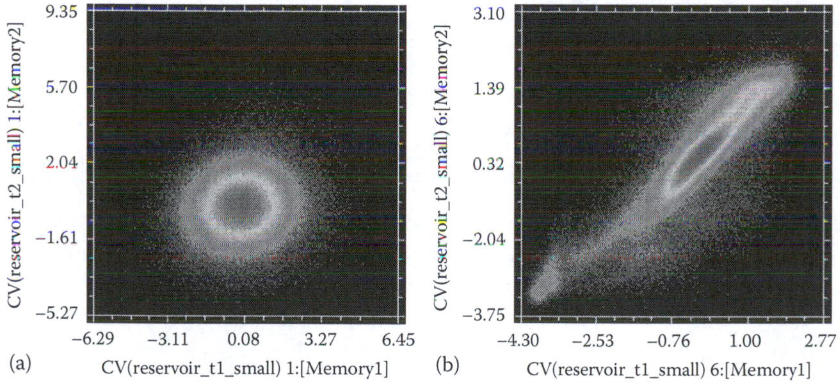

FIGURE 9.9
Scatterplots of (a) V_6 vs. U_6 and (b) V_1 vs. U_1 from Figure 9.8. The last two canonical variates are virtually uncorrelated, the first two are maximally correlated.

or, with $\boldsymbol{d} := \boldsymbol{L}^\top \boldsymbol{a}$ and the commutivity of inverse and transpose,

$$[\boldsymbol{L}^{-1}\boldsymbol{\Sigma}_1(\boldsymbol{L}^{-1})^\top]\boldsymbol{d} = \rho^2 \boldsymbol{d},$$

a standard eigenvalue problem for the symmetric matrix $\boldsymbol{L}^{-1}\boldsymbol{\Sigma}_1(\boldsymbol{L}^{-1})^\top$. Let its eigenvectors be $\boldsymbol{d}_i$. Since they are orthogonal and normalized, we have

$$\delta_{ij} = \boldsymbol{d}_i^\top \boldsymbol{d}_j = \boldsymbol{a}_i^\top \boldsymbol{L}\boldsymbol{L}^\top \boldsymbol{a}_j = \boldsymbol{a}_i^\top \boldsymbol{\Sigma} \boldsymbol{a}_j. \tag{9.14}$$

From Equations 9.4 and 9.14, taking $\boldsymbol{\Sigma} = \boldsymbol{\Sigma}_{11}$ and then $\boldsymbol{\Sigma} = \boldsymbol{\Sigma}_{22}$, it follows that

$$\begin{aligned} \text{cov}(U_i, U_j) &= \boldsymbol{a}_i^\top \boldsymbol{\Sigma}_{11} \boldsymbol{a}_j = \delta_{ij} \\ \text{cov}(V_i, V_j) &= \boldsymbol{b}_i^\top \boldsymbol{\Sigma}_{22} \boldsymbol{b}_j = \delta_{ij}. \end{aligned} \tag{9.15}$$

Furthermore, according to Equation 9.9,

$$\boldsymbol{b}_i = \frac{1}{\rho_i} \boldsymbol{\Sigma}_{22}^{-1} \boldsymbol{\Sigma}_{21} \boldsymbol{a}_i.$$

Therefore we have, with Equation 9.10,

$$\text{cov}(U_i, V_j) = \boldsymbol{a}_i^\top \boldsymbol{\Sigma}_{12} \boldsymbol{b}_j = \boldsymbol{a}_i^\top \frac{1}{\rho_j} \boldsymbol{\Sigma}_{12} \boldsymbol{\Sigma}_{22}^{-1} \boldsymbol{\Sigma}_{21} \boldsymbol{a}_j = \rho_j \, \boldsymbol{a}_i^\top \boldsymbol{\Sigma}_{11} \boldsymbol{a}_i = \rho_j \, \delta_{ij}. \tag{9.16}$$

Thus we see that the canonical variates are *all mutually uncorrelated* except for the pairs (U_i, V_i), and these are ordered by decreasing correlation. The MAD variates themselves are consequently also mutually uncorrelated, their covariances being given by (recalling the reversed order in Equation 9.12)

$$\text{cov}(M_i, M_j) = \text{cov}(U_{N-i+1} - V_{N-i+1}, U_{N-j+1} - V_{N-j+1}) = 0, \; i \neq j = 1 \ldots N, \tag{9.17}$$

and their variances by

$$\sigma_{M_i}^2 = \text{var}(U_{N-i+1} - V_{N-i+1}) = 2(1 - \rho_{N-i+1}), \quad i = 1 \ldots N. \tag{9.18}$$

The first MAD variate has maximum variance in its pixel intensities. The second MAD variate has maximum spread subject to the condition that its pixel intensities are statistically uncorrelated with those in the first variate, the third has maximum spread subject to being uncorrelated with the first two, and so on. Depending on the type of change present, any of the components may exhibit significant change information. Interesting small-scale anthropogenic changes, for instance, will generally be unrelated to dominating seasonal vegetation changes or stochastic image noise, so it is quite common that such changes will be concentrated in higher order MAD variates. This is apparently the situation in Figure 9.8, where the flooding shows up more strongly in M_6 than in M_1. In fact, one of the nicest aspects of the method is that it sorts different categories of change into different, uncorrelated image components.

9.4.3 Scale Invariance

An additional advantage of the MAD procedure stems from the fact that the calculations involved are invariant under linear affine transformations of the original image intensities. This implies insensitivity to linear differences in atmospheric conditions or sensor calibrations at the two acquisition times. Scale invariance also offers the possibility of using the MAD transformation to perform relative radiometric normalization of multitemporal imagery in a fully automatic way, as will be explained in Section 9.6.

We can see the invariance as follows. Suppose the second image $\boldsymbol{G}_2$ is transformed according to some linear transformation $\boldsymbol{H} = \boldsymbol{T}\boldsymbol{G}_2$. The relevant covariance matrices for the coupled generalized eigenvalue problems (Equations 9.10 and 9.11) are then

$$
\begin{aligned}
\boldsymbol{\Sigma}_{1h} &= \langle \boldsymbol{G}_1 \boldsymbol{H}^\top \rangle = \boldsymbol{\Sigma}_{12} \boldsymbol{T}^\top \\
\boldsymbol{\Sigma}_{11} &\quad \text{unchanged} \\
\boldsymbol{\Sigma}_{hh} &= \langle \boldsymbol{H}\boldsymbol{H}^\top \rangle = \boldsymbol{T}\boldsymbol{\Sigma}_{22}\boldsymbol{T}^\top.
\end{aligned}
$$

The coupled eigenvalue problems now read

$$
\begin{aligned}
\boldsymbol{\Sigma}_{12}\boldsymbol{T}^\top (\boldsymbol{T}\boldsymbol{\Sigma}_{22}\boldsymbol{T}^\top)^{-1} \boldsymbol{T}\boldsymbol{\Sigma}_{12}^\top \boldsymbol{a} &= \rho^2 \boldsymbol{\Sigma}_{11} \boldsymbol{a} \\
\boldsymbol{T}\boldsymbol{\Sigma}_{12}^\top \boldsymbol{\Sigma}_{11}^{-1} \boldsymbol{\Sigma}_{12} \boldsymbol{T}^\top \boldsymbol{c} &= \rho^2 \boldsymbol{T}\boldsymbol{\Sigma}_{22}\boldsymbol{T}^\top \boldsymbol{c},
\end{aligned}
$$

where $\boldsymbol{c}$ is the desired projection for the transformed image $\boldsymbol{H}$. These equations are easily seen to be equivalent to

$$
\begin{aligned}
\boldsymbol{\Sigma}_{12}\boldsymbol{\Sigma}_{22}^{-1}\boldsymbol{\Sigma}_{12}^\top \boldsymbol{a} &= \rho^2 \boldsymbol{\Sigma}_{11}\boldsymbol{a} \\
\boldsymbol{\Sigma}_{12}^\top \boldsymbol{\Sigma}_{11}^{-1} \boldsymbol{\Sigma}_{12} (\boldsymbol{T}^\top \boldsymbol{c}) &= \rho^2 \boldsymbol{\Sigma}_{22} (\boldsymbol{T}^\top \boldsymbol{c}),
\end{aligned}
$$

which are identical to Equations 9.10 and 9.11 with $\boldsymbol{b}$ replaced by $\boldsymbol{T}^\top \boldsymbol{c}$. Therefore the MAD components in the transformed situation are, as before,

$$
\boldsymbol{a}_i^\top \boldsymbol{G}_1 - \boldsymbol{c}_i^\top \boldsymbol{H} = \boldsymbol{a}_i^\top \boldsymbol{G}_1 - \boldsymbol{c}_i^\top \boldsymbol{T}\boldsymbol{G}_2 = \boldsymbol{a}_i^\top \boldsymbol{G}_1 - (\boldsymbol{T}^\top \boldsymbol{c}_i)^\top \boldsymbol{G}_2 = \boldsymbol{a}_i^\top \boldsymbol{G}_1 - \boldsymbol{b}_i^\top \boldsymbol{G}_2.
$$

9.4.4 Iteratively Reweighted MAD

Let us now imagine two images of the same scene, acquired at different times under similar conditions, but for which no ground reflectance changes have occurred whatsoever. Then the only differences between them will be due to random effects like instrument noise and atmospheric fluctuation. In such a case we would expect that the histogram of any difference component that we generate will be very nearly Gaussian. In particular, the MAD variates, being uncorrelated, should follow a multivariate, zero mean normal distribution with diagonal covariance matrix (see Chapter 2). Change observations would deviate more or less strongly from such a distribution. Just as for the iterated principal components method of Section 9.3.1, we might therefore expect an improvement in the sensitivity of the MAD transformation if we can establish an increasingly better background of no change against which to detect change (Nielsen, 2007). This can again be done in an iteration scheme in which, when calculating the means and covariance matrices for the next iteration of the MAD transformation, observations are weighted by the probability of no change determined on the preceding iteration.*

One way to determine the no-change probabilities is to perform an unsupervised classification of the MAD variates using, for example, the Gaussian mixture algorithm of Chapter 8 with K *a priori* clusters, one for no change and $K-1$ clusters for different categories of change. The pixel class membership probabilities for no change then provide the necessary weights for iteration. This method, being unsupervised, will not distinguish which clusters correspond to change and which to no change, but the no-change cluster can be identified by its compactness, for example, by its partition density (see Section 8.3.3 and also Figure 9.16). This strategy has, however, the disadvantage that the rather slow clustering procedure must be repeated for every iteration. Moreover one has to make a more or less arbitrary decision as to how many no-change clusters are present and tolerate the unpredictability of clustering results generally.

An alternative scheme is to examine the MAD variates directly. Let the random variable Z represent the sum of the squares of the standardized MAD variates,

$$Z = \sum_{i=1}^{N} \left(\frac{M_i}{\sigma_{M_i}} \right)^2, \tag{9.19}$$

where σ_{M_i} is given by Equation 9.18. Then, since the no-change observations are expected to be normally distributed and uncorrelated, the random variable Z should be chi-square distributed with N degrees of freedom (Section 2.4.1). For each iteration, the observations can thus be given weights determined by the chi-square distribution, namely

$$\Pr(\text{no change}) = 1 - P_{\chi^2;N}(z) \tag{9.20}$$

(see Equation 2.45). Here Pr(no change) is the probability that a sample z drawn from the chi-square distribution could be that large or larger. A small z implies a correspondingly large probability. Iteration of the MAD transformation continues until some stopping criterion is met, such as lack of significant change in the canonical correlations ρ_i, $i = 1 \ldots N$. Nielsen (2007) refers to this procedure as the *iteratively reweighted* MAD (IR-MAD) algorithm.

The ENVI/IDL extension given in Appendix C.2.22 includes the possibility of iteration of the MAD transformation on the basis of Equation 9.20. Figure 9.10 indicates the kind of improvement that can be achieved through iteration. The iterated canonical correlations are shown in Figure 9.11.

* This was in fact the main motivation for allowing for weights in the provisional means algorithm described in Chapter 2.

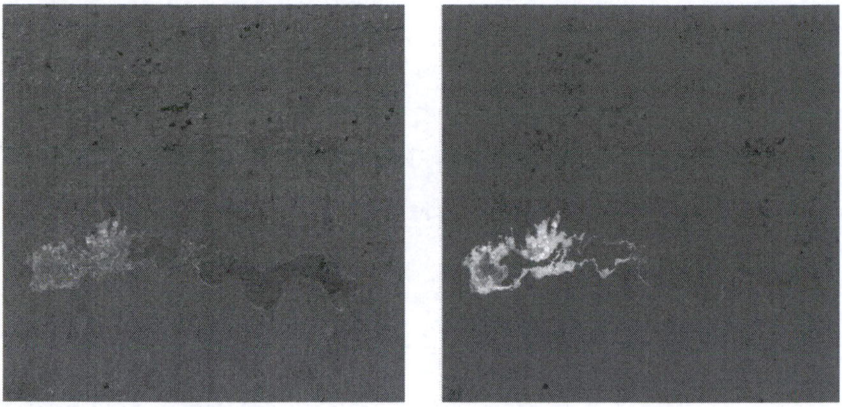

(a) (b)

FIGURE 9.10
The sixth MAD variate for the bitemporal image in Figure 9.1, shown in an unsaturated linear histogram stretch, (a) without iteration, (b) after iteration to convergence. Bright and dark pixels indicate change.

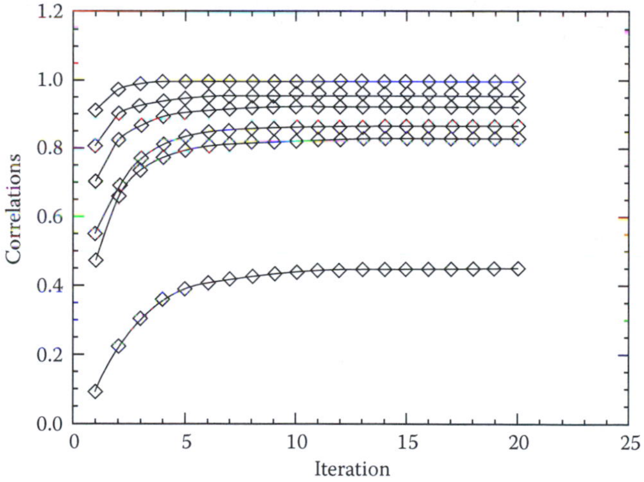

FIGURE 9.11
The canonical correlations ρ_i, $i = 1 \ldots 6$, under 20 iterations of the MAD transformation of the bitemporal image in Figure 9.1.

9.4.5 Correlation with the Original Observations

As in the case of PCA, the linear transformations involved in MAD mix the spectral bands of the images, making physical interpretation of observed

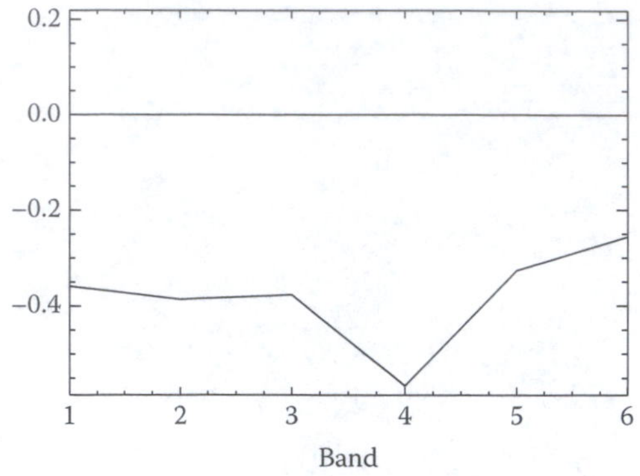

FIGURE 9.12
Correlation of the fourth iteratively reweighted MAD (IR-MAD) component with the six nonthermal spectral bands for the reservoir scene acquired on March 29, 1998.

changes more difficult. Interpretation can be facilitated by examining the correlation of the MAD variates with the original spectral bands (Nielsen et al., 1998). In a more compact matrix notation for the transformation coefficients, define

$$A = (\boldsymbol{a}_i \dots \boldsymbol{a}_N), \quad \boldsymbol{B} = (\boldsymbol{b}_i \dots \boldsymbol{b}_N),$$

so that $\boldsymbol{A}$ is an $N \times N$ matrix whose columns are the eigenvectors $\boldsymbol{a}_i$ of the MAD transformation, and similarly for $\boldsymbol{B}$. Furthermore let

$$\boldsymbol{M} = (M_N \dots M_1)^\top$$

be the random vector of MAD variates (in the same ordering as the CVs). Then the covariances with the original bands are given by

$$\begin{aligned}\langle G_1 \boldsymbol{M}^\top \rangle &= \langle G_1 (\boldsymbol{A}^\top G_1 - \boldsymbol{B}^\top G_2)^\top \rangle = \boldsymbol{\Sigma}_{11}\boldsymbol{A} - \boldsymbol{\Sigma}_{12}\boldsymbol{B} \\ \langle G_2 \boldsymbol{M}^\top \rangle &= \langle G_2 (\boldsymbol{A}^\top G_1 - \boldsymbol{B}^\top G_2)^\top \rangle = \boldsymbol{\Sigma}_{12}^\top \boldsymbol{A} - \boldsymbol{\Sigma}_{22}\boldsymbol{B},\end{aligned} \tag{9.21}$$

from which the correlations can be calculated by dividing by the square roots of the variances (Equation 2.33). Figure 9.12 shows the correlation of the fourth MAD variate with the spectral bands of the first acquisition of Figure 9.1. The strong negative correlation with spectral band 4 indicates that the change signal is due primarily to changes in vegetation. This is confirmed in Figure 9.13, showing the fourth IR-MAD component in which significant changes in the agricultural fields to the north of the reservoir are evident.

9.4.6 Regularization

To avoid possible near singularity problems in the solution of Equations 9.10 and 9.11 it is useful to introduce some form of regularization into the MAD algorithm (Nielsen, 2007). This may in particular be necessary when the number of spectral bands, N, is large, as is the case for change detection involving hyperspectral imagery. The concept of regularization by means

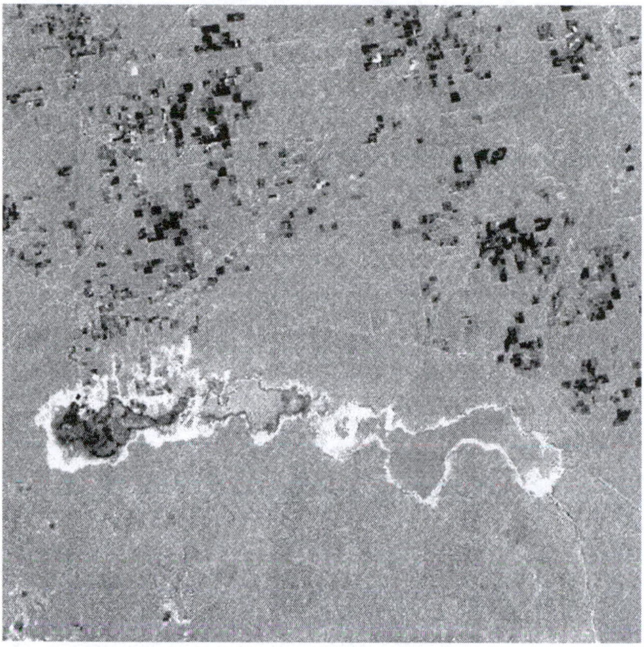

FIGURE 9.13
The fourth IR-MAD component, 2% saturated linear histogram stretch.

of length penalization was met in Section 2.6.3. Here we allow for length penalization of the eigenvectors $\boldsymbol{a}$ and $\boldsymbol{b}$ in CCA by replacing Equation 9.4 by

$$(1-\lambda)\text{var}(U) + \lambda\|\boldsymbol{a}\|^2 = (1-\lambda)\text{var}(V) + \lambda\|\boldsymbol{b}\|^2 = 1, \tag{9.22}$$

where
$\text{var}(U) = \boldsymbol{a}^\top \boldsymbol{\Sigma}_{11}\boldsymbol{a}$, $\text{var}(V) = \boldsymbol{b}^\top \boldsymbol{\Sigma}_{22}\boldsymbol{b}$
λ is a regularization parameter

To maximize the covariance $\text{cov}(U, V) = \boldsymbol{a}^\top \boldsymbol{\Sigma}_{12}\boldsymbol{b}$ under this constraint, we now maximize the unconstrained Lagrange function

$$L = \boldsymbol{a}^\top \boldsymbol{\Sigma}_{12}\boldsymbol{b} - \frac{\nu}{2}((1-\lambda)\boldsymbol{a}^\top \boldsymbol{\Sigma}_{11}\boldsymbol{a} + \lambda\|\boldsymbol{a}\|^2 - 1)$$
$$- \frac{\mu}{2}((1-\lambda)\boldsymbol{b}^\top \boldsymbol{\Sigma}_{22}\boldsymbol{b} + \lambda\|\boldsymbol{b}\|^2 - 1).$$

Setting the derivatives equal to zero, we obtain

$$\begin{aligned}\frac{\partial L}{\partial \boldsymbol{a}} &= \boldsymbol{\Sigma}_{12}\boldsymbol{b} - \nu(1-\lambda)\boldsymbol{\Sigma}_{11}\boldsymbol{a} - \nu\lambda\boldsymbol{a} = 0\\ \frac{\partial L}{\partial \boldsymbol{b}} &= \boldsymbol{\Sigma}_{12}^{\top}\boldsymbol{a} - \mu(1-\lambda)\boldsymbol{\Sigma}_{22}\boldsymbol{b} - \mu\lambda\boldsymbol{b} = 0.\end{aligned} \tag{9.23}$$

Multiplying the first equation above from the left with $\boldsymbol{a}^{\top}$ gives

$$\boldsymbol{a}^{\top}\boldsymbol{\Sigma}_{12}\boldsymbol{b} - \nu((1-\lambda)\boldsymbol{a}^{\top}\boldsymbol{\Sigma}_{11}\boldsymbol{a} + \lambda\|\boldsymbol{a}\|^2) = 0$$

and, from Equation 9.22, $\nu = \boldsymbol{a}^{\top}\boldsymbol{\Sigma}_{12}\boldsymbol{b}$. Similarly, multiplying the second equation from the left with $\boldsymbol{b}^{\top}$, we have $\mu = \boldsymbol{b}^{\top}\boldsymbol{\Sigma}_{12}^{\top}\boldsymbol{a} = \nu$. Equation 9.23 can now be written as the single generalized eigenvalue problem ($\boldsymbol{I}$ is the $N \times N$ identity matrix)

$$\begin{pmatrix} \boldsymbol{0} & \boldsymbol{\Sigma}_{12} \\ \boldsymbol{\Sigma}_{21} & \boldsymbol{0} \end{pmatrix}\begin{pmatrix} \boldsymbol{a} \\ \boldsymbol{b} \end{pmatrix} = \mu \begin{pmatrix} (1-\lambda)\boldsymbol{\Sigma}_{11} + \lambda\boldsymbol{I} & \boldsymbol{0} \\ \boldsymbol{0} & (1-\lambda)\boldsymbol{\Sigma}_{22} + \lambda\boldsymbol{I} \end{pmatrix}\begin{pmatrix} \boldsymbol{a} \\ \boldsymbol{b} \end{pmatrix} \tag{9.24}$$

(see also Exercise 3). Note that the eigenvalue μ is the covariance of U and V and not the correlation. Thus, with Equation 9.22,

$$\sigma^2_{MAD} = \mathrm{var}(U) + \mathrm{var}(V) - 2\mathrm{cov}(U,V) = \frac{2 - \lambda(\|\boldsymbol{a}\|^2 + \|b\|^2)}{1-\lambda} - 2\mu \tag{9.25}$$

and the correlation is

$$\rho = \frac{\mu}{\sqrt{\mathrm{var}(U)\mathrm{var}(V)}} = \frac{\mu(1-\lambda)}{\sqrt{(1-\lambda\|\boldsymbol{a}\|^2)(1-\lambda\|\boldsymbol{b}\|^2)}}. \tag{9.26}$$

Regularization is included optionally in the program `MAD_RUN` (see Appendix C.2.22).

Canonical correlation with regularization is illustrated in Listing 9.2. After collecting image statistics with the weighted provisional means method (Section 2.3.3) the covariance matrix and mean vector of the bitemporal image are extracted from the `COVPM` class instance in lines 162 and 163. The (regularized) variance–covariance matrices required for CCA are then constructed (lines 166–176) and the generalized eigenvalue problems (Equations 9.10 and 9.11) are solved in lines 178 and 179. In lines 181–185 the MAD standard deviations and canonical correlations are determined as given, respectively, by Equations 9.25 and 9.26.

9.4.7 Postprocessing

The MAD transformation can be augmented by subsequent application of the MAF transformation in order to improve the spatial coherence of the MAD variates (Nielsen et al., 1998). When image noise is estimated as the difference between intensities of neighboring pixels, the MAF transformation is

Listing 9.2
Excerpt from the program module IR_MAD.PRO.

```
; canonical correlation
      Sigma = covpm->Covariance()
      means = covpm->Means()
      Obj_Destroy, covpm

      S11 = Sigma[0:num_bands-1,0:num_bands-1]
      S11 = (1-lam)*S11 + lam*omega_L
      S22 = Sigma[num_bands:*,num_bands:*]
      S22 = (1-lam)*S22 + lam*omega_L
      S12 = Sigma[num_bands:*,0:num_bands-1]
      S21 = Sigma[0:num_bands-1,num_bands:*]

      C1 = S12 ## invert(S22,/double) ## S21
      B1 = S11
      C2 = S21 ## invert(S11,/double) ## S12
      B2 = S22

      gen_eigenproblem, C1, B1, A, mu2
      gen_eigenproblem, C2, B2, B, mu2

      mu = sqrt(mu2)
      a2=diag_matrix(transpose(A)##A) & PRINT,a2
      b2=diag_matrix(transpose(B)##B)
      sigma = sqrt( (2-lam*(a2+b2))/(1-lam)-2*mu )
      rho=mu*(1-lam)/sqrt( (1-lam*a2)*(1-lam*b2) )
```

equivalent to the MNF transformation, as was discussed in Chapter 3. MAD components postprocessed in this way will be referred to as MAD/MNF variates in the remainder of this chapter.

9.5 Decision Thresholds and Unsupervised Classification of Changes

Since the MAD (or MAD/MNF) variates are approximately normally distributed about zero and uncorrelated, thresholds for deciding between change and no change can be set in terms of standard deviations about the mean for each variate separately, as is done for example in spectral change vector analysis (Section 9.1). Thresholds may be chosen in an *ad hoc* fashion, for example by saying that all pixels in a MAD component M_i whose intensities are within $\pm 2\sigma_{M_i}$ of zero, where σ_{M_i} is given by Equation 9.18, are no-change pixels.

We can do better than this, however. Let us consider a simple, three-Gaussian mixture model for a random variable M representing one of the MAD components. The probability density of M is modeled as

$$p(m) = p(m \mid NC)\Pr(NC) + p(m \mid C-)\Pr(C-) + p(m \mid C+)\Pr(C+), \quad (9.27)$$

where

$C+$, $C-$, and NC denote positive change, negative change, and no change, respectively

$p(m \mid \cdot\,)$ is a normal density function

The set $S = \{m(\nu)\}$ of realizations of M may be partitioned into four disjoint subsets:

$$S_{NC}, \quad S_{C-}, \quad S_{C+}, \quad \text{and} \quad S_U = S \backslash (S_{NC} \cup S_{C-} \cup S_{C+}),$$

with S_U denoting a set of ambiguous pixels.* From the sample mean and sample variance, we can estimate initial values for the mean of the no-change distribution

$$\mu_{NC} = \frac{1}{|S_{NC}|} \cdot \sum_{\nu \in S_{NC}} m(\nu),$$

and for the variance:

$$(\sigma_{NC})^2 = \frac{1}{|S_{NC}|} \cdot \sum_{\nu \in S_{NC}} (m(\nu) - \mu_{NC})^2$$

($|S_{NC}|$ denotes set cardinality) and similarly for $C-$ and $C+$. Following a suggestion by Bruzzone and Prieto (2000), these estimates can be improved by applying the expectation maximization (EM) algorithm. The EM algorithm recalculates the parameters of the mixture model according to (see Chapter 8, Exercise 9)

$$\begin{aligned} \mu'_{NC} &= \sum_{\nu \in S} \Pr(NC \mid m(\nu)) \cdot m(\nu) \Big/ \sum_{\nu \in S} \Pr(NC \mid m(\nu)) \\ (\sigma'_{NC})^2 &= \sum_{\nu \in S} \Pr(NC \mid m(\nu)) \cdot (m(\nu) - \mu'_{NC})^2 \Big/ \sum_{\nu \in S} \Pr(NC \mid m(\nu)) \\ \Pr(NC)' &= \frac{1}{|S|} \cdot \sum_{\nu \in S} \Pr(NC \mid m(\nu)), \end{aligned} \quad (9.28)$$

where the primes denote improved estimates and the three equations are iterated to convergence. In Equation (9.28), $\Pr(NC \mid m(\nu))$ is the posterior

* The symbols $\cup$ and $\backslash$ denote set union and set difference, respectively. These sets can be determined in practice by setting generous, scene-independent thresholds for change and no-change pixel intensities (see Bruzzone and Prieto, 2000).

probability for a no-change pixel conditional on measurement $m(\nu)$. We have

1. $i \in S_{NC}$: $\Pr(NC \mid m(\nu)) = 1$
2. $i \in S_{C\pm}$: $\Pr(NC \mid m(\nu)) = 0$
3. $i \in S_U$: $\Pr(NC \mid m(\nu)) = p(m(\nu) \mid NC)\Pr(NC)/p(m(\nu))$ (Bayes' theorem), where

$$p(m(\nu) \mid NC) = \frac{1}{\sqrt{2\pi} \cdot \sigma_{NC}} \cdot \exp\left(\frac{-(m(\nu) - \mu_{NC})^2}{2\sigma_{NC}^2}\right).$$

Equation (9.28) can therefore be written in the form

$$\mu'_{NC} = \frac{\sum_{\nu \in S_U} \Pr(NC \mid m(\nu))x_i + \sum_{\nu \in S_{NC}} m(\nu)}{\sum_{\nu \in S_U} \Pr(NC \mid m(\nu)) + |S_{NC}|}$$

$$(\sigma'_{NC})^2 = \frac{\sum_{\nu \in S_U} \Pr(NC \mid m(\nu))(m(\nu) - \mu'_{NC})^2 + \sum_{\nu \in S_{NC}} (m(\nu) - \mu'_{NC})^2}{\sum_{\nu \in S_U} \Pr(NC \mid m(\nu)) + |S_{NC}|}$$

$$\Pr(NC)' = \frac{1}{|S|}\left(\sum_{\nu \in S_U} \Pr(NC \mid m(\nu)) + |S_{NC}|\right),$$

and a similar set of equations for $C+$ and $C-$. They can be iterated to obtain final estimates of the distributions.* A typical fit to a MAD/MNF variate is shown in Figure 9.14. Note the logarithmic scale.

Once a fit to the Gaussian mixture model has been obtained, one can determine the upper change threshold as the appropriate solution of

$$p(m \mid NC)\Pr(NC) = p(m \mid C+)\Pr(C+).$$

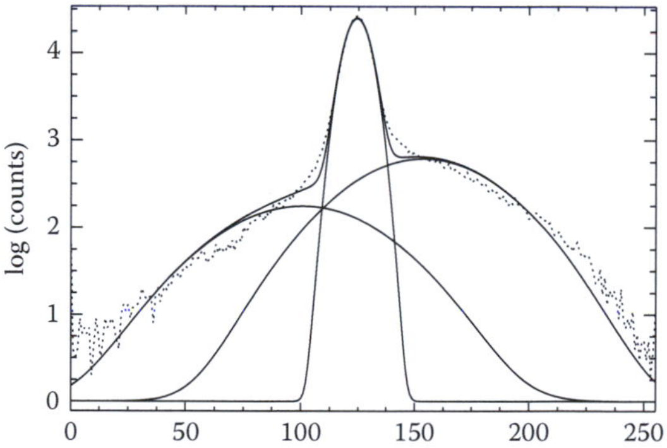

FIGURE 9.14
Gaussian mixture fit to a MAD/MNF component. Optimal decision thresholds correspond to the upper intersections of the central no-change Gaussian with the negative and positive change Gaussians.

* This is in fact just the Gaussian mixture algorithm of Chapter 8 for one-dimensional observations, including "frozen" memberships (Section 8.3.4).

Taking logarithms,

$$\frac{1}{2\sigma_{C+}^2}(m - \mu_{C+})^2 - \frac{1}{2\sigma_{NC}^2}(m - \mu_{NC})^2 = \log\left[\frac{\sigma_{NC}}{\sigma_{C+}} \cdot \frac{\Pr(C+)}{\Pr(NC)}\right] =: A$$

with solutions

$$m = \frac{\mu_{C+}\sigma_{NC}^2 - \mu_{NC}\sigma_{C+}^2 \pm \sigma_{NC}\sigma_{C+}\sqrt{(\mu_{NC} - \mu_{C+})^2 + 2A(\sigma_{NC}^2 - \sigma_{C+}^2)}}{\sigma_{NC}^2 - \sigma_{C+}^2}. \qquad (9.29)$$

A corresponding expression is obtained for the lower threshold. These thresholds minimize the overall error probabilities in deciding between change and no change, (see Exercise 6). A simple IDL GUI (Graphics User Interface), called `MAD_VIEW`, for determining change thresholds and displaying the thresholded MAD or MAD/MNF variates in various color combinations and histogram stretches is described in Appendix C.2.23, and

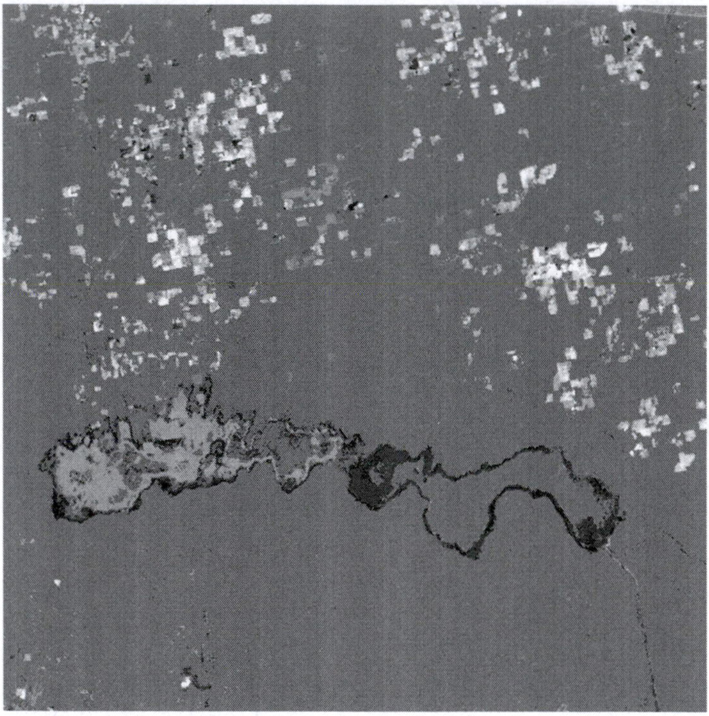

FIGURE 9.15
(See color insert following page 114.) Color composite of MAD/MNF components 1 (red), 2 (green), and 3 (blue) for the bitemporal image of Figure 9.1, generated with `MAD_VIEW` using decision thresholds obtained from the Gaussian mixture model fit, linear stretch over ±16 standard deviations of the no-change observations.

an example is shown in Figure 9.15. The image is stretched over ±16 standard deviations of the no-change observations. This very large dynamic range typifies the sensitivity of the IR-MAD method.

Figure 9.15 provides a rather nice visualization of the significant changes that have taken place. However, it constitutes a marginal analysis in the sense that the decision thresholds are determined band-wise. We might go a step further (Canty and Nielsen, 2006) and cluster the change and no-change pixels in multidimensional MAD or MAD/MNF feature space, again using the Gaussian mixture algorithm discussed in Chapter 8. Since the no-change cluster will tend to be very dominant, it is to be expected that this will only be sensible for situations in which a good deal of change has taken place and some *a priori* information about the number of change categories exists. The result of clustering the first three (i.e., least noisy) MAD/MNF variates for the reservoir scene is shown in Figure 9.16.

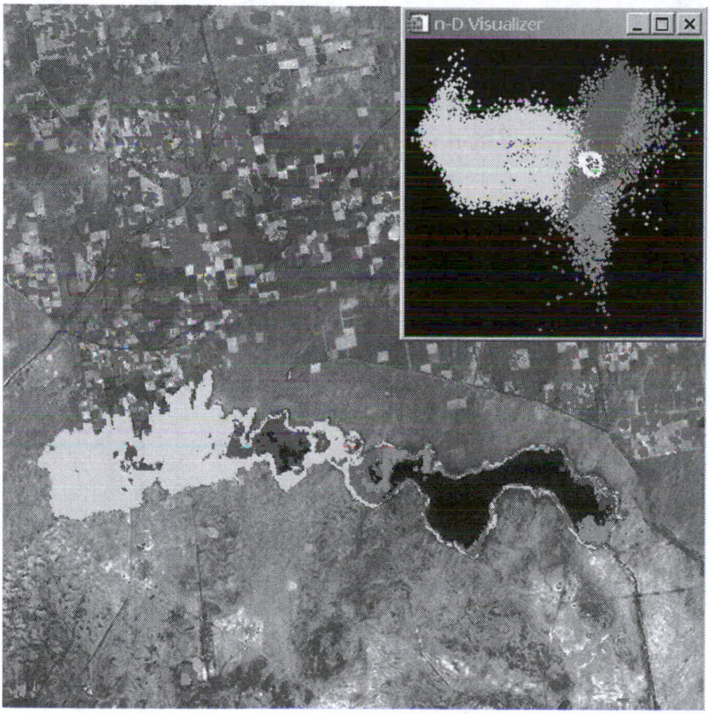

FIGURE 9.16
(See color insert following page 114.) Unsupervised classification of changes for the reservoir scenes of Figure 9.1. Five clusters were used, including no change. The four change clusters are shown superimposed onto spectral band 5 of the first image. The inset displays the clusters projected onto the MAD/MNF 1–2 plane using ENVI's n-dimensional visualizer. The partially obscured white cluster near the center is no change.

9.6 Radiometric Normalization

Ground reflectance determination from satellite imagery requires, among other things, an atmospheric correction algorithm and the associated atmospheric properties at the time of image acquisition. For most historical satellite scenes such data are not available, and even for planned acquisitions they may be difficult to obtain. A relative normalization based on the radiometric information intrinsic to the images themselves is an alternative whenever absolute surface reflectances are not required.

In performing relative radiometric normalization, one usually makes the assumption that the relationship between the at-sensor radiances recorded at two different times from regions of constant reflectance can be approximated by linear functions. The critical aspect is the determination of suitable time-invariant features upon which to base the normalization (Schott et al., 1988; Yang and Lo, 2000; Du et al., 2002).

As we have seen in Section 9.4.3, the MAD transformation is invariant under arbitrary linear transformations of the pixel intensities for the images involved. Thus, if one uses MAD for change detection applications, preprocessing by linear radiometric normalization is superfluous. However radiometric normalization of imagery is important for many other applications, such as mosaicking, tracking vegetation indices over time, comparison of supervised and unsupervised land cover classifications, etc. Furthermore, if some other, noninvariant change detection procedure is preferred, it must generally be preceded by radiometric normalization. Taking advantage of invariance, one can apply the MAD transformation to select the no-change pixels in unnormalized bitemporal images, and then use them for relative radiometric normalization. The procedure is simple, fast, and completely automatic, and compares very favorably with normalization using hand-selected, time-invariant features (Canty et al., 2004; Schroeder et al., 2006; Canty and Nielsen, 2007).

An ENVI/IDL extension `RADCAL_RUN` for radiometric normalization with the MAD transformation is given in Appendix C.2.24. The program reads the output from a previous IR-MAD transformation that has been performed on (overlapping portions of) the images to be normalized. Then Equations 9.19 and 9.20 are used to select pixels with a high no-change probability, typically ≥ 0.95. By regressing the reference image onto the target image at the no-change locations, slope and intercept parameters are obtained to perform the linear normalization. The preferred regression method is in this case orthogonal linear regression (see Appendix A) as both variables involved have similar uncertainties. Figure 9.17 shows a mosaic of two LANDSAT 7 ETM+ images taken over the same area. The October 2000 image (target) is to be normalized to the December 1999 scene (reference). The orthogonal regression coefficients obtained in a fit to 15,696 no-change pixels identified

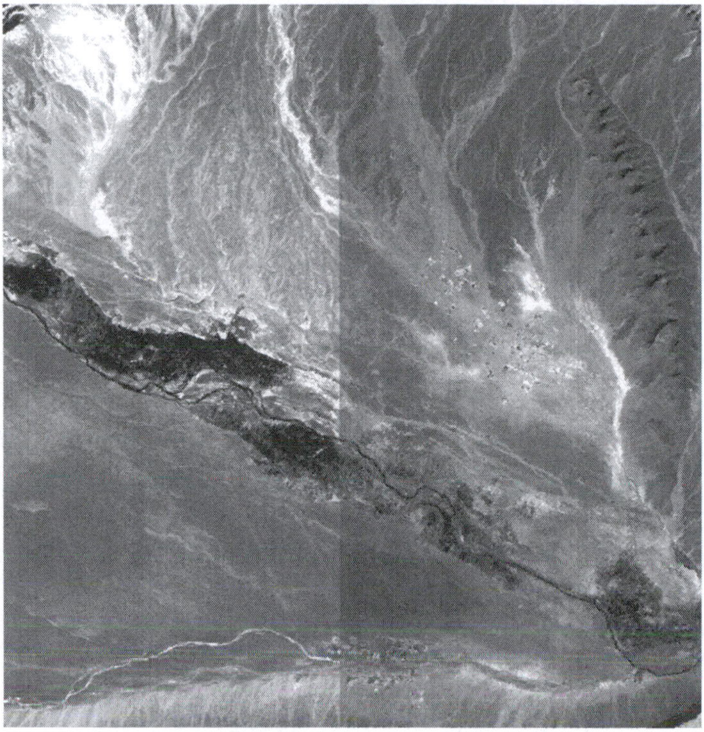

FIGURE 9.17
Mosaic of two LANDSAT 7 ETM+ images over Morocco, spectral band 5. The left side was acquired on December 19, 1999, the right side on October 18, 2000.

TABLE 9.1
Orthogonal Regression Coefficients and Statistics for Radiometric Normalization of the Images of Figure 9.17

Band	Intercept a	σ_a	Slope b	σ_b	Correlation	RMSE
1	−5.06	0.18	1.285	0.003	0.962	0.708
2	−6.84	0.13	1.226	0.002	0.976	0.917
3	−11.20	0.16	1.222	0.002	0.981	1.418
4	−9.69	0.11	1.285	0.002	0.985	1.012
5	−4.74	0.17	1.176	0.002	0.978	1.535
6	−4.30	0.17	1.182	0.002	0.973	1.357

by the IR-MAD procedure are given in Table 9.1. Figure 9.18 shows the regression lines obtained for the six nonthermal bands. The mosaic after radiometric normalization is shown in Figure 9.19.

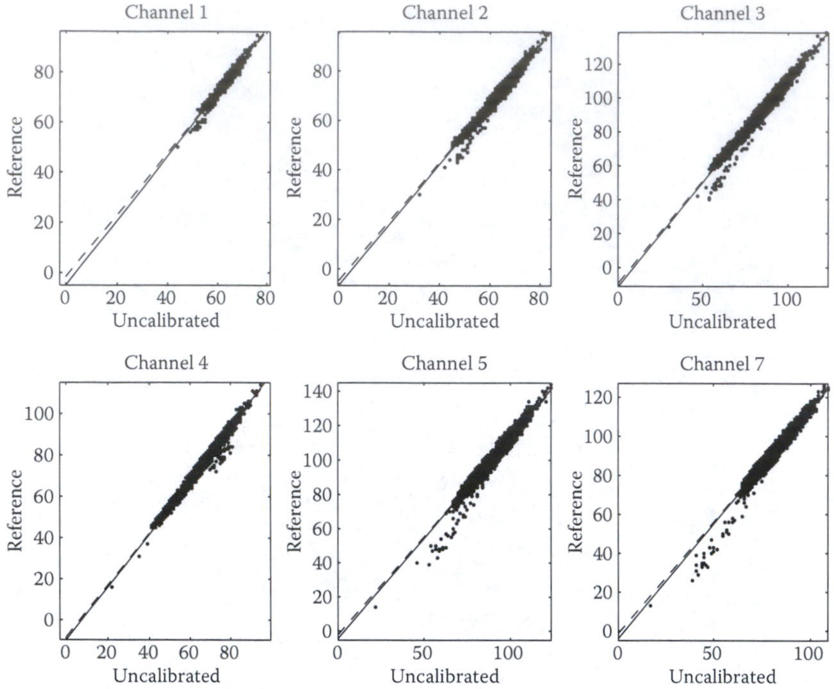

FIGURE 9.18
Regressions of the December 1999 reference scene on the October 2000 target (uncalibrated) scene. Solid line: orthogonal linear regression, dashed line: ordinary linear regression. (From Canty, M. J. et al., *Remote Sens. Environ.*, 91(3–4), 441, 2004.)

In order to evaluate the normalization procedure the program holds back one-third of the no-change pixels for testing purposes. These are used to calculate means and variances before and after normalization and also to perform statistical hypothesis tests for equal means and variances of the invariant pixels in the reference and normalized target images. These tests were discussed in Section 2.4. Results are given in Table 9.2 for 7849 no-change test pixels. The significance values (P-values) for the t-test for equal means and for the F-test for equal variances indicate that the hypotheses of equality cannot be rejected for any of the spectral bands.

9.7 Exercises

1. Use the ENVI BAND MATH facility to calculate NDVI (normalized difference vegetation index) images for two LANDSAT images and to perform change detection by subtracting one from the other.

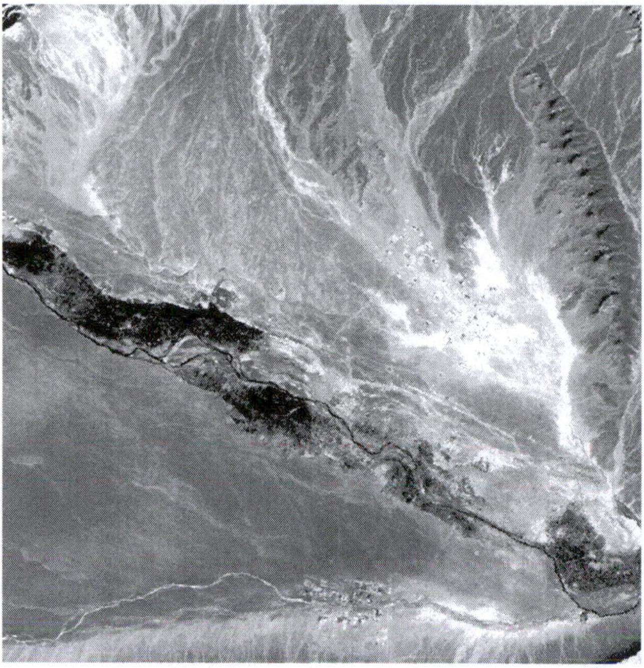

FIGURE 9.19
As Figure 9.17 after automatic radiometric normalization.

TABLE 9.2

Comparison of Means and Variances for Hold-Out Test Pixels, with t- and *F*-Tests for Unequal Means and Variances

	Band 1	Band 2	Band 3	Band 4	Band 5	Band 6
Target mean	62.684	61.460	83.761	64.462	88.000	80.009
Reference mean	75.509	68.501	91.131	73.193	98.775	90.335
Normalized mean	75.518	68.508	91.143	73.198	98.776	90.327
t-Statistic	−0.144	−0.0688	−0.0672	−0.036	−0.007	0.054
P-value	0.885	0.945	0.945	0.975	1.000	0.957
Target variance	10.61	28.83	87.78	55.46	95.24	59.63
Reference variance	17.10	42.72	129.64	90.83	129.82	81.98
Normalized variance	17.54	43.34	131.16	91.70	131.80	83.42
F-statistic	1.025	1.014	1.011	1.009	1.015	1.017
P-value	0.255	0.524	0.606	0.672	0.504	0.440

2. Show that, for one dimensional, zero mean images represented by random variables F and G, the MAD transformation (Equation 9.12) is just

$$MAD = F/\sigma_F - G/\sigma_G.$$

3. (a) Demonstrate the validity of Equation 9.11 for the MAD transformation vector $\boldsymbol{b}$.

   (b) Show that an equivalent formulation of the coupled generalized eigenvalue problems given by Equations 9.10 and 9.11 is the following single generalized eigenvalue problem:

$$\begin{bmatrix} \boldsymbol{0} & \boldsymbol{\Sigma}_{12} \\ \boldsymbol{\Sigma}_{21} & \boldsymbol{0} \end{bmatrix} \begin{bmatrix} \boldsymbol{a} \\ \boldsymbol{b} \end{bmatrix} = \rho \begin{bmatrix} \boldsymbol{\Sigma}_{11} & \boldsymbol{0} \\ \boldsymbol{0} & \boldsymbol{\Sigma}_{22} \end{bmatrix} \begin{bmatrix} \boldsymbol{a} \\ \boldsymbol{b} \end{bmatrix}. \tag{9.30}$$

4. The requirement that the correlations of the canonical variates be positive, namely

$$\boldsymbol{a}_i^\top \boldsymbol{\Sigma}_{12} \boldsymbol{b}_i > 0, \quad i = 1 \ldots N,$$

does not completely remove the ambiguity in the signs of the transformation vectors $\boldsymbol{a}_i$ and $\boldsymbol{b}_i$, since if we invert both their signs simultaneously the condition is still met. The ambiguity can be resolved by requiring that the sum of the correlations of the first image (represented here by $\boldsymbol{G}$) with each of the canonical variates $U_j = \boldsymbol{a}_i^\top \boldsymbol{G}, j = 1 \ldots N$, be positive:

$$\sum_{\nu=1}^{N} \mathrm{corr}(G_i, U_j) > 0, \quad j = 1 \ldots N. \tag{9.31}$$

This condition is implemented in the ENVI/IDL extension `MAD_RUN`.

   (a) Show that the matrix of correlations

$$\boldsymbol{C} = \begin{pmatrix} \mathrm{corr}(G_1, U_1) & \mathrm{corr}(G_1, U_2) & \cdots & \mathrm{corr}(G_1, U_N) \\ \mathrm{corr}(G_2, U_1) & \mathrm{corr}(G_2, U_2) & \cdots & \mathrm{corr}(G_2, U_N) \\ \vdots & \vdots & \ddots & \vdots \\ \mathrm{corr}(G_N, U_1) & \mathrm{corr}(G_N, U_2) & \cdots & \mathrm{corr}(G_N, U_N) \end{pmatrix}$$

   is given by

$$\boldsymbol{C} = \boldsymbol{D}\boldsymbol{\Sigma}_{11}\boldsymbol{A},$$

   where $\boldsymbol{A} = (\boldsymbol{a}_1, \boldsymbol{a}_2 \ldots \boldsymbol{a}_N)$, $\boldsymbol{\Sigma}_{11}$ is the covariance matrix for $\boldsymbol{G}$, and

$$\boldsymbol{D} = \begin{pmatrix} \frac{1}{\sqrt{\mathrm{var}(G_1)}} & 0 & \cdots & 0 \\ 0 & \frac{1}{\sqrt{\mathrm{var}(G_2)}} & \cdots & 0 \\ \vdots & \vdots & \ddots & \vdots \\ 0 & 0 & \cdots & \frac{1}{\sqrt{\mathrm{var}(G_N)}} \end{pmatrix}.$$

(b) Let $s_j = \sum_i C_{ij}$, $j = 1 \ldots N$, be the column sums of the correlation matrix. Show that Equation 9.31 is fulfilled by replacing $\boldsymbol{A}$ by $\boldsymbol{AS}$, where

$$\boldsymbol{S} = \begin{pmatrix} \frac{s_1}{|s_1|} & 0 & \cdots & 0 \\ 0 & \frac{s_2}{|s_2|} & \cdots & 0 \\ \vdots & \vdots & \ddots & \vdots \\ 0 & 0 & \cdots & \frac{s_N}{|s_N|} \end{pmatrix}.$$

5. (a) Consider the following experiment: The ENVI/IDL extension `MAD_RUN` is used to generate MAD variates from two coregistered multispectral images. Then a principal components transformation of one of the images is performed and the MAD transformation is repeated. Will the MAD variates have changed? Why(not)?

   (b) Demonstrate the invariance of the MAD transformation to linear affine transformations of the original images, that is, transformations of form

$$\boldsymbol{G} \rightarrow \boldsymbol{TG} + \boldsymbol{C}.$$

6. Given the validity of the three-Gaussian mixture model of Section 9.5, show that the choice of threshold given in Equation 9.29 minimizes the overall error (incorrectly interpreting a no-change observation as change and vice versa).

7. The program of Listing 9.3 simulates no-change pixels by copying a spatial subset of one image to another, and adding some Gaussian noise. Ideally, the iteratively reweighted MAD scheme should identify these pixels unambiguously. Experiment with a bitemporal image (in byte format) and different subset sizes to see the extent to which this is the case. Use both the `MAD_RUN` and `RADCAL_RUN` ENVI extensions.

8. The more recent versions of ENVI include, among the so-called SPEAR Tools, a change detection "wizard" which may be called from the ENVI main menu with

```
Spectral/SPEAR Tools/Change Detection
```

Experiment with the different algorithms offered by the wizard (image transform, subtractive, spectral angle), and compare them with the iteratively reweighted MAD method.

Listing 9.3

Copying a spatial subset between images.

```
PRO subset_copy

   dim = 256L
; get an image band for T1
   envi_select, title='Choose␣T1␣image', $
                fid=fid1, dims=dims1,pos=pos1
   IF (fid1 EQ -1) THEN RETURN
; get an image band for T2
   envi_select, title='Choose␣T2␣image', $
             fid=fid2, dims=dims2,pos=pos2
   IF (fid2 EQ -1) THEN RETURN
; read images
   num_cols = dims1[2]-dims1[1]+1
   num_rows = dims1[4]-dims1[3]+1
   num_bands = n_elements(pos1)
   num_pixels = num_cols*num_rows
   im1 = fltarr(num_cols,num_rows,num_bands)
   im2 = im1*0.0
   FOR i=0,num_bands-1 DO BEGIN
     im1[*,*,i] = envi_get_data(fid=fid1,dims=dims1,$
                    pos=pos1[i])
     im2[*,*,i] = envi_get_data(fid=fid2,dims=dims2, $
                    pos=pos2[i])
   ENDFOR
; copy subset with Gaussian noise
   im2[0:dim-1,0:dim-1,*] = im1[0:dim-1,0:dim-1,*] $
      + 0.1*randomu(seed,dim,dim,num_bands,/normal)
   envi_enter_data, im2

END
```

Appendix A: Mathematical Tools

A.1 Cholesky Decomposition

Cholesky decomposition is used in some of the IDL routines in this book to solve generalized eigenvalue problems associated with the maximum autocorrelation factor (MAF) and maximum noise fraction (MNF) transformations as well as with canonical correlation analysis. We sketch its justification in the following.

THEOREM A.1

If the $p \times p$ matrix $\boldsymbol{A}$ is symmetric positive definite and if the $p \times q$ matrix $\boldsymbol{B}$, where $q \le p$, has rank q, then $\boldsymbol{B}^\top \boldsymbol{A}\boldsymbol{B}$ is positive definite and symmetric.

Proof

Choose any q-dimensional vector $\boldsymbol{y} \ne \boldsymbol{0}$, and let $\boldsymbol{x} = \boldsymbol{B}\boldsymbol{y}$. We can write this as

$$\boldsymbol{x} = y_1 \boldsymbol{b}_1 + \cdots + y_q \boldsymbol{b}_q,$$

where $\boldsymbol{b}_i$ is the ith column of $\boldsymbol{B}$. Since $\boldsymbol{B}$ has rank q, we conclude that $\boldsymbol{x} \ne \boldsymbol{0}$ as well, for otherwise the column vectors would be linearly dependent. But

$$\boldsymbol{y}^\top (\boldsymbol{B}^\top \boldsymbol{A}\boldsymbol{B})\boldsymbol{y} = (\boldsymbol{B}\boldsymbol{y})^\top \boldsymbol{A}(\boldsymbol{B}\boldsymbol{y}) = \boldsymbol{x}^\top \boldsymbol{A}\boldsymbol{x} > 0,$$

since $\boldsymbol{A}$ is positive definite. So $\boldsymbol{B}^\top \boldsymbol{A}\boldsymbol{B}$ is positive definite (and clearly symmetric). □

A square matrix $\boldsymbol{A}$ is *diagonal* if $a_{ij} = 0$ for $i \ne j$. It is *lower triangular* if $a_{ij} = 0$ for $i < j$ and *upper triangular* if $a_{ij} = 0$ for $i > j$. The product of two lower(upper) triangular matrices is lower(upper) triangular. The inverse of a lower(upper) triangular matrix is lower(upper) triangular. If $\boldsymbol{A}$ is diagonal and positive definite, then all of its diagonal elements are positive. Otherwise if, say, $a_{ii} \le 0$ then, for $\boldsymbol{x} = (0\ldots,1,\ldots 0)^\top$ with 1 at the ith position,

$$\boldsymbol{x}^\top \boldsymbol{A}\boldsymbol{x} \le 0$$

contradicting the fact that $\boldsymbol{A}$ is positive definite. Now we state without proof the following theorem.

THEOREM A.2

If $\boldsymbol{A}$ is nonsingular, then there exists a nonsingular lower triangular matrix $\boldsymbol{F}$ such that $\boldsymbol{FA}$ is nonsingular upper triangular.

The proof is straightforward, but somewhat lengthy (see Anderson, 2003, Appendix A).

It follows directly that, if $\boldsymbol{A}$ is symmetric and positive definite, there exists a lower triangular matrix $\boldsymbol{F}$ such that $\boldsymbol{FAF}^\top$ is diagonal and positive definite. That is, from Theorem A.2, $\boldsymbol{FA}$ is upper triangular and nonsingular. But then, since $\boldsymbol{F}^\top$ is upper triangular, $\boldsymbol{FAF}^\top$ is also upper triangular. But it is clearly also symmetric, so it must be diagonal. Finally, since $\boldsymbol{F}$ is nonsingular it has full rank, so that $\boldsymbol{FAF}^\top$ is also positive definite by Theorem A.1.

One can now go one step further and claim that, if $\boldsymbol{A}$ is positive definite, then there exists a lower triangular matrix $\boldsymbol{G}$ such that $\boldsymbol{GAG}^\top = \boldsymbol{I}$. To show this, choose an $\boldsymbol{F}$ such that $\boldsymbol{FAF}^\top = \boldsymbol{D}$ is diagonal and positive definite. Let $\boldsymbol{D}'$ be the diagonal matrix whose diagonal elements are the positive square roots of the diagonal elements of $\boldsymbol{D}$. Choose $\boldsymbol{G} = \boldsymbol{D}'^{-1}\boldsymbol{F}$. Then $\boldsymbol{GAG}^\top = \boldsymbol{I}$.

THEOREM A.3 (Cholesky Decomposition)

If $\boldsymbol{A}$ is symmetric and positive definite, there exists a lower triangular matrix $\boldsymbol{L}$ such that $\boldsymbol{A} = \boldsymbol{LL}^\top$.

Proof

Since there exists a lower triangular matrix $\boldsymbol{G}$ such that $\boldsymbol{GAG}^\top = \boldsymbol{I}$, we must have

$$\boldsymbol{A} = \boldsymbol{G}^{-1}(\boldsymbol{G}^\top)^{-1} = \boldsymbol{G}^{-1}(\boldsymbol{G}^{-1})^\top = \boldsymbol{LL}^\top,$$

where $\boldsymbol{L} = \boldsymbol{G}^{-1}$ is lower triangular. □

Cholesky decomposition on a positive definite symmetric matrix is analogous to finding the square root of a positive real number. The IDL procedure `CHOLDC, A, D` performs Cholesky decomposition of a matrix `A` returning `L` in the lower diagonal of `A` except for its diagonal elements, which are returned in D. For example

```
PRO chol
   A = [[2,1],[1,3]]
   choldc, A, D, /double
   L = diag_matrix(D)
   L[0,1]=A[0,1]
   PRINT, L##transpose(L)
END
ENVI> chol
       2.0000000       1.0000000
       1.0000000       3.0000000
```

A.2 Vector and Inner Product Spaces

The real column vectors of dimension N, which were introduced in Chapter 1 to represent multispectral pixel intensities, provide the standard example of the more general concept of a *vector space*.

Definition A.1: A set S is a *(real) vector space* if the operations addition and scalar multiplication are defined on S so that, for $\boldsymbol{x}, \boldsymbol{y} \in S$ and $\alpha, \beta \in \mathbb{R}$,

$$\begin{aligned}
\boldsymbol{x} + \boldsymbol{y} &\in S \\
\alpha \boldsymbol{x} &\in S \\
1\boldsymbol{x} &= \boldsymbol{x} \\
0\boldsymbol{x} &= \boldsymbol{0} \\
\boldsymbol{x} + \boldsymbol{0} &= \boldsymbol{x} \\
\alpha(\boldsymbol{x} + \boldsymbol{y}) &= \alpha\boldsymbol{x} + \alpha\boldsymbol{y} \\
(\alpha + \beta)\boldsymbol{x} &= \alpha\boldsymbol{x} + \beta\boldsymbol{y}.
\end{aligned}$$

The elements of S are called *vectors*.

Another example of a vector space satisfying the above definition is the set of continuous functions $f(x)$ on the real interval $[a, b]$, which is encountered in Chapter 3 in connection with the discrete wavelet transform. Unlike the column vectors of Chapter 1, the elements of this vector space have infinite dimension. Both vector spaces are *inner product spaces* according to the following definition.

Definition A.2: A vector space S is an *inner product space* if there is a function mapping two elements $x, y \in S$ to a real number $\langle x, y\rangle$ such that

$$\langle x, y\rangle = \langle y, x\rangle$$
$$\langle x, x\rangle \geq 0$$
$$\langle x, x\rangle = 0 \text{ if and only if } x = 0.$$

In the case of the real column vectors, the inner product is defined by Equation 1.7, that is, $\langle x, y\rangle = x^\top y$. For the vector space of continuous functions, we define

$$\langle f, g\rangle = \int_a^b f(x)g(x)dx.$$

Definition A.3: Two elements x and y of an inner product space S are said to be *orthogonal* if $\langle x, y\rangle = 0$. The set of elements x_i, $i = 1 \dots n$ is *orthonormal* if $\langle x_i, x_j\rangle = \delta_{ij}$.

A finite set S of linearly independent vectors (see Definition 1.2) constitutes a *basis* for the vector space V comprising all vectors that can be expressed as a linear combination of the vectors in S. The number of vectors in the basis is called the *dimension* of V. An orthogonal basis for a finite dimensional inner product space can always be constructed by the *Gram–Schmidt orthogonalization procedure* (Press et al., 2002; Shawe-Taylor and Cristianini, 2004). If v_i, $i = 1 \dots N$, is an orthogonal basis for V, then for any $x \in V$,

$$x = \sum_{i=1}^{N} \frac{\langle x, v_i\rangle}{\langle v_i, v_i\rangle} v_i.$$

Let W be a subset of vector space V. Then it will have an orthogonal basis $\{w_1, w_2 \dots w_K\}$. For any $x \in V$, the *projection* y of x onto the subspace W is given by

$$y = \sum_{i=1}^{K} \frac{\langle x, w_i\rangle}{\langle w_i, w_i\rangle} w_i,$$

so that $y \in W$. We define the *orthogonal complement* $W^\perp$ of W as the set

$$W^\perp = \{x \in V \mid \langle x, y\rangle = 0 \text{ for all } y \in W\}.$$

It is then easy to show that the *residual vector* $\boldsymbol{y}_\perp = \boldsymbol{x} - \boldsymbol{y}$ is in $W^\perp$, that is, that $\langle \boldsymbol{y}_\perp, \boldsymbol{y} \rangle = 0$ for all $\boldsymbol{y} \in W$. Thus we can always write $\boldsymbol{x}$ as

$$\boldsymbol{x} = \boldsymbol{y} + \boldsymbol{y}_\perp,$$

where $\boldsymbol{y} \in W$ and $\boldsymbol{y}_\perp \in W^\perp$.

THEOREM A.4 (Orthogonal Decomposition Theorem)

If W is a finite-dimensional subspace of an inner product space V, then any $\boldsymbol{x} \in V$ can be written uniquely as $\boldsymbol{x} = \boldsymbol{y} + \boldsymbol{y}_\perp$, where $\boldsymbol{y} \in W$ and $\boldsymbol{y}_\perp \in W^\perp$.

A.3 Least Squares Procedures

In this section ordinary linear regression is extended to a recursive procedure for sequential data. This forms the basis of one of the neural network training algorithms derived in Appendix B. In addition, the orthogonal linear regression procedure used in Chapter 9 for radiometric normalization is explained.

A.3.1 Recursive Linear Regression

Consider the statistical model given by Equation 2.71, now in a slightly different notation:

$$Y(j) = \sum_{i=0}^{N} w_j x_i(j) + R(j), \quad j = 1 \ldots \nu. \tag{A.1}$$

This model relates the independent variables $\boldsymbol{x}(j) = (1, x_1(j) \ldots x_N(j))^\top$ to a measured quantity $Y(j)$ via the parameters $\boldsymbol{w} = (w_0, w_1 \ldots w_N)^\top$. The index ν is now intended to represent the number of measurements that have been made *so far*. The random variables $R(j)$ represent the measurement uncertainty in the realizations $y(j)$ of $Y(j)$. We assume that they are uncorrelated and normally distributed with zero mean and unit variance ($\sigma^2 = 1$), whereas the values $\boldsymbol{x}(j)$ are exact. We wish to determine the best values for parameters $\boldsymbol{w}$. Equation A.1 can be written in the terms of a design matrix $\boldsymbol{\mathcal{X}}_\nu$ as

$$\boldsymbol{Y}_\nu = \boldsymbol{\mathcal{X}}_\nu \boldsymbol{w} + \boldsymbol{R}_\nu, \tag{A.2}$$

where

$$\mathcal{X}_\nu = \begin{pmatrix} x(1)^\top \\ \vdots \\ x(\nu)^\top \end{pmatrix},$$

$$Y_\nu = (Y(1)\dots Y(\nu))^\top$$

$$R_\nu = (R(1)\dots R(\nu))^\top$$

As was shown in Chapter 2, the best solution in the least squares sense for the parameter vector w is given by

$$w(\nu) = \left[\left(\mathcal{X}_\nu^\top \mathcal{X}_\nu\right)^{-1} \mathcal{X}_\nu^\top\right] y_\nu = \Sigma(\nu)\mathcal{X}_\nu^\top y_\nu, \tag{A.3}$$

where the expression in square brackets is the pseudoinverse of $\mathcal{X}_\nu$ and where $\Sigma(\nu)$ is an estimate of the covariance matrix of w,

$$\Sigma(\nu) = \left(\mathcal{X}_\nu^\top \mathcal{X}_\nu\right)^{-1}. \tag{A.4}$$

Suppose a new observation $(x(\nu+1), y(\nu+1))$ becomes available. Now we must solve the least squares problem

$$\begin{pmatrix} Y_\nu \\ Y(\nu+1) \end{pmatrix} = \begin{pmatrix} \mathcal{X}_\nu \\ x(\nu+1)^\top \end{pmatrix} w + R_{\nu+1}. \tag{A.5}$$

With Equation A.3 the solution is

$$w(\nu+1) = \Sigma(\nu+1) \begin{pmatrix} \mathcal{X}_\nu \\ x(\nu+1)^\top \end{pmatrix}^\top \begin{pmatrix} y_\nu \\ y(\nu+1) \end{pmatrix}. \tag{A.6}$$

Inverting Equation A.4 with $\nu \to \nu+1$ we obtain a recursive formula for the new covariance matrix $\Sigma(\nu+1)$:

$$\Sigma(\nu+1)^{-1} = \begin{pmatrix} \mathcal{X}_\nu \\ x(\nu+1)^\top \end{pmatrix}^\top \begin{pmatrix} \mathcal{X}_\nu \\ x(\nu+1)^\top \end{pmatrix} = \mathcal{X}_\nu^\top \mathcal{X}_\nu + x(\nu+1)x(\nu+1^\top)$$

or

$$\Sigma(\nu+1)^{-1} = \Sigma(\nu)^{-1} + x(\nu+1)x(\nu+1)^\top. \tag{A.7}$$

To obtain a similar recursive formula for $w(\nu+1)$ we multiply Equation A.6 out, giving

$$w(\nu+1) = \Sigma(\nu+1)\left(\mathcal{X}_\nu^\top y_\nu + x(\nu+1)y(\nu+1)\right),$$

and replace y_ν with $\mathcal{X}_\nu w(\nu)$ to obtain

$$w(\nu+1) = \Sigma(\nu+1)\left(\mathcal{X}_\nu^\top \mathcal{X}_\nu w(\nu) + x(\nu+1)y(\nu+1)\right).$$

Using Equations A.4 and A.7,

$$\begin{aligned} w(\nu+1) &= \mathbf{\Sigma}(\nu+1)\Big(\mathbf{\Sigma}(\nu)^{-1}w(\nu) + x(\nu+1)y(\nu+1)\Big) \\ &= \mathbf{\Sigma}(\nu+1)\Big[\mathbf{\Sigma}(\nu+1)^{-1}w(\nu) - x(\nu+1)x(\nu+1)^{\top}w(\nu) \\ &\quad + x(\nu+1)y(\nu+1)\Big]. \end{aligned}$$

This simplifies to

$$w(\nu+1) = w(\nu) + K(\nu+1)\big[y(\nu+1) - x(\nu+1)^{\top}w(\nu)\big], \tag{A.8}$$

where the *Kalman gain* $K(\nu+1)$ is given by

$$K(\nu+1) = \mathbf{\Sigma}(\nu+1)x(\nu+1). \tag{A.9}$$

Equations A.7 through A.9 define a so-called *Kalman filter* for the least squares problem of Equation A.1. For observations

$$x(\nu+1) \quad \text{and} \quad y(\nu+1)$$

the *system response* $x(\nu+1)^{\top}w(\nu)$ is calculated and compared in Equation A.8 with the measurement $y(\nu+1)$. Then the *innovation*, that is to say the difference between the measurement and system response, is multiplied by the Kalman gain $K(\nu+1)$ determined by Equations A.7 and A.9 and the old estimate $w(\nu)$ for the parameter vector w is corrected to the new value $w(\nu+1)$.

Relation A.7 is inconvenient as it calculates the inverse of the covariance matrix $\mathbf{\Sigma}(\nu+1)$, whereas we require the noninverted form in order to determine the Kalman gain in Equation A.9. However Equations A.7 and A.9 can be reformed as follows:

$$\begin{aligned} \mathbf{\Sigma}(\nu+1) &= \Big[I - K(\nu+1)x(\nu+1)^{\top}\Big]\mathbf{\Sigma}(\nu) \\ K(\nu+1) &= \mathbf{\Sigma}(\nu)x(\nu+1)\Big[x(\nu+1)^{\top}\mathbf{\Sigma}(\nu)x(\nu+1) + 1\Big]^{-1}. \end{aligned} \tag{A.10}$$

To see this, first of all note that the second equation above is a consequence of the first equation and Equation A.9. Therefore it suffices to show that the first equation is indeed the inverse of Equation A.7:

$$\begin{aligned}\boldsymbol{\Sigma}(\nu+1)\boldsymbol{\Sigma}(\nu+1)^{-1} &= \left[\boldsymbol{I}-\boldsymbol{K}(\nu+1)\boldsymbol{x}(\nu+1)^\top\right]\boldsymbol{\Sigma}(\nu)\boldsymbol{\Sigma}(\nu+1)^{-1}\\ &= \boldsymbol{I}-\boldsymbol{K}(\nu+1)\boldsymbol{x}(\nu+1)^\top\\ &\quad+\left[\boldsymbol{I}-\boldsymbol{K}(\nu+1)\boldsymbol{x}(\nu+1)^\top\right]\boldsymbol{\Sigma}(\nu)\boldsymbol{x}(\nu+1)\boldsymbol{x}(\nu+1)^\top\\ &= \boldsymbol{I}-\boldsymbol{K}(\nu+1)\boldsymbol{x}(\nu+1)^\top+\boldsymbol{\Sigma}(\nu)\boldsymbol{x}(\nu+1)\boldsymbol{x}(\nu+1)^\top\\ &\quad-\boldsymbol{K}(\nu+1)\boldsymbol{x}(\nu+1)^\top\boldsymbol{\Sigma}(\nu)\boldsymbol{x}(\nu+1)\boldsymbol{x}(\nu+1)^\top.\end{aligned}$$

The second equality above follows from Equation A.7. But from the second of Equation A.10 we have

$$\boldsymbol{K}(\nu+1)\boldsymbol{x}(\nu+1)^\top\boldsymbol{\Sigma}(\nu)\boldsymbol{x}(\nu+1)=\boldsymbol{\Sigma}(\nu)\boldsymbol{x}(\nu+1)-\boldsymbol{K}(\nu+1)$$

and therefore

$$\begin{aligned}\boldsymbol{\Sigma}(\nu+1)\boldsymbol{\Sigma}(\nu+1)^{-1} &= \boldsymbol{I}-\boldsymbol{K}(\nu+1)\boldsymbol{x}(\nu+1)^\top+\boldsymbol{\Sigma}(\nu)\boldsymbol{x}(\nu+1)\boldsymbol{x}(\nu+1)^\top\\ &\quad-(\boldsymbol{\Sigma}(\nu)\boldsymbol{x}(\nu+1)-\boldsymbol{K}(\nu+1))\boldsymbol{x}(\nu+1)^\top=\boldsymbol{I}\end{aligned}$$

as required.

A.3.2 Orthogonal Linear Regression

In the model for ordinary linear regression described in Chapter 2 the independent variable x is assumed to be error free. If we are regressing one spectral band against another, for example, then this is manifestly not the case. If we impose the model

$$Y(\nu)-R(\nu)=a+b(X(\nu)-S(\nu)),\quad i=1\ldots m, \tag{A.11}$$

with $R(\nu)$ and $S(\nu)$ being uncorrelated, normally distributed random variables with mean zero and equal variances σ^2, then we might consider the analog of Equation 2.45 as a starting point:

$$z(a,b)=\sum_{i=1}^{m}\frac{(y(\nu)-a-bx(\nu))^2}{\sigma^2+b^2\sigma^2}. \tag{A.12}$$

Finding the minimum of Equation A.12 with respect to a and b is now more difficult because of the nonlinear dependence on b. Let us begin with the derivative with respect to a:

$$\frac{\partial z(a,b)}{\partial a}=0=-\frac{2}{\sigma^2(1+b^2)}\sum_{\nu}(y(\nu)-a-bx(\nu))$$

or

$$\hat{a} = \bar{y} - b\bar{x}. \tag{A.13}$$

Differentiating with respect to b, we obtain

$$0 = \frac{2b}{1+b^2}\sum_\nu (y(\nu) - a - bx(\nu))^2 + 2\sum_\nu (y(\nu) - a - bx(\nu))x(\nu),$$

which simplifies to

$$\sum_\nu (y(\nu) - a - bx(\nu))[b(y(\nu) - a) + x(\nu)] = 0.$$

Now substitute $\hat{a}$ for a using Equation A.13. This gives

$$\sum_\nu [y(\nu) - \bar{y} - b(x(\nu) - \bar{x})][b(y(\nu) - \bar{y} + b\bar{x}) + x(\nu)] = 0.$$

This equation is in fact only quadratic in b, since the cubic term is

$$-b^3\bar{x}\sum_\nu (x(\nu) - \bar{x}) = 0.$$

The quadratic term is, with Equation 2.47,

$$b^2\sum_\nu \left((y(\nu) - \bar{y})\bar{x} - (x(\nu) - \bar{x})(y(\nu) - \bar{y}) - (x(\nu) - \bar{x})\bar{x}\right) = -mb^2 s_{xy}.$$

The linear term is, with the definition

$$s_{yy} = \frac{1}{m}\sum_{\nu=1}^{m} (y(\nu) - \bar{y})^2,$$

given by

$$b\sum_\nu (y(\nu) - \bar{y})^2 - (x(\nu) - \bar{x})x(\nu) = mb(s_{yy} - s_{xx}),$$

since $\sum_\nu (x(\nu) - \bar{x})x(\nu) = \sum_\nu (x(\nu) - \bar{x})(x(\nu) - \bar{x})$. Similarly the constant term is

$$\sum_\nu (y(\nu) - \bar{y})x(\nu) = \sum_\nu (y(\nu) - \bar{y})(x(\nu) - \bar{x}) = ms_{xy}.$$

Thus b is a solution of the quadratic equation

$$b^2 s_{xy} + b(s_{xx} - s_{yy}) - s_{xy} = 0.$$

The solution (for positive slope) is

$$\hat{b} = \frac{(s_{yy} - s_{xx}) + \sqrt{(s_{yy} - s_{xx})^2 + 4s_{xy}^2}}{2s_{xy}}. \tag{A.14}$$

According to Patefield (1977) and Bilbo (1989) the variances in the regression parameters are given by

$$\sigma_a^2 = \frac{\sigma^2 \hat{b}(1+\hat{b}^2)}{m s_{xy}} \left(\bar{x}^2(1+\hat{\tau}) + \frac{s_{xy}}{\hat{b}} \right)$$

$$\sigma_b^2 = \frac{\sigma^2 \hat{b}(1+\hat{b}^2)}{m s_{xy}} (1+\hat{\tau}) \tag{A.15}$$

with

$$\hat{\tau} = \frac{\sigma^2 \hat{b}}{(1+\hat{b}^2) s_{xy}}. \tag{A.16}$$

If σ^2 is not known *a priori* it can be estimated by (Kendall and Stuart, 1979)

$$\hat{\sigma}^2 = \frac{m}{(m-2)(1+\hat{b}^2)} (s_{yy} - 2\hat{b} s_{xy} + \hat{b}^2 s_{xx}). \tag{A.17}$$

Listing A.1

A procedure for orthogonal linear regression.

```
PRO ortho_regress, X, Y, b, Xm, Ym, $
                   sigma_a, sigma_b, sigma, rank=rank
   m = n_elements(X)
   Xm = mean(X)
   Ym = mean(Y)
   S = correlate([X,Y],/covariance,/double)
   Sxx = S[0,0]
   Syy = S[1,1]
   Sxy = S[1,0]
   void = eigenql(S,eigenvectors=eigenvectors,/double)
; slope
   b = eigenvectors[1,0]/eigenvectors[0,0]
; standard errors
   sigma2 = m*(Syy-2*b*Sxy+b*b*Sxx)/((m-2)*(1+b^2))
   tau = sigma2*b/((1+b^2)*Sxy)
   sigma_b=sqrt( sigma2*b*(1+b^2)*(1+tau)/(m*Sxy))
   sigma_a=sqrt( sigma2*b*(1+b^2)*(Xm^2*(1+tau)+Sxy/b)$
                                          /(m*Sxy) )
   sigma = sqrt(sigma2)
   rank = r_correlate(X,Y)
END
```

It is easy to see that the estimate $\hat{b}$ (Equation A.14) can be calculated as the slope of the first principal component vector $\boldsymbol{u}$ of the covariance matrix

$$\mathbf{s} = \begin{pmatrix} s_{xx} & s_{xy} \\ s_{yx} & s_{yy} \end{pmatrix},$$

that is,

$$\hat{b} = u_2/u_1.$$

Thus orthogonal linear regression on one independent variable is equivalent to principal components analysis. This is the basis for the IDL procedure shown in Listing A.1, which performs orthogonal regression on the input arrays `X` and `Y`. The routine is used in some of the ENVI/IDL extensions described in Appendix C.

Appendix B: Efficient Neural Network Training Algorithms

The standard backpropagation algorithm introduced in Chapter 6 is notoriously slow to converge. In this appendix we will develop two additional training algorithms for the two-layer, feed-forward neural network of Figure 6.10. The first of these, *scaled conjugate gradient*, makes use of the second derivatives of the cost function with respect to the synaptic weights, that is, of the Hessian matrix. The second, the *extended Kalman filter* method, takes advantage of the statistical properties of the weight parameters themselves. Both techniques are considerably more efficient than backpropagation.

B.1 Hessian Matrix

We begin with a detailed discussion of the Hessian matrix and how to calculate it efficiently.

The Hessian matrix $\boldsymbol{H}$ for a neural network training cost function $E(\boldsymbol{w})$ is given by

$$(\boldsymbol{H})_{ij} = \frac{\partial^2 E(\boldsymbol{w})}{\partial w_i \partial w_j} \tag{B.1}$$

(see Equation 1.55). It is the (symmetric) matrix of second order partial derivatives of the cost function with respect to the synaptic weights, the latter being thought of as a single column vector

$$\boldsymbol{w} = \begin{pmatrix} \boldsymbol{w}_1^h \\ \vdots \\ \boldsymbol{w}_L^h \\ \boldsymbol{w}_1^o \\ \vdots \\ \boldsymbol{w}_K^o \end{pmatrix}$$

of length $n_w = L(N+1) + K(L+1)$ for the network architecture of Figure 6.10. Since $\boldsymbol{H}$ is symmetric, it is positive definite if and only if its eigenvalues are positive (see Section 1.3). Thus a good way to check if one is at or near a local minimum in the cost function is to examine the eigenvalues of the Hessian matrix.

The scaled conjugate gradient algorithm makes explicit use of the Hessian matrix for more efficient convergence to a minimum in the cost function. The disadvantage of using $\boldsymbol{H}$ is that it is difficult to compute. For example, for a typical classification problem with $N =$ three-dimensional input data, $L = 8$ hidden neurons, and $K = 12$ land use categories, there are

$$\big(L(N+1) + K(L+1)\big)\big(L(N+1) + K(L+1) + 1\big)/2 = 9870$$

matrix elements to determine at each iteration (allowing for symmetry). We develop in the following an efficient method (Bishop, 1995) not to calculate $\boldsymbol{H}$ directly, but rather the product $\boldsymbol{v}^\top \boldsymbol{H}$ for any vector $\boldsymbol{v}$ having n_w components.

B.1.1 *R*-Operator

Let us begin by summarizing some results from Chapter 6 for the two-layer, feed-forward network, changing the notation slightly to simplify what follows:

$$
\begin{aligned}
g' &= (g'_1 \dots g'_N)^\top && \text{input observation vector}\\
g &= \begin{pmatrix}1\\ g'\end{pmatrix} && \text{biased input observation}\\
\boldsymbol{\ell} &= (0\dots1\dots0)^\top && \text{class label}\\
\boldsymbol{I}^h &= \boldsymbol{W}^{h\top}\boldsymbol{g} && \text{activation vector for the hidden layer}\\
n'_j &= f(I^h_j),\ j = 1\dots L && \text{output signal vector from the hidden layer}\\
\boldsymbol{n} &= \begin{pmatrix}1\\ \boldsymbol{n}'\end{pmatrix} && \text{biased output signal vector}\\
\boldsymbol{I}^o &= \boldsymbol{W}^{o\top}\boldsymbol{n} && \text{activation vector for the output layer}\\
&m_k(\boldsymbol{I}^o),\ k = 1\dots K && \text{softmax output signal from } k\text{th output neuron.}
\end{aligned}
\tag{B.2}
$$

The corresponding activation functions are, for the hidden neurons,

$$f(I^h_j) = \frac{1}{1+e^{-I^h_j}}, \quad j = 1\dots L, \tag{B.3}$$

and for the output neurons,

$$m_k(\boldsymbol{I}^o) = \frac{e^{I^o_k}}{\sum_{k'=1}^{K} e^{I^o_{k'}}}, \quad k = 1\dots K. \tag{B.4}$$

The first derivatives of the local cross entropy cost function (Equation 6.34) with respect to the output and hidden weights (Equations 6.36 and 6.40) can be written concisely as

$$\begin{aligned}\frac{\partial E}{\partial \mathbf{W}^o} &= -\boldsymbol{n}\boldsymbol{\delta}^{o\top} \\ \frac{\partial E}{\partial \mathbf{W}^h} &= -\boldsymbol{g}\boldsymbol{\delta}^{h\top},\end{aligned} \tag{B.5}$$

where (see Equations 6.38 and 6.42)

$$\boldsymbol{\delta}^o = \boldsymbol{\ell} - \boldsymbol{m} \tag{B.6}$$

and

$$\begin{pmatrix} 0 \\ \boldsymbol{\delta}^h \end{pmatrix} = \boldsymbol{n} \cdot (1 - \boldsymbol{n}) \cdot \mathbf{W}^o\boldsymbol{\delta}^o. \tag{B.7}$$

(The dot denotes component-by-component multiplication.) Following Bishop (1995) we introduce the *R-operator* according to the definition

$$R_v\{\boldsymbol{x}\} := \boldsymbol{v}^\top \frac{\partial}{\partial \boldsymbol{w}}\boldsymbol{x}, \quad \boldsymbol{v}^\top = (v_1 \dots v_{n_w}).$$

We have

$$R_v\{\boldsymbol{w}\} = \boldsymbol{v}^\top \frac{\partial}{\partial \boldsymbol{w}}\boldsymbol{w} = \sum_j v_j \frac{\partial \boldsymbol{w}}{\partial w_j} = \boldsymbol{v}.$$

Note that we are now taking derivatives of vectors. This should not confuse us. For example the above result in two dimensions is

$$R_v\{\boldsymbol{w}\} = v_1\frac{\partial}{\partial w_1}(w_1\boldsymbol{i} + w_2\boldsymbol{j}) + v_2\frac{\partial}{\partial w_2}(w_1\boldsymbol{i} + w_2\boldsymbol{j}) = v_1\boldsymbol{i} + v_2\boldsymbol{j} = \boldsymbol{v}.$$

We adopt the convention that the result of applying the R-operator has the same structure as the argument to which it is applied. Thus for example

$$R_v\{\mathbf{W}^h\} = \mathbf{V}^h,$$

where $\mathbf{V}^h$, like $\mathbf{W}^h$, is an $(N+1) \times L$ matrix consisting of the first $(N+1) \times L$ components of the n_w-dimensional vector $\boldsymbol{v}$. Implicitly we set the last $n_w - (N+1)L$ components of $\boldsymbol{v}$ equal to zero.

Next we derive an expression for $\boldsymbol{v}^\top \boldsymbol{H}$ in terms of the R-operator.

$$(\boldsymbol{v}^\top\boldsymbol{H})_j = \sum_{i=1}^{n_w} v_i H_{ij} = \sum_{i=1}^{n_w} v_i \frac{\partial^2 E}{\partial w_i \partial w_j} = \sum_{i=1}^{n_w} v_i \frac{\partial}{\partial w_i}\left(\frac{\partial E}{\partial w_j}\right)$$

or

$$(v^\top H)_j = v^\top \frac{\partial}{\partial w}\left(\frac{\partial E}{\partial w_j}\right) = R_v\left\{\frac{\partial E}{\partial w_j}\right\}, \quad j = 1 \ldots n_w.$$

Since $v^\top H$ is a row vector, this can be written

$$v^\top H = R_v\left\{\frac{\partial E}{\partial w^\top}\right\} \cong \left(R_v\left\{\frac{\partial E}{\partial W^h}\right\}, R_v\left\{\frac{\partial E}{\partial W^o}\right\}\right). \tag{B.8}$$

Note the reorganization of the structure in the argument of R_v, namely $w^\top \to (W^h, W^o)$. This is merely for convenience of evaluation. Once the expressions on the right have been evaluated, the result must be "flattened" back to a row vector. Note also that Equation B.8 is understood to involve the local cost function. In order to complete the calculation we must sum over all training pairs (see Equation 6.33).

Applying R_v to Equation B.5,

$$\begin{aligned} R_v\left\{\frac{\partial E}{\partial W^o}\right\} &= -nR_v\{\delta^{o\top}\} - R_v\{n\}\delta^{o\top} \\ R_v\left\{\frac{\partial E}{\partial W^h}\right\} &= -gR_v\{\delta^{h\top}\}, \end{aligned} \tag{B.9}$$

so that, in order to evaluate Equation B.8, we need expressions for

$$R_v\{n\},\ R_v\{\delta^{o\top}\},\ \text{and } R_v\{\delta^{h\top}\}.$$

This is somewhat tedious, but well worth the effort.

B.1.1.1 Determination of $R_v\{n\}$

With the notation of Equation B.2 we can write

$$R_v\{n\} = \begin{pmatrix} 0 \\ R_v\{n'\} \end{pmatrix} \tag{B.10}$$

and, from the chain rule,

$$R_v\{n'\} = n' \cdot (1 - n') \cdot R_v\{I^h\} \tag{B.11}$$

where, by differentiation of I^h, we evaluate

$$R_v\{I^h\} = V^{h\top} g. \tag{B.12}$$

Note that, according to our convention, $V^{h\top}$ must be interpreted as an $L \times (N+1)$-dimensional matrix, since the argument I^h is a vector of length L and the result must have the same structure.

B.1.1.2 Determination of $R_v\{\delta^o\}$

With Equations B.6 and B.2 we get

$$R_v\{\boldsymbol{\delta}^o\} = -R_v\{\boldsymbol{m}\} = -v^\top \frac{\partial \boldsymbol{m}}{\partial w} = -v^\top \frac{\partial \boldsymbol{m}}{\partial \boldsymbol{I}^o} \cdot \frac{\partial \boldsymbol{I}^o}{\partial w} = -\frac{\partial \boldsymbol{m}}{\partial \boldsymbol{I}^o} \cdot R_v\{\boldsymbol{I}^o\}.$$

But from Equation B.4 it is easy to see that

$$\frac{\partial \boldsymbol{m}}{\partial \boldsymbol{I}^o} = \boldsymbol{m} \cdot (\mathbf{1} - \boldsymbol{m}),$$

and therefore

$$R_v\{\boldsymbol{\delta}^o\} = -\boldsymbol{m} \cdot (\mathbf{1} - \boldsymbol{m}) \cdot R_v\{\boldsymbol{I}^o\}. \tag{B.13}$$

Similarly with Equations B.2 we get

$$R_v\{\boldsymbol{I}^o\} = \boldsymbol{W}^{o\top} R_v\{\boldsymbol{n}\} + \boldsymbol{V}^{o\top} \boldsymbol{n}, \tag{B.14}$$

where $R_v\{\boldsymbol{n}\}$ is given by Equations B.10 through B.12.

B.1.1.3 Determination of $R_v\{\delta^h\}$

We begin by writing Equation B.7 in the form

$$\begin{pmatrix} 0 \\ \boldsymbol{\delta}^h \end{pmatrix} = \begin{pmatrix} 0 \\ f'(\boldsymbol{I}^h) \end{pmatrix} \cdot \boldsymbol{W}^o \boldsymbol{\delta}^o,$$

where

$$f'(\boldsymbol{I}^h) = (f'(I_1^h) \dots f'(I_L^h))^\top$$

and where the prime on f denotes differentiation with respect to its argument, $f'(x) = df(x)/dx$. Operating with $R_v\{\cdot\}$ and applying the chain rule, we obtain

$$\begin{pmatrix} 0 \\ R_v\{\boldsymbol{\delta}^h\} \end{pmatrix} = \begin{pmatrix} 0 \\ f''(\boldsymbol{I}^h) \end{pmatrix} \cdot \begin{pmatrix} 0 \\ R_v\{\boldsymbol{I}^h\} \end{pmatrix} \cdot \boldsymbol{W}^o \boldsymbol{\delta}^o$$
$$+ \begin{pmatrix} 0 \\ f'(\boldsymbol{I}^h) \end{pmatrix} \cdot \boldsymbol{V}^o \boldsymbol{\delta}^o + \begin{pmatrix} 0 \\ f'(\boldsymbol{I}^h) \end{pmatrix} \cdot \boldsymbol{W}^o R_v\{\boldsymbol{\delta}^o\}. \tag{B.15}$$

Finally, substitute the derivatives of the logistic function

$$f'(\boldsymbol{I}^h) = \boldsymbol{n}'(\mathbf{1} - \boldsymbol{n}')$$
$$f''(\boldsymbol{I}^h) = \boldsymbol{n}'(\mathbf{1} - \boldsymbol{n}')(\mathbf{1} - 2\boldsymbol{n}')$$

into Equation B.15 to obtain

$$\begin{pmatrix} 0 \\ R_v\{\delta^h\} \end{pmatrix} = \boldsymbol{n} \cdot (\boldsymbol{1} - \boldsymbol{n}) \cdot \left[(\boldsymbol{1} - 2\boldsymbol{n}) \cdot \begin{pmatrix} 0 \\ R_v\{\boldsymbol{I}^h\} \end{pmatrix} \cdot \boldsymbol{W}^o \delta^o + \boldsymbol{V}^o \delta^o + \boldsymbol{W}^o R_v\{\delta^o\} \right] \tag{B.16}$$

in which all of the terms on the right have now been determined. As already mentioned, we have done everything so far in terms of the local cost function. The final step in the calculation involves summing over all of the training examples. This concludes the evaluation of Equation B.8.

B.1.2 Calculating the Hessian

To calculate the Hessian matrix for the neural network, we evaluate Equation B.8 successively for the vectors

$$v_1^\top = (1, 0, 0 \ldots 0) \quad \ldots \quad v_{n_w}^\top = (0, 0, 0 \ldots 1)$$

and build up $\boldsymbol{H}$ row for row:

$$\boldsymbol{H} = \begin{pmatrix} v_1^\top \boldsymbol{H} \\ \vdots \\ v_{n_w}^\top \boldsymbol{H} \end{pmatrix}.$$

The excerpt from the IDL program `FFNCG__DEFINE` (see Appendix C.2.10) shown in Listing B.1 extends the object class `FFN` introduced in Chapter 6 and implements a vectorized version of the above determination of $v^\top \boldsymbol{H}$ (method `FFNCG::ROP`) and $\boldsymbol{H}$ (method `FFNCG::HESSIAN`). The eigenvalues of $\boldsymbol{H}$ are calculated in method `FFNCG::EIGENVALUES`.

B.2 Scaled Conjugate Gradient Training

The backpropagation algorithm of Chapter 6 attempts to minimize the cost function *locally*, that is, weight updates are made immediately after presentation of a single training pair to the network. We will now consider a *global* approach aimed at minimization of the full cost function (Equation 6.33) which we denote in the following by $E(w)$. The symbol w is, as before, the n_w-component vector of synaptic weights.

Listing B.1

Excerpt from the object class FFNCG.

```
FUNCTION FFNCG::Rop, V
  nw = self.LL*(self.NN+1)+ self.KK *(self.LL+1)
; reform V to dimensions of Wh and Wo and transpose
  VhT=transpose(reform(V[0:self.LL*(self.NN+1)-1], $
                 self.LL,self.NN+1))
  Vo=reform(V[self.LL*(self.NN+1):*],self.KK,self.LL+1)
  VoT = transpose(Vo)
; transpose the weights
  Wo  = *self.Wo
  WoT = transpose(Wo)
; vectorized forward pass
  M = self-> vForwardPass(*self.GTs,N)
; evaluation of v^T.H
  Zeroes = fltarr(self.ntp)
  D_o=*self.LTs-M                                 ;d^o
  RIh=VhT##(*self.GTs)                            ;Rv{I^h}
  RN=N*(1-N)*[[Zeroes],[RIh]]                     ;Rv{n}
  RIo=WoT##RN + VoT##N                            ;Rv{I^o}
  Rd_o=-M*(1-M)*RIo                               ;Rv{d^o}
  Rd_h=N*(1-N)*((1-2*N)*[[Zeroes],[RIh]]*(Wo##D_o) $
        + Vo##D_o + Wo##Rd_o)
  Rd_h=Rd_h[*,1:*]                                ;Rv{d^h}
  REo=-N##transpose(Rd_o)-RN##transpose(D_o);Rv{dE/dWo}
  REh=-*self.GTs##transpose(Rd_h)                 ;Rv{dE/dWh}
  RETURN, [REh[*],REo[*]]                         ;v^T.H
END

FUNCTION FFNCG::Hessian
   nw = self.LL*(self.NN+1)+self.KK*(self.LL+1)
   v = diag_matrix(fltarr(nw)+1.0)
   H = fltarr(nw,nw)
   FOR i=0,nw-1 DO H[*,i] = self -> Rop(v[*,i])
   RETURN, H
END

FUNCTION FFNCG::Eigenvalues
   t1 = systime(2)
   H = self-> Hessian()
   PRINT,'Hessian␣timer(CPU):␣',systime(2)-t1
   H = (H+transpose(H))/2
```

Let the gradient of the cost function at the point w be $\mathbf{g}(w)$,* that is,

$$\mathbf{g}(w) = \frac{\partial}{\partial w} E(w).$$

* The symbol $\mathbf{g}$ for gradient should not to be confused with the observation vector g.

The Hessian matrix

$$(\boldsymbol{H})_{ij} = \frac{\partial^2 E(\boldsymbol{w})}{\partial w_i \partial w_j}, \quad i,j = 1 \ldots n_w$$

can then be expressed conveniently as the outer product

$$\boldsymbol{H} = \frac{\partial}{\partial \boldsymbol{w}} \mathbf{g}(\boldsymbol{w})^\top. \tag{B.17}$$

B.2.1 Conjugate Directions

The search for a minimum in the cost function can be visualized as tracing out a series of points in the space of synaptic weight parameters,

$$\boldsymbol{w}^1, \boldsymbol{w}^2 \ldots \boldsymbol{w}^{k-1}, \boldsymbol{w}^k, \boldsymbol{w}^{k+1} \ldots,$$

where the point $\boldsymbol{w}^k$ is determined by minimizing $E(\boldsymbol{w})$ along some *search direction* $\boldsymbol{d}^{k-1}$ which originated at the preceding point $\boldsymbol{w}^{k-1}$. This is illustrated in Figure B.1 and corresponds to the vector equation

$$\boldsymbol{w}^k = \boldsymbol{w}^{k-1} + \alpha_{k-1}\boldsymbol{d}^{k-1}. \tag{B.18}$$

Here, $\boldsymbol{d}^{k-1}$ is a unit vector along the chosen search direction, and the scalar α_{k-1} minimizes the cost function along that direction:

$$\alpha_{k-1} = \arg\min_{\alpha} E\left(\boldsymbol{w}^{k-1} + \alpha \boldsymbol{d}^{k-1}\right).$$

If, starting from $\boldsymbol{w}^k$, we now wish to take the next minimizing step in the weight space, it is not efficient simply to choose, as in backpropagation, the direction of the local gradient $\mathbf{g}(\boldsymbol{w}_k)$ at the new starting point $\boldsymbol{w}^k$. Since the cost function has been minimized along the direction $\boldsymbol{d}^{k-1}$ at the point $\boldsymbol{w}^{k-1} + \alpha_{k-1}\boldsymbol{d}^{k-1}$, its gradient along that direction is zero,

$$\mathbf{g}(\boldsymbol{w}^k)^\top \boldsymbol{d}^{k-1} = 0, \tag{B.19}$$

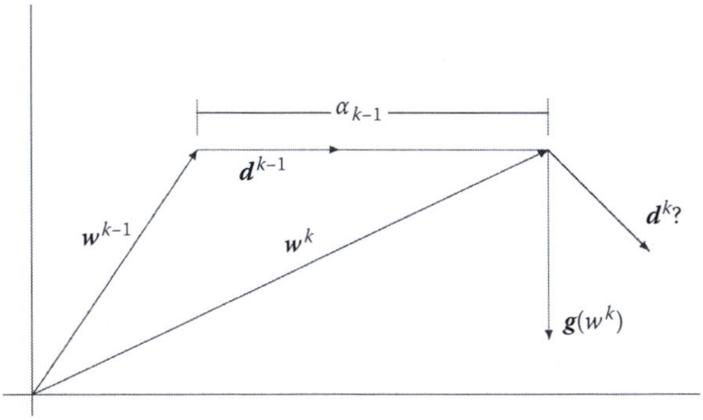

FIGURE B.1
Search directions in weight space.

as indicated in Figure B.1. Since the algorithm has just succeeded in reducing the gradient of the cost function along $\boldsymbol{d}^{k-1}$ to zero, we prefer to choose the new search direction $\boldsymbol{d}^k$ so that the component of the gradient along the old search direction remains as small as possible. Otherwise we are undoing what we have just accomplished. Therefore we choose $\boldsymbol{d}^k$ according to the condition

$$\mathbf{g}(\boldsymbol{w}^k + \alpha \boldsymbol{d}^k)^\top \boldsymbol{d}^{k-1} = 0. \tag{B.20}$$

But to first order in α we have, with Equation B.17,

$$\mathbf{g}(\boldsymbol{w}^k + \alpha \boldsymbol{d}^k)^\top = \mathbf{g}(\boldsymbol{w}^k)^\top + \alpha {\boldsymbol{d}^k}^\top \frac{\partial}{\partial \boldsymbol{w}} \mathbf{g}(\boldsymbol{w}^k)^\top = \mathbf{g}(\boldsymbol{w}^k)^\top + \alpha {\boldsymbol{d}^k}^\top \boldsymbol{H}$$

and Equation B.20 is, together with Equation B.19, equivalent to

$${\boldsymbol{d}^k}^\top \boldsymbol{H} \boldsymbol{d}^{k-1} = 0. \tag{B.21}$$

Directions that satisfy Equation B.21 are referred to as *conjugate directions*.

B.2.2 Minimizing a Quadratic Function

Although the neural network cost function is not quadratic in the synaptic weights, within a sufficiently small region of weight space it can be approximated as a quadratic function. We describe in the following an efficient procedure to find the global minimum of a quadratic function of $\boldsymbol{w}$ having the general form (see Equation 1.54)

$$E(\boldsymbol{w}) = E_0 + \boldsymbol{b}^\top \boldsymbol{w} + \frac{1}{2} \boldsymbol{w}^\top \boldsymbol{H} \boldsymbol{w}, \tag{B.22}$$

where $\boldsymbol{b}$ and $\boldsymbol{H}$ are constant and the matrix $\boldsymbol{H}$ is positive definite. This will form the basis of the neural network training algorithm presented in Appendix B.2.3.

The local gradient of $E(\boldsymbol{w})$ at the point $\boldsymbol{w}$ is given by

$$\mathbf{g}(\boldsymbol{w}) = \frac{\partial}{\partial \boldsymbol{w}} E(\boldsymbol{w}) = \boldsymbol{b} + \boldsymbol{H}\boldsymbol{w},$$

and it vanishes at the global minimum $\boldsymbol{w}^*$,

$$\boldsymbol{b} + \boldsymbol{H}\boldsymbol{w}^* = \boldsymbol{0}. \tag{B.23}$$

Now let $\{\boldsymbol{d}^k \mid k = 1 \ldots n_w\}$ be a set of n_w conjugate directions satisfying Equation B.21,*

$${\boldsymbol{d}^k}^\top \boldsymbol{H} \boldsymbol{d}^\ell = 0 \quad \text{for } k \neq \ell,\ k, \ell = 1 \ldots n_w. \tag{B.24}$$

* It can be shown that such a set always exists (see Bishop, 1995).

The search directions, d^k, are in fact linearly independent. In order to demonstrate this let us assume the contrary, that is, that there exists an index k and constants $\alpha_{k'}$, $k' \neq k$, not all of which are zero, such that

$$d^k = \sum_{\substack{k'=1 \\ k' \neq k}}^{n_w} \alpha_{k'} d^{k'}.$$

Substituting this into Equation B.24, we have at once

$$\alpha_{k'} d^{k'^\top} H d^{k'} = 0 \quad \text{for } k' \neq k$$

and, since H is positive definite,

$$\alpha_{k'} = 0 \quad \text{for } k' \neq k.$$

The assumption leads to a contradiction, hence the d^k are indeed linearly independent. The conjugate directions thus constitute a (nonorthogonal) vector basis for the entire weight space.

In the search for the global minimum, suppose we begin at an arbitrary point w^1 and express the vector $w^* - w^1$ spanning the distance to the global minimum as a linear combination of the basis vectors d^k:

$$w^* - w^1 = \sum_{k=1}^{n_w} \alpha_k d^k. \tag{B.25}$$

Further, define

$$w^k = w^1 + \sum_{\ell=1}^{k-1} \alpha_\ell d^\ell \tag{B.26}$$

and split Equation B.25 up into n_w steps

$$w^{k+1} = w^k + \alpha_k d^k, \quad k = 1 \ldots n_w. \tag{B.27}$$

At the kth step the search starts at the point w^k and proceeds a distance α_k along the conjugate direction d^k. After n_w such steps the global minimum w^* is reached, since from Equations B.25 through B.27 it follows that

$$w^* = w^1 + \sum_{k=1}^{n_w} \alpha_k d^k = w^2 + \sum_{k=2}^{n_w} \alpha_k d^k = \cdots = w^{n_w} + \alpha_{n_w} d^{n_w} = w^{n_w+1}.$$

We get the necessary step sizes α_k from Equation B.25 by multiplying from the left with $d^{\ell\top} H$,

$$d^{\ell\top} H w^* - d^{\ell\top} H w^1 = \sum_{k=1}^{n_w} \alpha_k d^{\ell\top} H d^k.$$

From Equations B.23 and B.24 we can write this as

$$-\boldsymbol{d}^{\ell^\top}(\boldsymbol{b} + \boldsymbol{H}\boldsymbol{w}^1) = \alpha_\ell \boldsymbol{d}^{\ell^\top}\boldsymbol{H}\boldsymbol{d}^\ell,$$

so that an explicit formula for the step sizes is given by

$$\alpha_\ell = -\frac{\boldsymbol{d}^{\ell^\top}(\boldsymbol{b} + \boldsymbol{H}\boldsymbol{w}^1)}{\boldsymbol{d}^{\ell^\top}\boldsymbol{H}\boldsymbol{d}^\ell}, \quad \ell = 1 \dots n_w.$$

But with Equations B.24 and B.26,

$$\boldsymbol{d}^{k^\top}\boldsymbol{H}\boldsymbol{w}^k = \boldsymbol{d}^{k^\top}\boldsymbol{H}\boldsymbol{w}^1 + 0,$$

and therefore, replacing index k by ℓ,

$$\boldsymbol{d}^{\ell^\top}\boldsymbol{H}\boldsymbol{w}^\ell = \boldsymbol{d}^{\ell^\top}\boldsymbol{H}\boldsymbol{w}^1.$$

The step lengths are thus

$$\alpha_\ell = -\frac{\boldsymbol{d}^{\ell^\top}(\boldsymbol{b} + \boldsymbol{H}\boldsymbol{w}^\ell)}{\boldsymbol{d}^{\ell^\top}\boldsymbol{H}\boldsymbol{d}^\ell}, \quad \ell = 1 \dots n_w.$$

Finally, using the notation $\mathbf{g}^\ell = \mathbf{g}(\boldsymbol{w}^\ell) = \boldsymbol{b} + \boldsymbol{H}\boldsymbol{w}^\ell$ and substituting $\ell \rightarrow k$,

$$\alpha_k = -\frac{\boldsymbol{d}^{k^\top}\mathbf{g}^k}{\boldsymbol{d}^{k^\top}\boldsymbol{H}\boldsymbol{d}^k}, \quad k = 1 \dots n_w. \tag{B.28}$$

For want of a better alternative we can choose the first search direction along the negative local gradient

$$\boldsymbol{d}^1 = -\mathbf{g}^1 = -\frac{\partial}{\partial \boldsymbol{w}}E(\boldsymbol{w}^1).$$

(Note that $\boldsymbol{d}^1$ is not a unit vector.) We move, according to Equation B.28, a distance

$$\alpha_1 = \frac{\boldsymbol{d}^{1^\top}\boldsymbol{d}^1}{\boldsymbol{d}^{1^\top}\boldsymbol{H}\boldsymbol{d}^1}$$

along this direction to the point $\boldsymbol{w}^2$, at which the local gradient $\mathbf{g}^2$ is orthogonal to $\boldsymbol{d}^1$. We then choose the new conjugate search direction $\boldsymbol{d}^2$ as a linear combination of the two:

$$\boldsymbol{d}^2 = -\mathbf{g}^2 + \beta_1 \boldsymbol{d}^1$$

or, at the kth step,

$$\boldsymbol{d}^{k+1} = -\mathbf{g}^{k+1} + \beta_k \boldsymbol{d}^k. \tag{B.29}$$

We get the coefficient β_k from Equations B.29 and B.21 by multiplication on the left with ${d^k}^\top H$:

$$0 = -{d^k}^\top H g^{k+1} + \beta_k {d^k}^\top H d^k,$$

from which follows

$$\beta_k = \frac{{g^{k+1}}^\top H d^k}{{d^k}^\top H d^k}. \tag{B.30}$$

Equations B.27 through B.30 constitute a recipe with which, starting at an arbitrary point w^1 in weight space, the global minimum of the quadratic function (Equation B.22) is found in precisely n_w steps.

B.2.3 Algorithm

Returning now to the nonquadratic neural net cost function $E(w)$ we will apply the above method to minimize it. We must take two things into consideration.

First of all, the Hessian matrix H is neither constant nor everywhere positive definite. When H is not positive definite it can happen that Equation B.28 leads to a step along the wrong direction—the denominator might turn out to be negative. Therefore, we replace Equation B.28 with*

$$\alpha_k = -\frac{{d^k}^\top g^k}{{d^k}^\top H d^k + \lambda_k \|d^k\|^2}, \quad k = 1 \dots n_w. \tag{B.31}$$

The constant λ_k is supposed to ensure that the denominator in Equation B.31 is always positive. It is initialized for $k = 1$ with a small numerical value. If, at the kth iteration, it is determined that

$$\delta_k := {d^k}^\top H d^k + \lambda_k \|d^k\|^2 < 0,$$

then λ_k is replaced by the larger value $\bar{\lambda}_k$ given by

$$\bar{\lambda}_k = 2\left(\lambda_k - \frac{\delta_k}{\|d^k\|^2}\right). \tag{B.32}$$

This ensures that the denominator in Equation B.31 becomes positive again. Note that this increase in λ_k has the effect of *decreasing* the step size α_k, as is apparent from Equation B.31.

Second, we must take into account any deviation of the cost function from its local quadratic approximation. Such deviations are to be expected

* This corresponds to the substitution $H \rightarrow H + \lambda_k I$, where I is the identity matrix.

for large step sizes α_k. As a measure of the *quadricity* of $E(w)$ along the chosen step length we can use the ratio

$$\Delta_k = -\frac{2\big(E(w^k) - E(w^k + \alpha_k d^k)\big)}{\alpha_k {d^k}^\top \mathbf{g}^k}. \tag{B.33}$$

With $\boldsymbol{\alpha}_k$ given by Equation B.28, this quantity is precisely 1 for a strictly quadratic function like Equation B.22. Therefore we can use the following heuristic: For the $(k+1)$st iteration

$$\begin{aligned} \text{if } \Delta_k > 3/4, &\quad \lambda_{k+1} := \lambda_k/2 \\ \text{if } \Delta_k < 1/4, &\quad \lambda_{k+1} := 4\lambda_k \\ \text{else,} &\quad \lambda_{k+1} := \lambda_k. \end{aligned}$$

In other words, if the local quadratic approximation looks good according to criterion of Equation B.33, then the step size can be increased (λ_{k+1} is reduced relative to λ_k). If this is not the case then the step size is decreased (λ_{k+1} is made larger).

All of these leads us, at last, to the following algorithm (Moeller, 1993):

Algorithm (Scaled Conjugate Gradient)

1. Initialize the synaptic weights w with random numbers, set $k = 0$, $\lambda = 0.001$ and $d = -\mathbf{g} = -\partial E(w)/\partial w$.
2. Set $\delta = d^\top H d + \lambda \|d\|^2$. If $\delta < 0$, set $\lambda = 2(\lambda - \delta/\|d\|^2)$ and $\delta = -d^\top H d$. Save the current cost function $E1 = E(w)$.
3. Determine the step size $\alpha = -d^\top \mathbf{g}/\delta$ and new synaptic weights $w = w + \alpha d$.
4. Calculate the quadricity $\Delta = -2(E1 - E(w))/(\alpha d^\top \mathbf{g})$. If $\Delta < 1/4$, restore the old weights: $w = w - \alpha d$, set $\lambda = 4\lambda$, $d = -\mathbf{g}$ and go to 2.
5. Set $k = k + 1$. If $\Delta > 3/4$ set $\lambda = \lambda/2$.
6. Determine the new local gradient $\mathbf{g} = \partial E(w)/\partial w$ and the new search direction $d = -\mathbf{g} + \beta d$, whereby, if $k \bmod n_w \neq 0$ then $\beta = \mathbf{g}^\top H d/(d^\top H d)$ else $\beta = 0$.
7. If $E(w)$ is small enough stop, else go to 2.

A few remarks on this algorithm:

1. The integer k counts the total number of iterations. Whenever $k \bmod n_w = 0$, then exactly n_w weight updates have been carried out and the minimum of a truly quadratic function would have been reached. This is taken as a good stage at which to restart the search along the negative local gradient $-g$ rather than continuing

along the current conjugate direction $\boldsymbol{d}$. One expects that approximation errors will gradually corrupt the determination of the conjugate directions and the "fresh start" is intended to counter this.

2. Whenever the quadricity condition is not filled, that is, whenever $\Delta < 1/4$, the last weight update is canceled and the search again restarted along $-\mathbf{g}$.
3. Since the Hessian only occurs in the forms $\boldsymbol{d}^\top \boldsymbol{H}$, and $\mathbf{g}^\top \boldsymbol{H}$, these quantities can be determined efficiently with the R-operator method.

Listing B.2 shows an excerpt from the object class `FFNCG` (see Appendix C.2.10) extending `FFN`. It lists the training method that implements the scaled conjugate gradient algorithm.

B.3 Kalman Filter Training

In this section the recursive linear regression method described in Appendix A will be applied to train the feed-forward neural network of Figure 6.10. The appropriate cost function in this case is the quadratic function (Equation 6.30) or, more specifically, its local version

$$E(\nu) = \frac{1}{2}\|\boldsymbol{\ell}(\nu) - \boldsymbol{m}(\nu)\|^2. \tag{B.34}$$

However our algorithm will also minimize the cross entropy cost function, as will be mentioned later.

We begin with consideration of the training process of an isolated neuron. Figure B.2 depicts an output neuron in the network during presentation of the νth training pair $(\boldsymbol{g}(\nu), \boldsymbol{\ell}(\nu))$. The neuron receives its input from the hidden layer (input vector $\boldsymbol{n}(\nu)$) and generates the softmax output signal

$$m_k(\nu) = \frac{e^{\boldsymbol{w}_k^{o\top}(\nu)\boldsymbol{n}(\nu)}}{\sum_{k'=1}^{K} e^{\boldsymbol{w}_{k'}^{o\top}(\nu)\boldsymbol{n}(\nu)}},$$

which is compared to the desired output $\ell_k(\nu)$. It is easy to show that differentiation of m_k with respect to $\boldsymbol{w}_k^o$ yields

$$\frac{\partial}{\partial \boldsymbol{w}_k^o} m_k(\nu) = m_k(\nu)(1 - m_k(\nu))\boldsymbol{n}(\nu) \tag{B.35}$$

and with respect to $\boldsymbol{n}$,

$$\frac{\partial}{\partial \boldsymbol{n}} m_k(\nu) = m_k(\nu)(1 - m_k(\nu))\boldsymbol{w}_k^o(\nu). \tag{B.36}$$

Listing B.2
Excerpt from the object class FFNCG.

```
PRO FFNCG::Train, key=key
   IF n_elements(key) EQ 0 THEN key=0
; if validation option then train only on odd examples
   IF key THEN void = $
   where((lindgen(self.np) MOD 2) EQ 0, $
                      complement=indices) $
   ELSE indices = lindgen(self.np)
   self.ntp = n_elements(indices)
   self.GTs = ptr_new((*self.Gs)[indices,*])
   self.LTs = ptr_new((*self.Ls)[indices,*])
   w = [(*self.Wh)[*],(*self.Wo)[*]]
   nw = n_elements(w)
   g = self->gradient()
   d = -g     ; search direction, row vector
   k = 0L
   lambda = 0.001
   window,12,xsize=600,ysize=400, $
      title='FFN(scaled␣conjugate␣gradient)'
   wset,12
   progressbar = Obj_New('progressbar', $
     Color='blue', Text='0', $
     title='Training:␣epoch␣No...',xsize=250,ysize=20)
   progressbar->start
   eivminmax = '?'
   REPEAT BEGIN
      IF progressbar->CheckCancel() THEN BEGIN
         PRINT,'Training␣interrupted'
         progressbar->Destroy
         RETURN
      ENDIF
      d2 = total(d*d)                  ; d^2
      dTHd = total(self->Rop(d)*d)     ; d^T.H.d
      delta = dTHd+lambda*d2
      IF delta LT 0 THEN BEGIN
         lambda = 2*(lambda-delta/d2)
         delta = -dTHd
      ENDIF
      E1 = self->cost(key)             ; E(w)
      (*self.cost_array)[k] = E1
      IF key THEN $
          (*self.valid_cost_array)[k]=self->cost(2)
      dTg = total(d*g)                 ; d^T.g
      alpha = -dTg/delta
      dw = alpha*d
```

Listing B.2
Excerpt from the object class FFNCG (continued).

```
      w = w+dw
      *self.Wh = reform(w[0:self.LL*(self.NN+1)-1], $
         self.LL,self.NN+1)
      *self.Wo = reform(w[self.LL*(self.NN+1):*], $
         self.KK,self.LL+1)
      E2 = self->cost(key)          ; E(w+dw)
      Ddelta = -2*(E1-E2)/(alpha*dTg); quadricity
      IF Ddelta LT 0.25 THEN BEGIN
         w = w - dw    ; undo weight change
         *self.Wh = reform(w[0:self.LL*(self.NN+1)-1],$
            self.LL,self.NN+1)
         *self.Wo = reform(w[self.LL*(self.NN+1):*], $
            self.KK,self.LL+1)
         lambda = 4*lambda        ; decrease step size
         IF lambda GT 1e20 THEN $; if step too small
           k=self.iterations   $ ;  then give up
         ELSE d = -g              ;  else restart
      END ELSE BEGIN
         k++
         IF Ddelta GT 0.75 THEN lambda = lambda/2
         g = self->gradient()
         IF k MOD nw EQ 0 THEN BEGIN
             beta = 0
             eivs = self->eigenvalues()
             eivminmax = string(min(eivs)/max(eivs),$
                format='(F10.6)')
         END ELSE beta = total(self->Rop(g)*d)/dTHd
         d = beta*d-g
         PLOT,*self.cost_array,xrange=[0,k>100L], $
            color=0, background='FFFFFF'XL, $
            ytitle='cross␣entropy', $
            xtitle='Epoch␣['+ $
             'min(lambda)/max(lambda)='+ $
             eivminmax+']'
         IF key THEN oplot, $
            *self.valid_cost_array,color=0,linestyle=2
      ENDELSE
      progressbar->Update,k*100/self.iterations, $
         text=strtrim(k,2)
   ENDREP UNTIL k GE self.iterations
   progressbar->Destroy
END
```

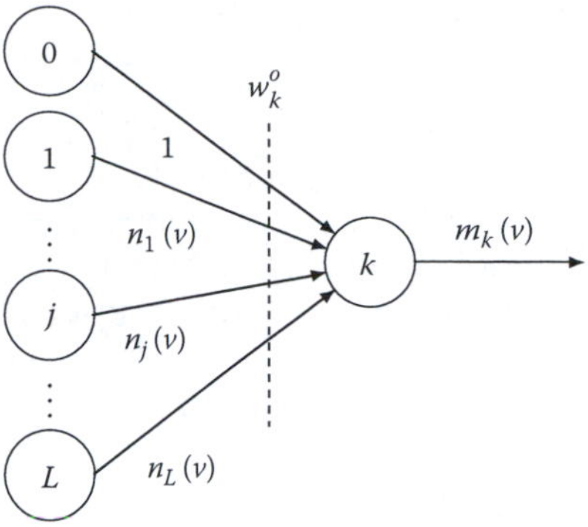

FIGURE B.2
An isolated output neuron.

B.3.1 Linearization

We shall drop for the time being the indices on $\boldsymbol{w}_k^o$, m_k, and ℓ_k, writing them simply as $\boldsymbol{w}$, m, and ℓ. The weight vectors for the other $K-1$ output neurons are considered to be frozen, so that m can be thought of as being a function of $\boldsymbol{w}$ only:

$$m(\nu) = m(\boldsymbol{w}(\nu)^\top \boldsymbol{n}(\nu)).$$

The weight vector $\boldsymbol{w}(\nu)$ is an approximation to the desired synaptic weight vector for our isolated output neuron, one which has been achieved so far in the training process, after presentation of the first ν labeled training observations. A linear approximation to $m(\nu+1)$ can be obtained by expanding in a first order Taylor series about the point $\boldsymbol{w}(\nu)$,

$$m(\nu+1) \approx m(\boldsymbol{w}(\nu)^\top \boldsymbol{n}(\nu+1)) + \left(\frac{\partial}{\partial \boldsymbol{w}} m(\boldsymbol{w}(\nu)^\top \boldsymbol{n}(\nu+1))\right)^\top (\boldsymbol{w} - \boldsymbol{w}(\nu)).$$

From Equation B.35 we then have

$$m(\nu+1) \approx \hat{m}(\nu+1) + \hat{m}(\nu+1)(1-\hat{m}(\nu+1))\boldsymbol{n}(\nu+1)^\top(\boldsymbol{w} - \boldsymbol{w}(\nu)), \quad \text{(B.37)}$$

where $\hat{m}(\nu+1)$ is given by

$$\hat{m}(\nu+1) = m(\boldsymbol{w}(\nu)^\top \boldsymbol{n}(\nu+1)).$$

The caret indicates that the signal is calculated from the next (i.e., the $(\nu+1)$st) training input, but using the current (i.e., the νth) weights. With the definition of the *linearized input*

$$\boldsymbol{a}(\nu) = \hat{m}(\nu)(1-\hat{m}(\nu))\boldsymbol{n}(\nu)^\top \quad \text{(B.38)}$$

we can write Equation B.37 in the form

$$m(\nu+1) \approx \boldsymbol{a}(\nu+1)\boldsymbol{w} + [\hat{m}(\nu+1) - \boldsymbol{a}\nu+1)\boldsymbol{w}(\nu)].$$

The term in square brackets is — to first order — the error that arises from the fact that the neuron's output signal is *not* simply linear in $\boldsymbol{w}$. If we neglect it altogether, then we get the *linearized* neuron output signal

$$m(\nu+1) = \boldsymbol{a}(\nu+1)\boldsymbol{w}.$$

Note that $\boldsymbol{a}$ has been defined in Equation B.38 as a row vector. In order to calculate the synaptic weight vector $\boldsymbol{w}$, we can now apply the theory of recursive linear regression developed in Appendix A. We simply identify the parameter vector $\boldsymbol{w}$ with the synaptic weight vector, y with the desired output ℓ, and observation $\boldsymbol{x}(\nu+1)^T$ with $\boldsymbol{a}(\nu+1)$. We then have the least squares problem

$$\begin{pmatrix} \boldsymbol{\ell}_\nu \\ \ell(\nu+1) \end{pmatrix} = \begin{pmatrix} \mathcal{A}_\nu \\ \boldsymbol{a}(\nu+1) \end{pmatrix} \boldsymbol{w} + \boldsymbol{R}_{\nu+1}$$

(see Equation A.5). The Kalman filter equations for the recursive solution of this problem are unchanged:

$$\begin{aligned} \boldsymbol{\Sigma}(\nu+1) &= \left[\boldsymbol{I} - \boldsymbol{K}(\nu+1)\boldsymbol{a}(\nu+1)\right]\boldsymbol{\Sigma}(\nu) \\ \boldsymbol{K}(\nu+1) &= \boldsymbol{\Sigma}(\nu)\boldsymbol{a}(\nu+1)^\top\left[\boldsymbol{a}(\nu+1)\boldsymbol{\Sigma}(\nu)\boldsymbol{a}(\nu+1)^\top + 1\right]^{-1}, \end{aligned} \tag{B.39}$$

while the recursive expression for the parameter vector (Equation A.8) can be improved somewhat by replacing the linear approximation to the neuron output $\boldsymbol{a}(\nu+1)\boldsymbol{w}(\nu)$ by the actual output for the $(\nu+1)$st training observation, namely $\hat{m}(\nu+1)$, so we have

$$\boldsymbol{w}(\nu+1) = \boldsymbol{w}(\nu) + \boldsymbol{K}(\nu+1)\left[\ell(\nu+1) - \hat{m}(\nu+1)\right]. \tag{B.40}$$

B.3.2 Algorithm

The recursive calculation of $\boldsymbol{w}$ is depicted in Figure B.3. The input is the current weight vector $\boldsymbol{w}(\nu)$, its covariance matrix $\boldsymbol{\Sigma}(\nu)$, and the output vector of the hidden layer $\boldsymbol{n}(\nu+1)$ obtained by propagating the next input observation $\boldsymbol{g}(\nu+1)$ through the network. After determining the linearized input $\boldsymbol{a}(\nu+1)$ from Equation B.38, the Kalman gain $\boldsymbol{K}(\nu+1)$ and the new covariance matrix $\boldsymbol{\Sigma}(\nu+1)$ are calculated with Equation B.39. Finally, the weights are updated according to Equation B.40 to give $\boldsymbol{w}(\nu+1)$ and the procedure is repeated.

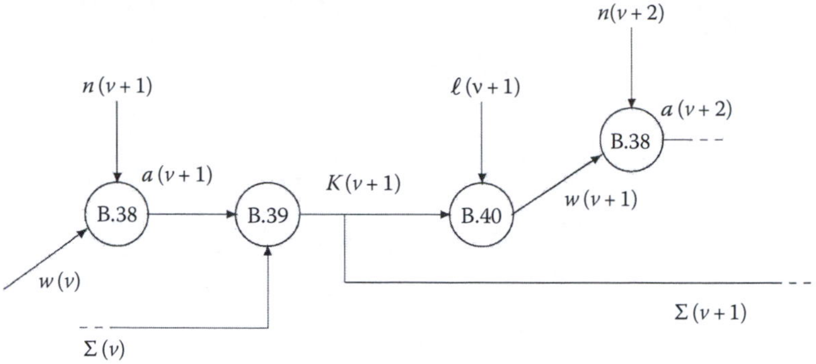

FIGURE B.3
Determination of the synaptic weights for an isolated neuron with the Kalman filter.

To make our notation explicit for the output neurons, we substitute

$$
\begin{aligned}
\ell(\nu) &\to \ell_k(\nu) \\
\boldsymbol{w}(\nu) &\to \boldsymbol{w}_k^o(\nu) \\
\hat{m}(\nu+1) &\to m_k\left(\boldsymbol{w}_k^{o\top}(\nu)\boldsymbol{n}(\nu+1)\right) \\
\boldsymbol{a}(\nu+1) &\to \boldsymbol{a}_k^o(\nu+1) = \hat{m}_k(\nu+1)(1-\hat{m}_k(\nu+1))\boldsymbol{n}(\nu+1)^\top \\
\boldsymbol{K}(\nu) &\to \boldsymbol{K}_k^o(\nu) \\
\boldsymbol{\Sigma}(\nu) &\to \boldsymbol{\Sigma}_k^o(\nu),
\end{aligned}
$$

for $k = 1 \ldots K$. Then Equation B.40 becomes

$$\boldsymbol{w}_k^o(\nu+1) = \boldsymbol{w}_k^o(\nu) + \boldsymbol{K}_k^o(\nu+1)\big[\ell_k(\nu+1) - \hat{m}_k(\nu+1)\big], \quad k = 1\ldots K. \quad \text{(B.41)}$$

Recalling that we wish to minimize the local quadratic cost function (Equation B.34) note that the expression in square brackets in Equation B.41 is in fact the negative derivative of $E(\nu)$ with respect to the output signal of the neuron, that is,

$$\ell_k(\nu+1) - m_k(\nu+1) = -\frac{\partial E(\nu+1)}{\partial m_k(\nu+1)},$$

so that Equation B.41 can be expressed in the form

$$\boldsymbol{w}_k^o(\nu+1) = \boldsymbol{w}_k^o(\nu) - \boldsymbol{K}_k^o(\nu+1)\left[\frac{\partial E(\nu+1)}{\partial m_k(\nu+1)}\right]_{\hat{m}_k(\nu+1)}. \quad \text{(B.42)}$$

With this observation, we can turn consideration to the hidden neurons, making the substitutions

$$
\begin{aligned}
w(\nu) &\rightarrow w_j^h(\nu) \\
\hat{m}(\nu+1) &\rightarrow \hat{n}_j(\nu+1) = f\left(w_j^{h\top}(\nu)g(\nu+1)\right) \\
a(\nu+1) &\rightarrow a_j^h(\nu+1) = \hat{n}_j(\nu+1)(1-\hat{n}_j(\nu+1))g(\nu+1)^\top \\
K(\nu) &\rightarrow K_j^h(\nu) \\
\Sigma(\nu) &\rightarrow \Sigma_j^h(\nu),
\end{aligned}
$$

for $j = 1 \ldots L$. Then, analogously to Equation B.42, the update equation for the weight vector of the jth hidden neuron is

$$w_j^h(\nu+1) = w_j^h(\nu) - K_j^h(\nu+1)\left[\frac{\partial E(\nu+1)}{\partial n_j(\nu+1)}\right]_{\hat{n}_j(\nu+1)}. \tag{B.43}$$

To obtain the partial derivative in Equation B.43, we differentiate the cost function (Equation B.34) applying the chain rule:

$$\frac{\partial E(\nu+1)}{\partial n_j(\nu+1)} = -\sum_{k=1}^{K}(\ell_k(\nu+1) - m_k(\nu+1))\frac{\partial m_k(\nu+1)}{\partial n_j(\nu+1)}.$$

From Equation B.36, noting that $(w_k^o)_j = W_{jk}^o$, we have

$$\frac{\partial m_k(\nu+1)}{\partial n_j(\nu+1)} = m_k(\nu+1)(1-m_k(\nu+1))W_{jk}^o(\nu+1).$$

Combining the last two equations,

$$\frac{\partial E(\nu+1)}{\partial n_j(\nu+1)} = -\sum_{k=1}^{K}(\ell_k(\nu+1) - m_k(\nu+1))m_k(\nu+1)(1-m_k(\nu+1))W_{jk}^o(\nu+1),$$

which we can write more compactly as

$$\frac{\partial E(\nu+1)}{\partial n_j(\nu+1)} = -W_{j\cdot}^o(\nu+1)\beta^o(\nu+1), \tag{B.44}$$

where $W_{j\cdot}^o$ is the jth *row* (!) of the output layer weight matrix, and where

$$\beta^o(\nu+1) = (\ell(\nu+1) - m(\nu+1)) \cdot m(\nu+1) \cdot (1 - m(\nu+1)).$$

As usual, the dot indicates component-by-component multiplication. The correct update relation for the weights of the jth hidden neuron is therefore

$$w_j^h(\nu+1) = w_j^h(\nu) + K_j^h(\nu+1)\left[W_{j\cdot}^o(\nu+1)\beta^o(\nu+1)\right]. \tag{B.45}$$

Apart from initialization of the covariance matrices $\mathbf{\Sigma}_j^h(0)$, $\mathbf{\Sigma}^o(0)$, the Kalman training procedure has no adjustable parameters whatsoever. The initial covariance matrices are simply taken to be proportional to the corresponding identity matrices:

$$\mathbf{\Sigma}_j^h(0) = Z\mathbf{I}^h, \quad \mathbf{\Sigma}_k^o(0) = Z\mathbf{I}^o, \quad Z \gg 1,\ j = 1 \ldots L,\ k = 1 \ldots K,$$

where
$\mathbf{I}^h$ is the $(N+1) \times (N+1)$ identity matrix
$\mathbf{I}^o$ is the $(L+1) \times (L+1)$ identity matrix

We choose $Z = 100$ and obtain the following algorithm:

Algorithm (Kalman Filter Training)

1. Set $\nu = 0$, $\mathbf{\Sigma}_j^h(0) = 100\mathbf{I}^h, j = 1 \ldots L$, $\mathbf{\Sigma}_k^o(0) = 100\mathbf{I}^o$, $k = 1 \ldots K$ and initialize the synaptic weight matrices $\mathbf{W}^h(0)$ and $\mathbf{W}^o(0)$ with random numbers.
2. Choose a training pair $(\mathbf{g}(\nu+1), \boldsymbol{\ell}(\nu+1))$ and determine the hidden layer output vector

$$\hat{\mathbf{n}}(\nu+1) = \begin{pmatrix} 1 \\ f\left(\mathbf{W}^h(\nu)^\top \mathbf{g}(\nu+1)\right) \end{pmatrix}$$

and with it the quantities

$$\mathbf{a}_j^h(\nu+1) = \hat{n}_j(\nu+1)(1 - \hat{n}_j(\nu+1))\mathbf{g}(\nu+1)^\top, \quad j = 1 \ldots L,$$

$$\hat{m}_k(\nu+1) = m_k\left(\mathbf{w}_k^{o\top}(\nu)\hat{\mathbf{n}}(\nu+1)\right)$$

$$\mathbf{a}_k^o(\nu+1) = \hat{m}_k(\nu+1)(1 - \hat{m}_k(\nu+1))\hat{\mathbf{n}}(\nu+1)^\top, \quad k = 1 \ldots K$$

and

$$\boldsymbol{\beta}^o(\nu+1) = (\boldsymbol{\ell}(\nu+1) - \hat{\mathbf{m}}(\nu+1)) \cdot \hat{\mathbf{m}}(\nu+1) \cdot (\mathbf{1} - \hat{\mathbf{m}}(\nu+1)).$$

3. Determine the Kalman gains for all of the neurons according to

$$\mathbf{K}_k^o(\nu+1) = \mathbf{\Sigma}_k^o(\nu)\mathbf{a}_k^o(\nu+1)^\top \left[\mathbf{a}_k^o(\nu+1)\mathbf{\Sigma}_k^o(\nu)\mathbf{a}_k^o(\nu+1)^\top + 1\right]^{-1},$$
$$k = 1 \ldots K$$

$$\mathbf{K}_j^h(\nu+1) = \mathbf{\Sigma}_j^h(\nu)\mathbf{a}_j^h(\nu+1)^\top \left[\mathbf{a}_j^h(\nu+1)\mathbf{\Sigma}_j^h(\nu)\mathbf{a}_j^h(\nu+1)^\top + 1\right]^{-1},$$
$$j = 1 \ldots L$$

Listing B.3
Excerpt from the object class FFNKAL.

```
PRO FFNKAL:: train, key=key
   IF n_elements(key) EQ 0 THEN key=0
; define update matrices for Wh and Wo
   dWh = dblarr(self.LL,self.NN+1)
   dWo = dblarr(self.KK,self.LL+1)
   iter = 0L
   iter100 = 0L
   progressbar = Obj_New('progressbar', $
     Color='blue', Text='0',$
     title='Training:␣example␣number...', $
     xsize=250,ysize=20)
   progressbar->start
   window,12,xsize=600,ysize=400, $
     title='FFN(Kalman␣filter)'
   wset,12
   REPEAT BEGIN
      IF progressbar->CheckCancel() THEN BEGIN
         PRINT,'Training␣interrupted'
         progressbar->Destroy
         RETURN
      ENDIF
; select trainig pair at random
      ell = long(self.np*randomu(seed))
; use only odd ones if validation option is set
      IF key EQ 1 THEN ell = (ell*2 MOD (self.np-2))+1
      x=(*self.Gs)[ell,*]
      y=(*self.Ls)[ell,*]
; send it through the network
      m=self->forwardPass(x)
; error at output
      e=y-m
; loop over the output neurons
      FOR k=0,self.KK-1 DO BEGIN
;    linearized input (column vector)
         Ao = m[k]*(1-m[k])*(*self.N)
;    Kalman gain
         So = (*self.So)[*,*,k]
         SA = So##Ao
         Ko = SA/((transpose(Ao)##SA)[0]+1)
;    determine delta for this neuron
         dWo[k,*] = Ko*e[k]
;    update its covariance matrix
         So = So - Ko##transpose(Ao)##So
         (*self.So)[*,*,k] = So
      ENDFOR
```

Listing B.3

Excerpt from the object class FFNKAL (continued).

```
113 ; update the output weights
114       *self.Wo = *self.Wo + dWo
115 ; backpropagated error
116       beta_o =e*m*(1-m)
117 ; loop over the hidden neurons
118       FOR j=0,self.LL-1 DO BEGIN
119 ;    linearized input (column vector)
120          Ah = X*(*self.N)[j+1]*(1-(*self.N)[j+1])
121 ;    Kalman gain
122          Sh = (*self.Sh)[*,*,j]
123          SA = Sh##Ah
124          Kh = SA/((transpose(Ah)##SA)[0]+1)
125 ;    determine delta for this neuron
126          dWh[j,*] = Kh*((*self.Wo)[*,j+1]##beta_o)[0]
127 ;    update its covariance matrix
128          Sh = Sh - Kh##transpose(Ah)##Sh
129          (*self.Sh)[*,*,j] = Sh
130       ENDFOR
131 ; update the hidden weights
132       *self.Wh = *self.Wh + dWh
133 ; record cost history
134       IF iter MOD 100 EQ 0 THEN BEGIN
135          IF key THEN BEGIN
136             (*self.cost_array)[iter100]= $
137              alog10(self->cost(1)) ; training cost
138             (*self.valid_cost_array)[iter100]= $
139              alog10(self->cost(2)) ; validation cost
140          END ELSE (*self.cost_array)[iter100]= $
141              alog10(self->cost(0)) ; full training cost
142          iter100 = iter100+1
143          progressbar->Update,iter*100/self.iterations,$
144                text=strtrim(iter,2)
145          PLOT,*self.cost_array,xrange=[0,iter100], $
146             color=0,background='FFFFFF'XL,$
147             xtitle='Iterations/100)', $
148             ytitle='log(cross_entropy)'
149          IF key THEN oplot, *self.valid_cost_array, $
150             color=0,linestyle=2
151       END
152       iter=iter+1
153    ENDREP UNTIL iter EQ self.iterations
154    progressbar->Destroy
155 END
```

4. Update the synaptic weight matrices:

$$\boldsymbol{w}_k^o(\nu+1) = \boldsymbol{w}_k^o(\nu) + \boldsymbol{K}_k^o(\nu+1)[\ell_k(\nu+1) - \hat{m}_k(\nu+1)], \quad k = 1 \ldots K$$

$$\boldsymbol{w}_j^h(\nu+1) = \boldsymbol{w}_j^h(\nu) + \boldsymbol{K}_j^h(\nu+1)\left[\boldsymbol{W}_{j\boldsymbol{\cdot}}^o(\nu+1)\boldsymbol{\beta}^o(\nu+1)\right], \quad j = 1 \ldots L$$

5. Determine the new covariance matrices:

$$\boldsymbol{\Sigma}_k^o(\nu+1) = \left[\boldsymbol{I}^o - \boldsymbol{K}_k^o(\nu+1)\boldsymbol{a}_k^o(\nu+1)\right]\boldsymbol{\Sigma}_k^o(\nu), \quad k = 1 \ldots K$$

$$\boldsymbol{\Sigma}_j^h(\nu+1) = \left[\boldsymbol{I}^h - \boldsymbol{K}_j^h(\nu+1)\boldsymbol{a}_j^h(\nu+1)\right]\boldsymbol{\Sigma}_j^h(\nu), \quad j = 1 \ldots L$$

6. If the overall cost function (Equation 6.30) is sufficiently small, stop, else set $\nu = \nu + 1$ and go to 2.

This method was originally suggested by Shah and Palmieri (1990), who called it the *multiple extended Kalman algorithm* (MEKA),* and explained in

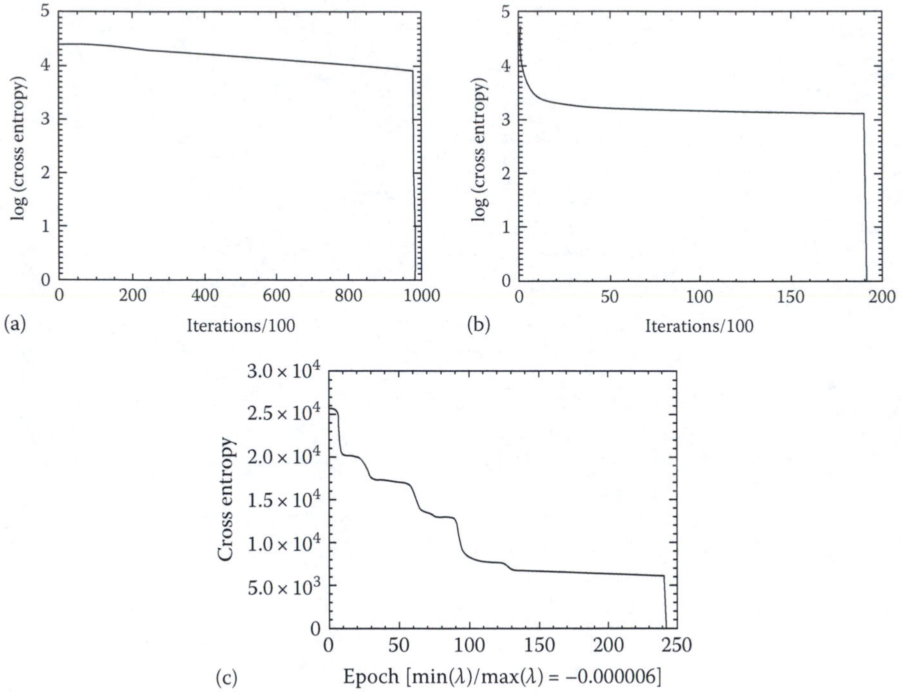

FIGURE B.4
(a) Training with backpropagation, (b) with Kalman filter, and (c) with scaled conjugate gradient. Note that the upper two plots are logarithmic.

---

* *Multiple*, because the algorithm is applied to multiple neurons. The adjective *extended* characterizes Kalman filter methods that linearize nonlinear models by using a first order Taylor expansion.

detail by Hayken (1994). Its IDL implementation is given in Listing B.3 showing an excerpt from the object class `FFNKAL` which extends `FFN` (see Appendix C.2.10).

B.4 A Neural Network Classifier with Hybrid Training

The Kalman filter algorithm is designed for minimization of the quadratic cost function of Equation 6.30 rather than of the cross entropy, Equation 6.33, which is more appropriate to the softmax output signal generated by the neural network. However when one is minimized, so is the other. The ENVI/IDL extension `FFN_RUN` described in Appendix C.2.10 implements both the Kalman filter and scaled conjugate gradient training algorithms, beginning with the former in order to approach a minimum quickly, and then using the latter to refine the synaptic weights. (The scaled conjugate gradient algorithm is guaranteed to reduce the overall cost function at each iteration.) The switch is achieved with the `GET_WEIGHTS()` and `PUT_WEIGHTS()` functions defined in the parent object class `FFN`. Convergence is extremely fast when compared to backpropagation (see Figure B.4).

Appendix C: ENVI Extensions in IDL

This appendix gives installation instructions and documentation for all of the ENVI/IDL extensions described in the text and their associated routines. The programs are available from the author's website: `http://mcanty.homepage.t-online.de/software.html`

C.1 Installation

To install the complete extension package

1. Place the program files (extension `.PRO`) anywhere in the IDL `!PATH`.
2. Place the file `MADVIEWHELP.PDF` anywhere in the IDL `!HELP_PATH`.
3. Place the dynamic link library (DLL) file `PROV_MEANS.DLL` in the operating system path (e.g., the `WINDOWS` directory for Windows systems) or in the `!DLM_PATH`. The compiled DLL is provided for Windows only. Unix flavors and Mac OS will need to create it from the C source code `PROV_MEANS.C` provided. See the IDL documentation for `MAKE_DLL.PRO`.
4. Most of the programs with filenames of the form `<program name>_RUN.PRO` will create their own menu items automatically at startup, so if you wish to call an extension directly from the ENVI menu system, copy it additionally to the `SAVE_ADD` directory before invoking ENVI.
5. (This step is only necessary for the calculation of structure heights with `CALCHEIGHT_RUN.PRO` and can be skipped if desired.) Place the file `MJC_CURSOR_MOTION.PRO` in the `SAVE_ADD` directory. In `File/Preferences/User Defined Motion Routine` in the ENVI main menu enter the string `MJC_CURSOR_MOTION`.
6. Programs with the prefix `GPU` (e.g., `GPUKPCA_RUN.PRO`) can take advantage of NVIDIA's *Compute Unified Device Architecture* (CUDA), and require the GPULIB API available from

   `http://www.txcorp.com/products/GPULib/`

   in order to run, irrespective of whether CUDA is available or not. See the above Web site for installation instructions.
7. Many of the routines make use of David Fanning's object class `PROGRESS-BAR__DEFINE.PRO` and associated dependencies, so his

COYOTE library must also be in the IDL !PATH. The library may be downloaded from

http://www.dfanning.com/documents/programs.html

C.2 Extensions

Table C.1 lists all of the ENVI extensions discussed in the text together with the auxiliary programs upon which they depend. If copied to the SAVE_ADD

TABLE C.1

ENVI Extensions and Their Required Subroutines

Extension	Dependency	Sub-Dependency
kpca_run	center	
	gausskernel_matrix	
	gpugausskernel_matrix	**gpulib**
	coyote	
	gpulib	
dwt_run	dwt__define	
	ortho_regress	
atwt_run	atwt__define	
	ortho_regress	
quality_index_run	quality_index	
	phase_corr	
calcheight_run	mjc_cursor_motion	
c_correction_run		
contour_match_run	ci__define	
	coyote	
maxlike_run	difference	remove_duplicates
gausskernel_run	minf_bracket	
	minf_parabolic	
	difference	remove_duplicates
	gausskernel_matrix	
	gpugausskernel_matrix	**gpulib**
	gpulib	
ffn_run	ffnkal__define	ffn__define
		coyote
	ffncg__define	ffn__define
		coyote
	difference	
	coyote	

TABLE C.1 (continued)
ENVI Extensions and Their Required Subroutines

Extension	Dependency	Sub-Dependency
svm_run	difference	remove_duplicates
plr_run	**coyote**	
plr_reclass_run	**coyote**	
ct_run		
mcnemar_run		
ffn3ab_run	ffn3kal__run	ffn3__define
		coyote
	difference	remove_duplicates
	coyote	
gpukkmeans_run	gausskernel_matrix	
	gpugausskernel_matrix	**gpulib**
	class_lookup_table	
	coyote	
	gpulib	
hcl_run	hcl	**coyote**
	class_lookup_table	
fkm_run	fkm	**coyote**
	cluster_fkm	
	class_lookup_table	
em_run	em	**coyote**
		intersection
	dwt__define	
	class_lookup_table	
	coyote	
som_run	**coyote**	
segment_class_run	segment_class	
mean_shift_run	class_lookup_table	
	mean_shift	
	coyote	
mad_run	mad_iter	gen_eigenproblem
		covpm__define
		coyote
	coyote	
mad_view_run	em	**coyote**
	coyote	
radcal_run	ortho_regress	
	wishart	
	coyote	

directory, most of the extensions may be called from the ENVI menu system. They are described in detail below. In addition, the file headers of all extensions and associated programs are given in reStructuredText (rst) format.

C.2.1 Kernel Principal Components Analysis

The ENVI extension `GPUKPCA_RUN` for kernel principal components analysis (Section 4.4.2) is invoked from the ENVI main menu as shown in Figure C.1. The user is queried for an input file, a training sample size, the number of kernel principal components to retain, and the kernel parameter `NSCALE`. The Gaussian kernel parameter, γ, is calculated from the latter as

$$\gamma = \frac{1}{2(\text{NSCALE}\,\sigma)^2}, \tag{C.1}$$

where $\sigma = \langle \|g(\nu) - g(\nu')\| \rangle$ is the average Euclidean distance between the training observations. The value of γ is printed to the IDL console. Finally, the output destination (memory or file) must be selected.

After centering and diagonalizing the kernel matrix, a plot of the logarithm of the eigenvalues is displayed (see Figure 4.12). Then the projection is carried out in two passes through the image. On the first pass the matrix column, row and overall sums required for centering are accumulated. These are applied on the second pass as each image pixel is projected. If CUDA is available then both centering and projection are performed on the GPU, resulting in a considerable speedup.

This extension does not make use of the ENVI tiling facility and is not intended to be used with large images.

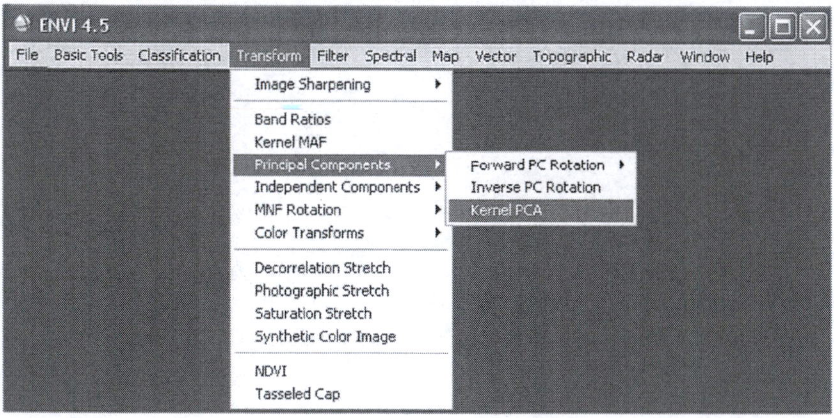

FIGURE C.1
Kernel principal component analysis from the ENVI menu.

Listing C.1
HEADER: GPUKPCA_RUN.PRO.

```
; :Description:
;       Performs kernel principal components analysis
;       using a Gaussian kernel::
;          Nielsen, A. A. and Canty, M. J. (2008).
;          Kernel principal component analysis
;          for change detections. In SPIE Europe
;          Remote Sensing Conference, Cardiff,
;          Great Britain, 15-18 September, volume 7109.
; :Params:
;       event:  in, required
;          if called from the ENVI menu
; :Uses:
;       ENVI::
;       CENTER::
;       GAUSSKERNEL_MATRIX::
;       GPUGAUSSKERNEL_MATRIX::
;       COYOTE::
;       GPULIB
```

Listing C.2
HEADER: CENTER.PRO.

```
; :Description:
;        center a kernel matrix::
;           Shawe-Taylor, J. and Cristianini, N. (2004).
;           Kernel Methods for Pattern Analysis.
;           Cambridge University Press.
; :Params:
;        K:  in, required, type = array of float
;           kernel matrix to be centered
; :Returns:
;        array of float
```

Listing C.3
HEADER: GAUSSKERNEL_MATRIX.PRO.

```
; :Description:
;        Returns array of Gaussian kernel functions
;        for data matrices G1 and G2.
;        G1 is an N x m array,
;        G2 is an N x n array.
;        Returned array is n x m.
;        If G2 not present then returns a symmetric,
;        positive definite Guassian kernel matrix
; :Params:
;        G1:  in, required
;           data matrix
;        G2:  in, optional
;           data matrix
```

```
; :Keywords:
;        gma: in,out,optional,type=float
;             if not given, calculated from the data:
;             GMA=1/(2*(nscale*scale)^2) where
;             scale = average distance between observations
;             in the feature space
;        nscale: in, optional,type=float
;             multiple of scale for GMA (default 1.0)
```

Listing C.4
HEADER: GPUGAUSSKERNEL_MATRIX.PRO.

```
; :Description:
;        Returns array of Gaussian kernel functions
;        for data matrices G1_gpu and G2_gpu.
;        G1_gpu is an N x m array,
;        G2_gpu is an N x n array.
;        If G2_gpu not present then returns a symmetric,
;        positive definite Guassian kernel matrix
; :Returns:
;        {GPUHANDLE} to n x m array
; :Params:
;        G1_gpu:  in, required, type={GPUHANDLE}
;           data matrix
;        G2_gpu:  in, optional, type={GPUHANDLE}
;           data matrix
; :Keywords:
;        gma: in,out,optional,type=float
;             if not given, calculated from the data:
;             GMA=1/(2*(nscale*scale)^2) where
;             scale = average distance between observations
;             in the feature space
;        nscale: in, optional,type=float
;             multiple of scale for GMA (default 1.0)
; :Uses:
;        GPULib
```

C.2.2 Discrete Wavelet Transform Fusion

`DWT_RUN` is an ENVI extension for pan-sharpening using image fusion with the discrete wavelet transform as discussed in Section 5.3.4. Its invocation from the ENVI main menu is shown in Figure C.2.

The low and high resolution images must be georeferenced (often they are acquired simultaneously from the same platform). The user is queried for the (spatial/spectral subset of the) image to be sharpened and then for the corresponding panchromatic or high resolution image. This image should overlap the low resolution image completely. Then the output destination (memory or file) must be entered.

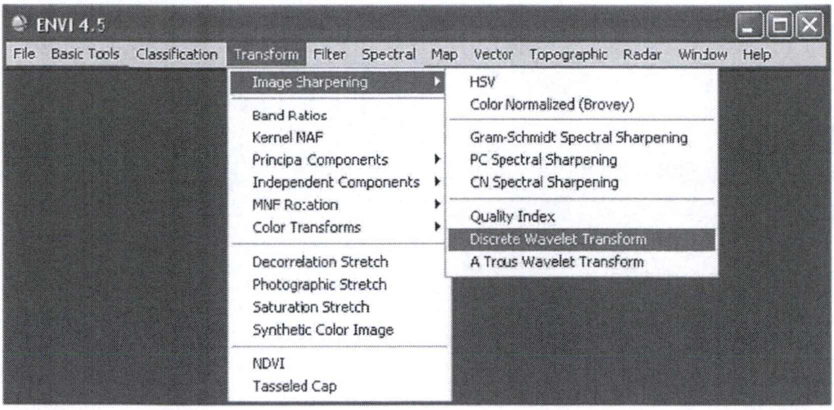

FIGURE C.2
Wavelet fusion in the ENVI menu.

During the calculation scatter plots are shown of the wavelet coefficients for the low and high resolution bands.

Listing C.5

HEADER: DWT_RUN.PRO.

```
; :Description:
;       ENVI extension for panchromatic sharpening
;       under ARSIS model with Mallat's discrete wavelet
;       transform and Daubechies wavelets
; :Params:
;      event:  in, required
;         if called from the ENVI menu
; :Uses:
;       ENVI::
;       DWT__DEFINE::
;       ORTHO_REGRESS
```

Listing C.6

HEADER: DWT__DEFINE.PRO.

```
; :Description:
;       Discrete wavelet transform object class using
;       Daubechies wavelets for construction of pyramid
;       representations of images::
;         Ranchin, T. and Wald, L. (2000).
;         Fusion of high spatial and spectral resolution
;         images: the ARSIS concept and its implementation.
;         Photogrammetric Engineering and Remote Sensing,
;         66(1), 49   61.
; :Params:
;       image: in,required
```

```
;          grayscale image to be transformed
; :Examples:
;        dwt = Obj_New("DWT",image)
```

Listing C.7
HEADER: ORTHO_REGRESS.PRO.

```
; :Description:
;      Orthogonal regression between two vectors::
;        Canty, M. J., Nielsen, A. A., and Schmidt, M.
;        (2004). Automatic radiometric normalization
;        of multitemporal satellite imagery.
;        Remote Sensing of Environment,
;        91(3-4), 441   451. I
; :Params:
;       X: in,required
;       Y: in,required
;       b: out, required
;       Xm: out, required
;       Ym: out, required
;       sigma_a: out, required
;       sigma_b: out, required
;       sigma(RMSE): out, required
; :Keywords:
;       rank: out,optional
;           Spearman rank order correlation coefficient
;           (two-element vector containing the rank
;           correlation coefficient and the two-sided
;           significance of its deviation from zero)
```

C.2.3 *À Trous* Wavelet Transform Fusion

ATWT_RUN is an ENVI extension for pan-sharpening using image fusion with the à trous wavelet transform as discussed in Section 5.3.5. Its location in the ENVI main menu is also shown in Figure C.2 and the input requirements are the same as for DWT_RUN.

Listing C.8
HEADER: ATWT_RUN.PRO.

```
; :Description:
;       ENVI extension for panchromatic sharpening under
;       ARSIS model with "a trous" wavelet transform.
; :Params:
;      event:  in, required
;         if called from the ENVI menu
; :Uses:
;       ENVI::
;       ATWT__DEFINE::
;       ORTHO_REGRESS
```

Listing C.9
HEADER: ATWT_DEFINE.PRO.

```
; :Description:
;       A trous wavelet transform object class using
;       cubic spline wavelet for panchromatic
;       sharpening of multispectral images::
;           Aiazzi, B., Alparone, L., Baronti, S., and
;           Garzelli, A. (2002). Context-driven
;           fusion of high spatial and spectral resolution
;           images based on oversampled ;multiresolution
;           analysis. IEEE Transactions on Geoscience and
;           Remote ;Sensing, 40(10), 2300âĂŞ2312.
; :Params:
;       image: in,required
;          grayscale image to be transformed
; :Examples:
;        dwt = Obj_New("ATWT",image)
```

C.2.4 Quality Index

QUALITY_INDEX_RUN is an ENVI extension to determine the Wang–Bovik index for radiometric fidelity of a pan-sharpened image (see Section 5.3.6 and Figure C.2). The user is first prompted for the reference multispectral image to which the sharpened image is to be compared and then for the image whose quality is to be determined. The reference image must completely overlap the pan-sharpened image. The latter is downsampled to the same resolution as the reference image and then the image bands are matched to the nearest pixel using phase correlation (Section 5.5.1), plots of which are shown during the calculation. The calculated indices are displayed in an IDL widget.

Listing C.10
HEADER: QUALITY_INDEX_RUN.PRO.

```
; :Description:
;       ENVI extension for radiometric comparison of two
;       multispectral images
; :Params:
;      event:  in, required
;         if called from the ENVI menu
; :Uses:
;       ENVI::
;       PHASE_CORR::
;       QUALITY_INDEX
```

Listing C.11
HEADER: QUALITY_INDEX.PRO.

```
; :Description:
;       Determine the Wang-Bovik quality index for a
```

```
;        pan-sharpened image band::
;          Ref: Wang and Bovik, IEEE Signal Processing
;          Letters9(3) 2002, 81-84
; :Params:
;       band1:  in, required
;          reference spectral band
;       band2:  in, required
;          degraded pan-sharpend spectral band
; :Keywords:
;       blocksize: in, optional
;           size of image blocks to calculate
;           index (default 8, i.e., 8 x 8)
```

Listing C.12

HEADER: PHASE_CORR.PRO.

```
; :Description:
;        Returns relative offset [xoff,yoff] of two images
;        using phase correlation.
;        Maximum offset should not exceed +- 5 pixels
;        in each dimension
;        Returns -1 if dimensions are not equal
;
;        Ref: H, Shekarforoush et al. (1995) INRIA 2707
; :Params:
;        im1: input,required
;        im2: input,required
; :Keywords:
;        display:  input,optional
;           show a surface plot if the
;           correlation in window with display
;           number display
;        subpixel: input, optional
;           returns result to subpixel accuracy if
;           set, otherwise nearest integer (default)
```

C.2.5 Calculating Heights of Man-Made Structures in High-Resolution Imagery

`CALC_HEIGHT_RUN` is an ENVI extension to determine heights of vertical structures in high-resolution images (such as QuickBird and IKONOS). It makes use of rational function models (RFMs) accompanying ortho-ready imagery distributed by the image providers as discussed in Section 5.4.3. It is implemented as a small graphical user interface (GUI) using IDL widgets (see Figure 5.13) and may be called from the active ENVI display menu as shown in Figure C.3.

After displaying a high resolution image and starting the GUI, an RFM file must be loaded from the GUI menu

```
File/Load RPC File
```

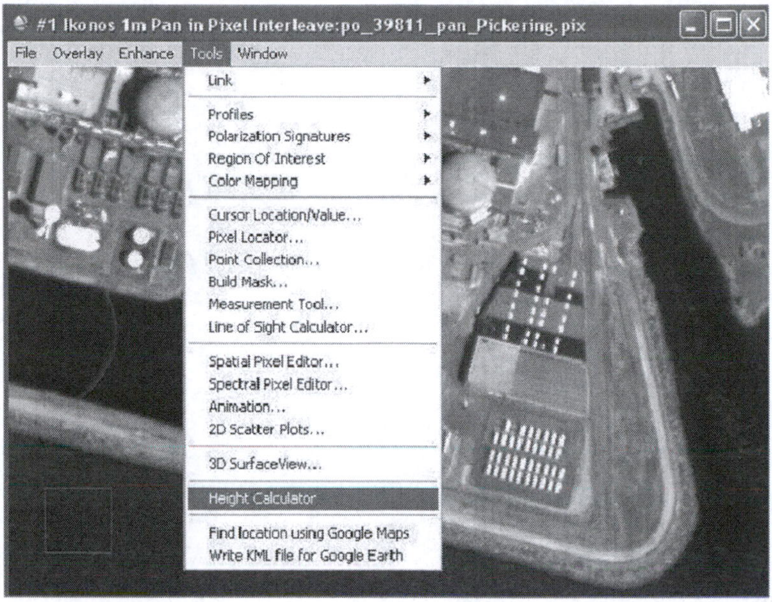

FIGURE C.3
Height calculation in the ENVI display menu.

(extension RPC or RPB). If a digital elevation model (DEM) is available for the scene, this can also be entered with

`File/Load DEM File`

A DEM is not required, however. Click on the bottom of a vertical structure to set the base height and then shift-click on the top of the structure. Press the CALC button to display the structure's height, latitude, longitude, and base elevation. The number in brackets next to the height is the minimum distance (in pixels) between the top pixel and a (projected) vertical line through the bottom pixel. It should be of the order of 1 or less. If no DEM is supplied, the base elevation is the average value for the whole scene. If a DEM is used, the base elevation is taken from it. The latitude and longitude are then orthorectified values.

Listing C.13
HEADER: CALCHEIGHT_RUN.PRO.

```
; :Description:
;       ENVI extention to determine height of vertical
;       structures in high resolution images using RFMs
; :Params:
;       event:  in, required
;          if called from the ENVI menu
```

```
; :Uses:
;       ENVI::
;       MJC_CURSOR_MOTION
```

Listing C.14
HEADER: MJC_CURSOR_MOTION.PRO.

```
; :Description:
;       Cursor communication with ENVI image windows
; :Params:
;       dn: in, required
;          display number
;       xloc,yloc: out, required
;          mouse position
; :Keywords:
;       xstart, ystart: out, optional
;          display origin
;       event: in, required
;          mouse event
; :Uses:
;       ENVI
```

C.2.6 Illumination Correction

C_CORRECTION_RUN is an ENVI extension for performing local solar incidence angle corrections for multispectral images over rough terrain (Section 5.4.6). Its location in the ENVI menu is shown in Figure C.4.

Select the (spectral/spatial subset of the) image to be corrected and (optionally) a mask. Then in the edit box enter the solar elevation and

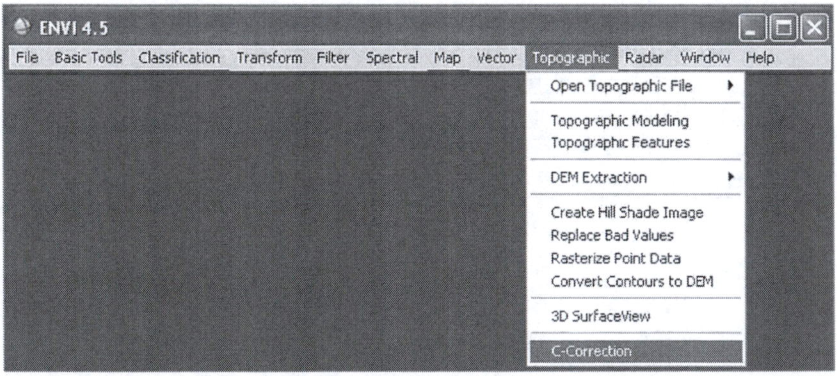

FIGURE C.4
Solar illumination correction in the ENVI menu.

azimuth in degrees and a size for the kernel used for slope/aspect determination (default 9×9). Then select the corresponding DEM file and the output destination (file or memory).

Listing C.15
HEADER: C_CORRECTION_RUN.PRO.

```
;       c-correction for solar
;       illuminatin in rough terrain::
;          Teillet, P. M., Guindon, B., and Goodenough,
;          D. G. (1982). On the slope aspect ;correction
;          of multispectral scanner data. Canadian Journal
;          of Remote Sensing, 8(2), 84   106.
; :Params:
;       event:  in, required
;          if called from the ENVI menu
; :Uses:
;       ENVI
```

C.2.7 Image Registration

CONTOUR_MATCH_RUN is an ENVI extension for extracting tie-points for image–image registration (see Section 5.5.2). It is primarily intended for fine adjustment, assuming that the images are already georeferenced and, if necessary, resampled to the same ground sample distance (GSD). Its location in the ENVI menu is shown in Figure C.5.

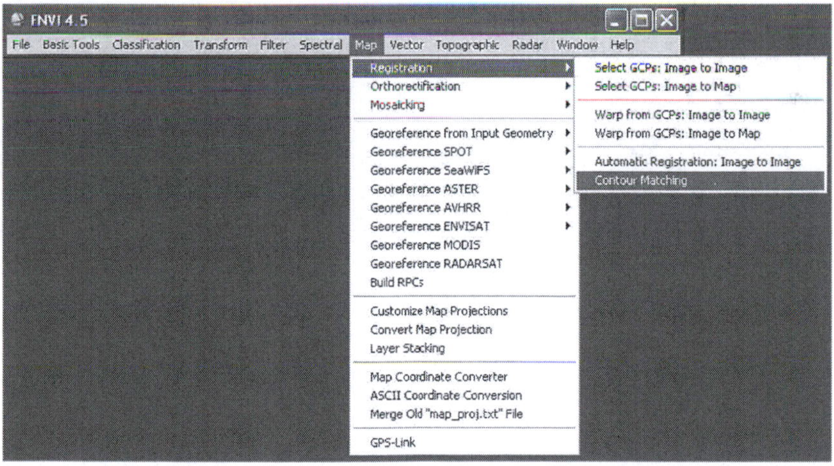

FIGURE C.5
Image-image registration with contour matching in the ENVI menu.

From the ensuing dialog windows choose the base and warp image bands. Then enter the σ parameter for the LoG filter (default value is 2.5 pixels). Finally give a filename (extension .PTS) for the tie-points (ground control points). If tie-points are found, they can be read in from this file into ENVI's `Ground Control Points Selection` dialog. It may be necessary to edit them further to eliminate outliers.

Listing C.16

HEADER: CONTOUR_MATCH_RUN.PRO.

```
; :Description:
;       ENVI extension for extraction of tie points for
;       image-image registration.Images may be already
;       georeferenced, in which case GCPs are for "fine
;       adjustement". Uses Laplacian of Gaussian and Sobel
;       filter and contour tracing to match contours::
;          Li, H., Manjunath, B. S., and Mitra, S. K. (1995).
;          A contour-based approach ;to multisensor image
;          registration. IEEE Transactions on Image Processing,
;          4(3), 320   334.
; :Params:
;      event:  in, required
;         if called from the ENVI menu
; :Uses:
;       ENVI::
;       CI__DEFINE::
;       COYOTE
```

Listing C.17

HEADER: CI__DEFINE.PRO.

```
; :Description:
;       Object class to find thin closed contours in an
;       image band with combined Sobel-LoG filtering
; :Params:
;       image: in, required
;           image to be searched for contours
;       sigma: in, required
;           radius for LoG filter
; :Examples:
;        ci = Obj_New("CI",image,sigma)
```

C.2.8 Maximum Likelihood Classification

`MAXLIKE_RUN` is a wrapper for ENVI's maximum likelihood classifier, discussed in Section 6.3.2, which reserves test data from the training regions of interest (ROIs) in the ratio 2:1 for training:test. The routine can be called from the ENVI command prompt with `MaxLike_Run`.

The user is prompted for an input file, the training ROIs (which must be present beforehand) and output destinations for the classified image, the rule image (optional) and the test results. The test result file can be processed with the procedures `CT_RUN` and `McNEMAR_RUN` discussed below. The routine `RULE_CONVERT`, also callable only from the ENVI command prompt, may be used to convert the ENVI rule image to probabilities (see Section 6.3.2).

Listing C.18
HEADER: MAXLIKE_RUN.PRO.

```
; :Description:
;       ENVI extension for classification of
;       a multispectral image with maximum likelihood
;       with option for reserving a random sample
;       of training data for subsequent evaluation
; :Uses:
;      ENVI::
;      DIFFERENCE
```

Listing C.19
HEADER: DIFFERENCE.PRO.

```
; :Description:
;       integer set difference
; :Params:
;       set1, set2:  in, required, type = array of integer
;          sets to be differenced
; :Returns:
;       array of integer
; :Uses:
;       REMOVE_DUPLICATES
```

Listing C.20
HEADER: REMOVE_DUPLICATES.PRO.

```
; :Description:
;       remove duplicates in an integer array
; :Params:
;       array:  in, required, type = array of integer
; :Returns:
;       array of integer
```

Listing C.21
HEADER: RULE_CONVERT.PRO.

```
; :Description:
;       Convert an ENVI rule image to class
;       membership probabilities
;  :Uses:
;       ENVI
```

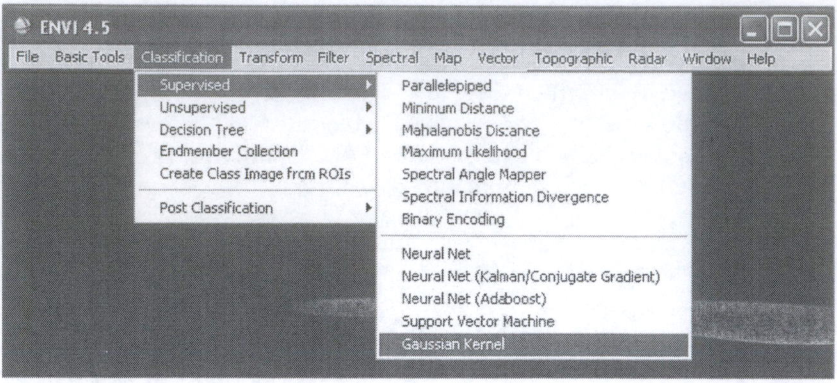

FIGURE C.6
Extensions for supervised classification in the ENVI menu.

C.2.9 Gaussian Kernel Classification

GPUKERNEL_RUN performs nonparametric classification of a multispectral image using the Gaussian kernel algorithm described in Section 6.4. The routine is invoked from the ENVI main menu as shown in Figure C.6. The program will make use of CUDA, if it is installed, to calculate the Gauss kernel matrices.

The user is prompted for an input file, the training ROIs (which must be present beforehand) and output destinations for the classified image, the probability image (optional) and the test results (optional). During calculation a plot window shows the minimization of the misclassification rate with respect to the parameter σ as shown in Figure 6.5. The program uses spectral tiling and may be used for large images. The test result file can be processed with the procedures CT_RUN and McNEMAR_RUN discussed below.

Listing C.22
HEADER: GPUKERNEL_RUN.PRO.

```
; :Description:
;       ENVI extension for classification of
;       a multispectral image with a Gassian
;       kernel-based classifier::
;           Masters, T. (1995). Advanced Algorithms
;           for Neural Networks, A C++ ;Sourcebook.
;           J. Wiley and Sons
; :Params:
;      event:  in, required
;         if called from the ENVI menu
; :Uses:
;       ENVI::
;       DIFFERENCE::
;       MINF_BRACKET::
```

```
;        MINF_PARABOLIC::
;        GAUSSKERNEL_MATRIX::
;        GPUGAUSSKERNEL_MATRIX::
;        GPULIB
```

C.2.10 Neural Network Classification

FFN_RUN classifies a multispectral image with a two-layer, feed-forward neural network trained with a combination of the Kalman filter algorithm and the scaled conjugate gradient algorithm described in Appendix B. The routine can be called from the ENVI main menu with

`Classification/Supervised/Neural Net(Kalman/Conjugate Gradient)`

(see Figure C.6). The user is prompted for an input file, the training ROIs (which must be present beforehand), output destinations for the classified image, the probability image (optional), and the test results (optional), the number of hidden neurons, and, finally, whether or not to use validation. During the calculation the cross entropy cost function is displayed. Training can be interrupted at any time, whereupon processing continues to the next phase. After training the Hessian matrix may be calculated, in which case its eigenvalues are written to a text window.

The program uses spectral tiling and may be used for large images. The test result file can be processed with the procedures CT_RUN and McNEMAR_RUN discussed below.

Listing C.23
HEADER: FFN_RUN.PRO.

```
; :Description:
;        ENVI extension for classification of a
;        multispectral image  with a feed forward
;        neural network using Kalman filter
;        plus scaled conjugate gradient training
; :Params:
;        event:  in, required
;           if called from ENVI
; :Uses:
;        ENVI::
;        DIFFERENCE::
;        FFKAL__DEFINE::
;        FFNCG__DEFINE::
;        COYOTE
```

Listing C.24
HEADER: FFNKAL__DEFINE.PRO.

```
; :Description:
;      Object class for implementation of a
;      two-layer, feed-forward neural network classifier.
;      Implements Kalman filter training::
```

```
;         Shah, S. and Palmieri, F. (1990). Meka     A fast,
;         local algorithm for training feed forward neural
;         networks. Proceedings of the International Joint
;         Conference on Neural Networks, San Diego, I(3),
;         41   46.
; :Inherits:
;         FFN__DEFINE
; :Params:
;         Gs: in, required
;            array of observation column vectors
;         Ls: in, required
;             array of class label column vectors
;             of form (0,0,1,0,0,...0)^T
;         L: in, required
;            number of hidden neurons
; :Examples:
;      ffn = Obj_New("FFNKAL",Gs,Ls,L)
; :Uses:
;        COYOTE
```

Listing C.25
HEADER: FFNCG__DEFINE.PRO.

```
; :Description:
;      Object class for implementation of a
;      two-layer, feed-forward neural network classifier.
;        Implements scaled conjugate gradient training::
;            Bishop, C. M. (1995). Neural Networks for
;            Pattern Recognition. Oxford ;University Press.
; :Inherits:
;         FFN__DEFINE
; :Params:
;         Gs: in, required
;            array of observation column vectors
;         Ls: in, required
;             array of class label column vectors
;             of form (0,0,1,0,0,...0)^T
;         L: in, required
;            number of hidden neurons
; :Examples:
;      ffn = Obj_New("FFNCG",Gs,Ls,L)
; :Uses:
;        COYOTE
```

Listing C.26
HEADER: FFN__DEFINE.PRO.

```
; :Description:
;      Object class for implementation of a
;      two-layer, feed-forward neural network classifier.
```

```
;      This is a generic class with no training method
; :Params:
;         Gs: in, required
;            array of observation column vectors
;         Ls: in, required
;             array of class label column vectors
;             of form (0,0,1,0,0,...0)^T
;         L: in, required
;            number of hidden neurons
```

C.2.11 Support Vector Machine Classification

SVM_RUN is a wrapper for ENVI's support vector machine classifier (discussed in Section 6.6.6) which reserves test data from the training ROIs in the ratio 2:1 for training:test. The routine can be called from the ENVI command prompt with SVM_Run. The user is prompted for an input file and the training ROIs (which must be present beforehand) and output destination for the test results. Classification and rule images are written to memory only. SVM-specific parameters (kernel type, kernel parameters, pyramid depth) must be set programmatically. The test result file can be processed with the procedures CT_RUN and McNEMAR_RUN discussed below.

Listing C.27
HEADER: SVM_RUN.PRO.

```
; :Description:
;        ENVI extension for running and testing
;        classification of a multispectral image with SVM.
;        Kernel parameters can only be set programmatically.
;        Classification and rule images are written to memory.
; :Uses:
;       ENVI
;       DIFFERENCE
```

C.2.12 Probabilistic Label Relaxation

PLR_RUN is an ENVI extension for probabilistic label relaxation (PLR) post-processing of supervised or unsupervised classification images (see Section 7.1.2). It takes as input a class probability vector image generated by any of the supervised classification extensions described in this appendix, as well as from the clustering routine EM_RUN described below, and generates a modified class probability vector image. The routine is invoked from the ENVI main menu as shown in Figure C.7. At the prompt, choose a class membership probabilities image and the number of iterations (default = 3). Then choose a destination either in memory or as a named file.

The generated rule image can then be processed with the ENVI extension PLR_RECLASS to generate an improved classification image with better spatial coherence. The menu item is PLR Reclassify (see Figure C.7). At

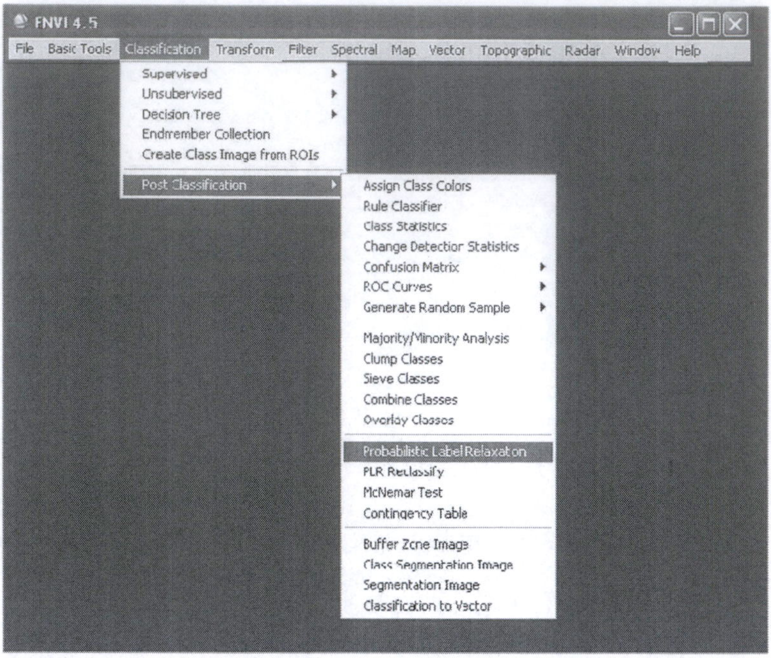

FIGURE C.7
Extensions for postclassification and evaluation in the ENVI menu.

the prompt, choose a previous classification image if available (to synchronize the class colors) and then a class membership probabilities image. At next prompt for a dummy unclassified class reply "Yes." Then choose a destination either in memory or as a named file.

Listing C.28
HEADER: PLR_RUN.PRO.

```
; :Description:
;       ENVI extension for postclassification with
;       Probabilistic Label Relaxation.
;       Processes a rule image (class membership
;       probabilities), outputs a new rule image::
;          Richards, J. A. and Jia, X. (2006). Remote
;          Sensing Digital Image Analysis ;(4th Ed.).
;          Springer.
; :Params:
;      event:  in, required
;         if called from the ENVI menu
; :Uses:
;      ENVI::
;      COYOTE
```

Listing C.29
HEADER: PLR_RECLASS_RUN.PRO.

```
; :Description:
;       ENVI extension for postclassification with
;       Probabilistic Label Relaxation
;       Processes a rule image (class membership
;       probabilities), outputs a new classification
;       file
; :Params:
;      event:  in, required
;         if called from the ENVI menu
; :Uses:
;      ENVI::
;      COYOTE
```

C.2.13 Classifier Evaluation and Comparison

The procedures `CT_RUN` and `McNEMAR_RUN` are used to evaluate and compare test results generated by the supervised classifiers `MAXLIKE_RUN`, `GPUKERNEL_RUN`, `FFN_RUN`, `FFNAB_RUN`, and `SVM_RUN` (see Section 7.2).

Both routines may be called from the ENVI menu as shown in Figure C.7. They request input files with extension `TST` and generate their outputs (contingency tables, accuracies, test statistics, etc.) in the IDL Log window.

Listing C.30
HEADER: CT_RUN.PRO.

```
; :Description:
;       ENVI extention to determine contingency
;       table (confusion matrix) and classification
;       accuracies from test classifcation results::
;          Richards, J. A. and Jia, X. (2006). Remote
;          Sensing Digital Image Analysis ;(4th Ed.).
;          Springer.
; :Params:
;       event:  in, required
;          if called from ENVI
; :Uses:
;       ENVI
```

Listing C.31
HEADER: CT_RUN.PRO.

```
; :Description:
;       ENVI extention to compare classifiers
;       on the basis of misclassifications
;       using McNemar's statistic::
;          Dietterich, T. G. (1998). Approximate
;          statistical  tests for comparing supervised
```

```
;           classification learning algorithms.
;           Neural Computation, 10(7), 1895   ;1923.
; :Params:
;        event:  in, required
;           if called from ENVI
; :Uses:
;        ENVI
```

C.2.14 Adaptive Boosting a Neural Network Classifier

FFN3AB_RUN is an ENVI extension for applying the AdaBoost.M1 algorithm to a series of three-layer neural network classifiers (see Section 7.3). The routine can be called from the ENVI main menu with

Classification/Supervised/Neural Net(AdaBoost)

(see Figure C.6). The user is prompted for an input file, the training ROIs (which must be present beforehand), output destinations for the classified image, the probability image (optional), and the test results (required). Then the number of training epochs for each network in the sequence (default 5) and the numbers of neurons in each hidden layer (default 10) are prompted for. During the calculation the cross entropy cost function for the current network is displayed as well as the overall training and generalization errors of the sequence. Training can be interrupted at any time, whereupon the boosted sequence is terminated and the image is classified.

The program uses spectral tiling and may be used for large images. The test result file can be processed with the procedures CT_RUN and McNEMAR_RUN discussed above.

Listing C.32
HEADER: FFN3AB_RUN.PRO.

```
; :Description:
;        ENVI extension for classification of a
;        multispectral image  with a 3-layer feed forward
;        neural network trained with Kalman filter
;        and using adaptive boosting AdaBoost.M1::
;          Canty, M. J. (2009). Boosting a fast neural
;          network for supervised land cover classification.
;          Computers and Geosciences, accepted.
; :Params:
;        event:  in, required
;           if called from ENVI
; :Uses:
;        ENVI::
;        DIFFERENCE::
;        FFN3KAL::
;        COYOTE
```

Listing C.33
HEADER: FFN3KAL__DEFINE.PRO.

```
; :Description:
;     Object class for implementation of a
;     three-layer, feed-forward neural network classifier.
;     Implements Kalman filter training::
;        Shah, S. and Palmieri, F. (1990). Meka âĂŤ A fast,
;        local algorithm for training feed forward neural
;        networks. Proceedings of the International Joint
;        Conference on Neural Networks, San Diego, I(3),
;        41   46.
; :Inherits:
;        FFN3__DEFINE
; :Params:
;        Gs: in, required
;           array of observation column vectors
;        Ls: in, required
;            array of class label column vectors
;            of form (0,0,1,0,0,...0)^T
;        L1, L2: in, required
;           number of hidden neurons
; :Keywords:
;        epochs: in, optional
;           training epochs per network in emsemble
;        D: in, required
;           training sample distribution
; :Examples:
;     ffn = Obj_New("FFN3KAL",Gs,Ls,L1,L2)
; :Uses:
;       COYOTE
```

Listing C.34
HEADER: FFN3__DEFINE.PRO.

```
; :Description:
;     Object class for implementation of a
;     three-layer, feed-forward neural network classifier.
;     This is a generic class with no training method
; :Inherits:
;        FFN3__DEFINE
; :Params:
;        Gs: in, required
;           array of observation column vectors
;        Ls: in, required
;            array of class label column vectors
;            of form (0,0,1,0,0,...0)^T
;        L1, L2: in, required
;           number of hidden neurons
```

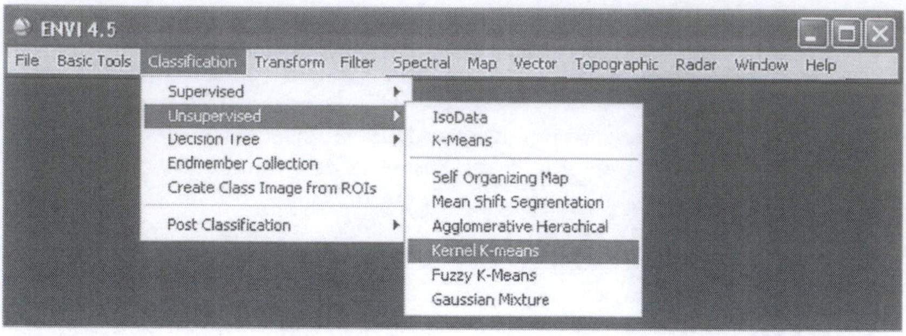

FIGURE C.8
Extensions for unsupervised classification in the ENVI menu.

C.2.15 Kernel K-Means Clustering

GPUKKMEANS_RUN is an ENVI/IDL extension for kernel K-means clustering of multispectral imagery. The algorithm is discussed in Section 8.2.2. Figure C.8 shows the corresponding menu location. Choose (a spatial/spectral subset of) a multispectral image in the first dialog. Then enter the number of random training samples and number of clusters in the corresponding prompts. The next prompt is for the kernel parameter NSCALE. The Gaussian kernel parameter, γ, is calculated as in Equation C.1. Finally, decide whether or not to save the clustered training data as ROIs and choose the output destination (memory or file). If CUDA is available then both training and generalization (classification) are performed on the GPU, resulting in a considerable speedup.

This extension does not make use of the ENVI tiling facility and is not intended to be used with large images.

Listing C.35
HEADER: GPUKKMEANS_RUN.PRO.

```
; :Description:
;       Performs kernel k-means clustering
;       using a Gaussian kernel::
;           Shawe-Taylor, J. and Cristianini, N. (2004).
;           Kernel Methods for Pattern Analysis.
;           Cambridge University Press.
; :Params:
;       event:  in, optional
;          required if called from ENVI
; :Uses:
;       ENVI::
;       GAUSSKERNEL_MATRIX::
;       GPUGAUSSKERNEL_MATRIX::
;       CLASS_LOOKUP_TABLE::
;       COYOTE::
;       GPULIB
```

Listing C.36
HEADER: CLASS_LOOKUP_TABLE.PRO.

```
; :Description:
;       returns colors from a 16-color lookup table
; :Params:
;       ptr:  in, required
;          color indices
```

C.2.16 Agglomerative Hierarchical Clustering

HCL_RUN is an ENVI/IDL extension for agglomerative hierarchical clustering (HCL) of multispectral imagery as discussed in Section 8.2.4. Clustering is performed on a small random sample of image pixels ($\leq$2000), and generalized to the entire image by using ENVI's built-in maximum likelihood supervised classification algorithm. The routine is invoked from the ENVI main menu as shown in Figure C.8. Choose (a spatial/spectral subset of) a multispectral image in the first dialog. Then enter the number of random samples and number of clusters in the corresponding prompts. Finally, indicate whether the classified image is to be saved to disk or memory. During computation, the current number of clusters is displayed in the progress bar.

This extension does not make use of the ENVI tiling facility, and is not intended to be used with large images.

Listing C.37
HEADER: HCL_RUN.PRO.

```
; :Description:
;       ENVI extension for agglomerative hierarchical
;       clustering
; :Params:
;       event:  in, optional
;          required if called from ENVI
; :Uses:
;       ENVI::
;       HCL::
;       CLASS_LOOKUP_TABLE::
```

Listing C.38
HEADER: HCL.PRO.

```
; :Description:
;       Agglomerative hierarchic clustering with
;       sum of squares cost function::
;          Fraley, C. (1996). Algorithms for model-based
;          Gaussian hierarchical clustering.
;          Technical report 311, Department of Statistics,
;          University of Washington, Seattle.
; :Params:
;        G: in, required
;           data matrix
```

```
;          K: in, required
;             number of clusters
;          Ls: out, required
;             Cluster labels of observations
; :Uses:
;        COYOTE
```

C.2.17 Fuzzy K-Means Clustering

FKM_RUN is an ENVI/IDL extension for fuzzy K-means (FKM) clustering of multispectral imagery (Section 8.2.5). Clustering is performed on a random sample of image pixels ($\leq 10^5$), and generalized to the entire image by using a modification of the IDL distance classifier CLUSTER_FKM. The routine is invoked from the ENVI main menu as shown in Figure C.8. Choose (a spatial/spectral subset of) a multispectral image at the prompt. Then enter the desired number of clusters and whether or not they should be saved as ROIs at the corresponding prompts. Finally, indicate whether the class image is to be saved to disk or memory.

This extension does not make use of the ENVI tiling facility, and is not intended to be used with large images.

Listing C.39
HEADER: FKM_RUN.PRO.

```
; :Description:
;        ENVI extension for fuzzy K-means clustering
;        with sampled data
; :Params:
;        event:  in, optional
;           required if called from ENVI
; :Uses:
;        ENVI::
;        FMK::
;        CLUSTER_FKM (Modified distance clusterer
;        from IDL library)::
;        CLASS_LOOKUP_TABLE
```

Listing C.40
HEADER: FKM.PRO.

```
; :Description:
;        Fuzzy Kmeans clustering algorithm::
;           Dunn, J. C. (1973). A fuzzy relative of the
;           isodata process and its use in detecting
;           compact well-separated clusters.
;           Journal of Cybernetics, PAM1-1, 32   57.
; :Params:
;          G: in, required
;             input data matrix
```

```
;         U: in, out, required
;            class probability membership matrix
;         Ms: out, required
;            cluster means (output)
; :Uses:
;        COYOTE
```

C.2.18 Gaussian Mixture Clustering

`EM_RUN` is an ENVI/IDL extension for Gaussian mixture clustering using the expectation maximization (EM) algorithm, also referred to as the fuzzy maximum likelihood estimation (FMLE) algorithm. It is discussed in Section 8.3. Clustering occurs optionally at different scales, and both simulated annealing and spatial memberships can be included if desired. The program will also optionally generate class membership probability images that can be post-processed with PLR (see Appendix C.2.12). The routine is invoked from the ENVI main menu as shown in Figure C.8.

Choose (a spatial/spectral subset of) a multispectral image at the prompt. Then enter the desired number of clusters and compressions (pyramid depth) at the corresponding prompts. In the `Clustering parameters` dialog enter the initial annealing temperature (zero for no annealing) and spatial membership parameter `Beta` (zero for no spatial memberships). At the next prompt indicate whether or not to save the clusters as ROIs. Finally, say whether the class image is to be saved to disk or memory and similarly for the membership probability image (optional).

This extension does not make use of the ENVI tiling facility, and is not intended to be used with large images.

Listing C.41
HEADER: EM_RUN.PRO.

```
; :Description:
;        ENVI extension for Gaussian mixture clustering
;        at different scales using DWT compression,
;        simulated annealing and the EM algorithm
; :Params:
;        event:  in, optional
;           required if called from ENVI
; :Uses:
;        ENVI::
;        EM::
;        DWT__DEFINE::
;        CLASS_LOOKUP_TABLE
```

Listing C.42
HEADER: EM.PRO.

```
; :Description:
;       Clustering with Gaussian mixtures, EM algorithm::
;          Gath, I. and Geva, A. B. (1989). Unsupervised
;          optimal fuzzy clustering.
;          IEEE Transactions on Pattern Analysis and
;          Machine Intelligence, 3(3), 773   781.
; :Params:
;       G: in, required
;          observations data matrix
;       U: in, out, required
;          initial class probability membership matrix
;          (column vectors)
;       Ms: out, required
;          cluster means
;       Ps: out, required
;          cluster priors
;       Cs: out, required
;          cluster covariance matrices
; :Keywords::
;       unfrozen: in, optional
;          indices of the observations which
;          take part in the iteration (default all)
;       wnd: in, optional
;          window for displaying the log likelihood
;       mask: in, optional
;          defines the region to be included in
;          clustering
;       maxinter: in, optional
;          maximum iterations
;       mininter: in, optional
;          minimum iterations
;       pdens: out, optional
;          partition density
;       fhv: out, optional
;          fuzzy hypervolume
;       T0: in, optional
;          initial annealing temperature (default 1.0)
;       verbose: in, optional
;          set to print output info to IDL log
;          (default 0)
;       pb_msg: in, optional
;          string for progressbar
;       beta: in, optional
;          spatial field parameter
;       num_cols, num_rows: in, required
```

```
;            if beta>0 image dimensions
;        status: out, optional
;            0 if successful else 1
; :Uses:
;      INTERSECTION::
;      COYOTE
```

Listing C.43
HEADER: INTERSECTION.PRO.

```
; :Description:
;       integer set intersection
; :Params:
;       set1, set2:  in, required, type = array of integer
;          sets to be intersected
; :Returns:
;       array of integer
```

C.2.19 Kohonen Self-Organizing Map

SOM_RUN is an ENVI extension for image cluster visualization using the Kohonen self-organizing map (see Section 8.6). The routine is invoked from the ENVI main menu (Figure C.8) with the command

```
Classification/Unsupervised/Self Organizing Map
```

At the prompts enter the (spectral/spatial subset of the) image to be visualized, the dimension of the neural network cube (default $6 \times 6 \times 6$), and the output destination for the map. The network is trained with a random sample of 10^4 pixel vectors. A three-dimensional plot of the neuron cube projected onto the space of the first three image bands is generated, as shown in Figure 8.7.

The routine does not use the ENVI tiling facility for the final classification, and is not intended to be used with large images.

Listing C.44
HEADER: SOM_RUN.PRO.

```
; :Description:
;       ENVI extension for Kohonen Self
;       Organizing Map::
;           Kohonen, T. (1989). Self-Organization
;           and Associative Memory. Springer.
; :Params:
;       event:  in, optional
;          required if called from ENVI
; :Uses:
;       ENVI::
;       COYOTE
```

C.2.20 Classified Image Segmentation

`SEGMENT_CLASS_RUN` is an ENVI extension for segmenting a classified image by finding connected components for each separate class. It is intended as an alternative to the ENVI `.../Segmentation Image` command that ignores class labels (see the discussion in Section 8.7.1). The routine is invoked from the ENVI main menu with the command (Figure C.7)

```
Classification/Post Classification/Class Segmentation Image
```

At the prompts enter the (spatial subset of the) classification image to be segmented and select the classes to include in segmentation. Then choose the minimum segment size to accept, the connectivity (four or eight) for blobbing and, finally, the output destination (memory or file).

For completeness, the header for the Hu-moments function described in Section 5.2.2 is also listed below.

Listing C.45

HEADER: SEGMENT_CLASS_RUN.PRO.

```
; :Description:
;        ENVI extension for segmentation of a classified
;        image. Replaces ENVI built-in
; :Params:
;       event:  in, required
;          if called from the ENVI menu
; :Uses:
;        ENVI::
;        SEGMENT_CLASS
```

Listing C.46

HEADER: SEGMENT_CLASS.PRO.

```
; :Description:
;       perform blobbing on a segmented or classified image
; :Params:
;       climg: in, required
;          classification image to be segmented
;       cl: in, required, type=array of integer
;          classes to be included in segmentation
; :Keywords:
;       minsize: in, optional
;          minimum segment size (default 0)
;       all_neighbours: in, optional
;          set for 8-neighborhood
;          (default4-neighborhood)
; :Uses:
;       COYOTE
```

Listing C.47
HEADER: HU_MOMENTS.PRO.

```
; :Description:
;     Takes a rectangular array A of gray values and returns
;     the 7 Hu invariant moments
; :Params:
;       A:  in, required
;           array of gray-scale values
; : Keywords:
;      log: in, optional
;           natural logarithms of the moments
;           are returned if set
```

C.2.21 Mean Shift Segmentation

The mean shift algorithm described in Section 8.7.3 is implemented as the ENVI extension MEAN_SHIFT_RUN, and its location in the ENVI menu can be seen in Figure C.8.

The user is prompted for the (spectral/spatial subset of the) multispectral (or grayscale) image to be segmented, and then, in an edit widget, for the spatial and spectral bandwidths (defaults 15) and minimum segment size (default 30). Finally the user must choose a destination (memory or file) for the output. Three images are generated:

1. The segmented image (ENVI standard)
2. The segment boundaries (ENVI classification file)
3. The segment labels (ENVI standard)

The boundaries can be overlayed onto the segmented image (or segment labels) with the Overlay/Classification command in the ENVI display menu.

As explained in Section 8.7.3, the implementation associates pixels along the current mean shift path with the final mode reached on that path, leading to an unavoidable fragmentation of some of the segments. To counter this effect to some extent, the segmented image is postprocessed with a 3×3 median filter.

Listing C.48
HEADER: MEAN_SHIFT_RUN.PRO.

```
; :Description:
;        ENVI extension for mean shift segmentation::
;           Comaniciu, D. and Meer, P. (2002). Mean shift:
;           a robust approach toward feature space analysis.
;           IEEE Transactions on Pattern Analysis and
;           Machine Intelligence, 24(5), 603   619.
```

```
; :Params:
;       event:  in, optional
;          required if called from ENVI
; :Uses:
;       ENVI::
;       MEAN_SHIFT::
;       CLASS_LOOKUP_TABLE::
;       COYOTE
```

Listing C.49
HEADER: MEAN_SHIFT.PRO.

```
; :Description:
;       Determines the mode of the indexed vector.
;       The vectors are in the common block
;       variable DATA. The bandwidth is in the
;       commen block variable HS
; :Params:
;        idx: in, required
;           pixel index
;        cpts: in, required
;           labeled pixels within distance HS/2
;           along path to mode and
;           within distance HS of mode
;        cpts_max: in, required
;           maximum labeled pixel
```

C.2.22 Multivariate Alteration Detection

The ENVI extension `MAD_RUN` performs multivariate alteration detection (MAD) on bitemporal imagery as discussed in Section 9.4, allowing for iterative re-weighting. It (optionally) generates two canonical variate images, and it (optionally) generates the MAD variates together with the associated chi-square image. The latter can be used for choosing invariant pixels for radiometric normalization with `RADCAL_RUN` discussed in Appendix C.2.24 below.

The routine is invoked from the ENVI main menu as illustrated in Figure C.9. Choose (spatial/spectral subsets of) the two multispectral images at the prompts. When entering the first image, a mask band can be optionally specified. Then enter the number of iterations (minimum zero for uniterated MAD, maximum 100) and the output destinations for the MAD and canonical variates (CVs). If the latter are to be output to a file, the program will append `_1` and `_2` to the file names in order to distinguish the two CV files that are created. If the images are not of the same spectral/spatial dimension, the program will abort. If an image is not in band interleaved by pixel (BIP) or band interleaved by line (BIL) format, the user will be prompted to allow it to be converted to BIP in place (the image must reside on disk in this case).

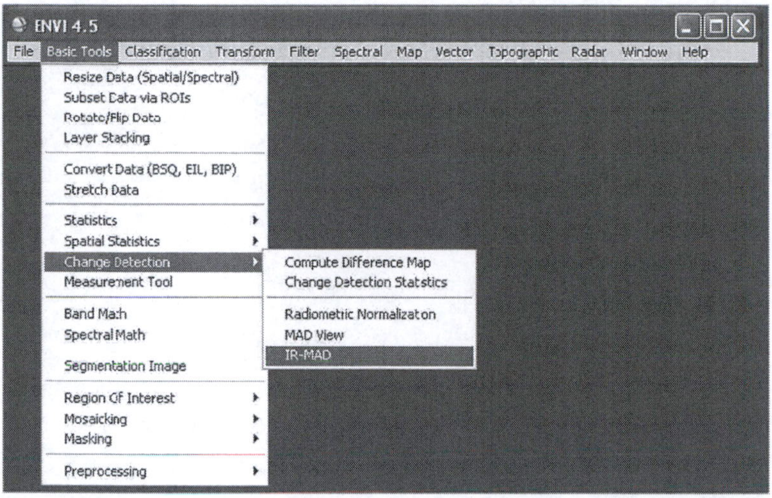

FIGURE C.9
Extensions for change detection in the ENVI menu.

This extension uses ENVI spectral tiling and can be run on large datasets, for example, Landsat TM full scenes.

Listing C.50
HEADER: MAD_RUN.PRO.

```
; :Description:
;        ENVI extension for Iteratively Re-weighted
;        Multivariate Alteration Detection (IR-MAD)::
; :Params:
;        event:  in, optional
;           required if called from ENVI
; :Uses:
;        ENVI::
;        MAD_ITER::
;        COYOTE
```

Listing C.51
HEADER: MAD_ITER.PRO.

```
; :Description:
;        Function for Iteratively Re-weighted Multivariate
;        Alteration Detection (IR-MAD)::
;           Nielsen, A. A. (2007). The regularized
;           iteratively reweighted MAD method for change
;           detection in multi- and hyperspectral data.
;           IEEE Transactions on Image Processing,
;           16(2), 463   478.
```

```
; :Params:
;       fid1,fid2: in,required
;          image file IDs
;       dims1,dims2: in,required
;          image spatial subsets
;       pos1,pos2: in,required
;          image spectral subsets
;       m_fid: in,required
;          mask band ID
; :KEYWORDS:
;       A, B: out, optional
;          arrays of transformation eigenvectors in columns,
;          smallest correlation first
;       means1,means2: out,optional
;          weighted mean values, row-replicated
;       rho: out,optional
;         canonical correlations in decreasing order
;       sigma: out,optional
;         standard deviations of MAD variates
;         in increasing order
;       lam: in,optional
;           length penalization parameter
;           (default 0.0)
;       niter: in,optional
;           max number of iterations,
;           (default is 50)
;       verbose: in,optional
;           print intermediate results to log
;           (default 0);
; :Uses:
;       ENVI::
;       COVPM_DEFINE::
;       GEN_EIGENPROBLEM::
;       COYOTE
```

Listing C.52

HEADER: COVPM__DEFINE.PRO.

```
; :Description:
;       Object class for sequential variance-
;       covariance matrix calculation using
;       the method of provisional means
; :Params:
;     p: in,required
;         dimension of observations
; :Examples:
;         covpm = Obj_New("COVPM",p)
; :Uses:
;     prov_means.dll
```

Listing C.53
HEADER: GEN_EIGENPROBLEM.PRO.

```
; :Description:
;       Solve the generalized eigenproblem
;       using Cholesky factorization
; :Params:
;       C:  in, required
;       B:  in, required
;       A:  out, required
;          matrix of eigenvectors (columns)
;       lambda: out, required
;            eigenvalues, largest to smallest
```

C.2.23 Viewing Changes

`MAD_VIEW_RUN` (Figure C.10) is an ENVI/IDL GUI for viewing MAD or MAD/MNF variates (see Section 9.5). It may be called from the ENVI menu

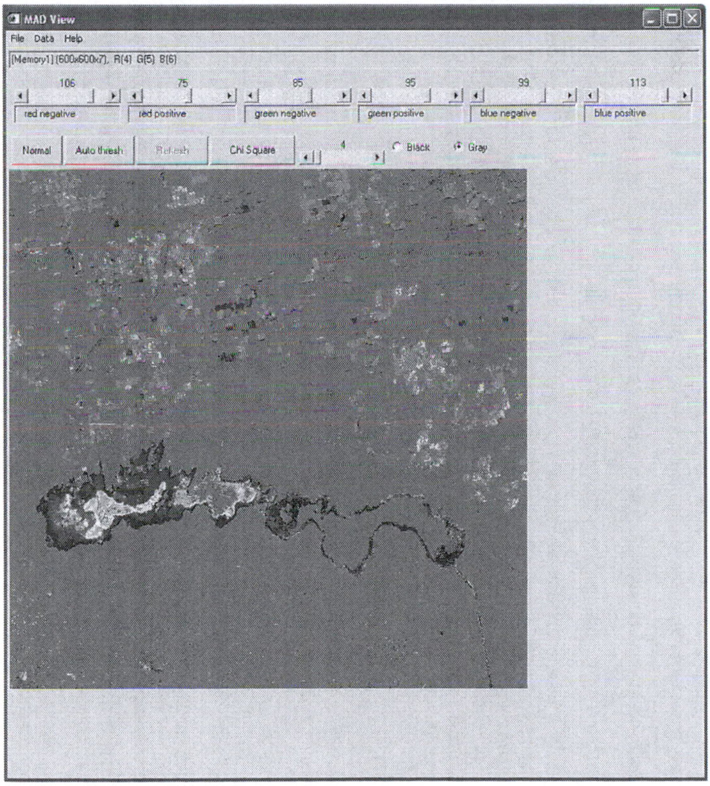

FIGURE C.10
GUI for viewing IR-MAD images.

as indicated in Figure C.9, and is provided with a rudimentary online help PDF file.

Listing C.54
HEADER: MAD_VIEW_RUN.PRO.

```
; :Description:
;       GUI for viewing and thresholding MAD images
; :Params:
;       event:  in, optional
;          required if called from ENVI
; :Uses:
;       ENVI::
;       EM::
;       COYOTE
```

C.2.24 Radiometric Normalization

RADCAL_RUN is an ENVI/IDL extension for radiometric normalization of two multispectral images using iteratively reweighted MAD to determine time invariant pixels (see Section 9.6). The routine is invoked from the ENVI main menu as shown in Figure C.9.

Choose (spatial/spectral subsets of) the two multispectral images (reference and target) at the prompts. Then enter the output destination. If the images are not of the same spectral/spatial dimension, the program will abort. Then the chi-square image generated previously by MAD_RUN on the same spatial/spectral subsets is requested. The histogram of the

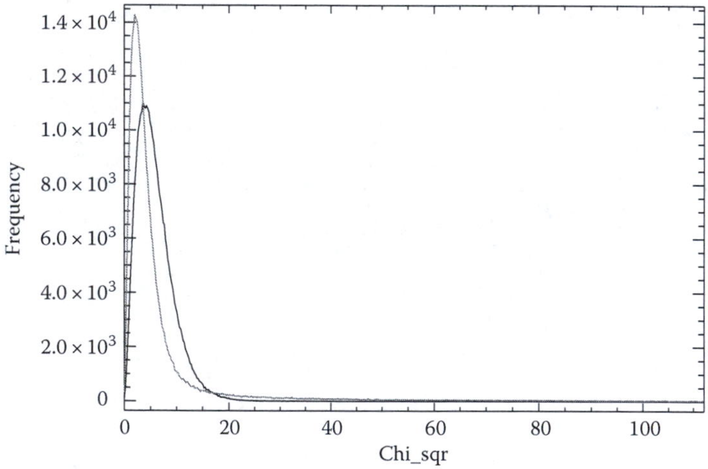

FIGURE C.11
Histogram of the chi-square image (upper curve) and theoretical histogram for uncorrelated, normally distributed MAD components (lower curve).

chi-square image is displayed (Figure C.11, upper curve) and compared with the theoretical histogram for purely no-change observations, that is, the chi-square distribution for uncorrelated, normally distributed MAD variates (lower curve). Over a slider widget the minimum probability to use to identify no-change pixels must then be chosen and, finally, the output destination (file or memory) for the normalized target image. During the calculation, orthogonal regressions are plotted in separate plot windows, once for each spectral band. After completion, another spatial subset (e.g., a full scene) may be normalized with the regression coefficients that were determined. The output is then to file only. Regression statistics and results of statistical tests calculated with the invariant pixels are written to a text window (see Tables 9.1 and 9.2).

Listing C.55
HEADER: RADCAL_RUN.PRO.

```
; :Description:
;     Radiometric calibration using IR-MAD::
;        Canty, M. J. and Nielsen, A. A. (2007).
;        Automatic radiometric normalization of multitemporal
;        satellite imagery with the iteratively re-weighted MAD
;        transformation. International Journal of Remote Sensing,
;        112(3), 1025   1036.
; :Params:
;       event:  in, optional
;          required if called from ENVI
; :Uses:
;       ENVI::
;       ORTHO_REGRESS::
;       WISHART::
;       COYOTE
```

Listing C.56
HEADER: WISHART.PRO.

```
; :Description:
;       Test two multivariate distributions for
;       equal means and covariance matrices
;       assuming both parameter sets are estimated
;       with the same sample size.
;       Returns p-value::
;          Anderson, T. W. (2003). An Introduction
;          to Multivariate Statistical Analysis.
;          Wiley Series in Probability and Statistics,
;          third edition.
; :Params:
;       M1: in,required
;          mean of first distribution
;       M2: in,required
```

```
;           mean of second distribution
;        S1: in,required
;           variance of first distribution
;        S2: in,required
;           variance of second distribution
;        n: in,required
;           common number of samples
; :Keywords:
;        statistic: out,optional
;           test statistic -2*rho*log(lambda)
```

Appendix D: Mathematical Notation

G	random variable
g	realization of G (observation, gray-value)
$\boldsymbol{G}$	random (column) vector
$\boldsymbol{g}$	realization of $\boldsymbol{G}$ (vector observation, gray-values)
$\boldsymbol{\mathcal{G}}$	data design matrix for $\boldsymbol{G}$
$\boldsymbol{g}^\top$	transposed form of the vector $\boldsymbol{g}$ (row vector)
$\Vert\boldsymbol{g}\Vert$	length or (2-)norm of $\boldsymbol{g}$
$\boldsymbol{f}^\top\boldsymbol{g}$	inner product of two vectors (scalar)
$\boldsymbol{f}\cdot\boldsymbol{g}$	Hadamard (component-by-component) product
$\boldsymbol{f}\boldsymbol{g}^\top$	outer product of two vectors (matrix)
$\boldsymbol{C}$	matrix
$\boldsymbol{f}^\top\boldsymbol{C}\boldsymbol{g}$	quadratic form (scalar)
$\vert\boldsymbol{C}\vert$	determinant of matrix $\boldsymbol{C}$
$\mathrm{tr}(\boldsymbol{C})$	trace of $\boldsymbol{C}$
$\boldsymbol{I}$	identity matrix
$\boldsymbol{0}$	column vector of zeroes
$\boldsymbol{1}$	column vector of ones
$\boldsymbol{\Lambda}$	$\mathrm{Diag}(\lambda_1 \ \ldots \ \lambda_N)$ (diagonal matrix of eigenvalues)
$\frac{\partial}{\partial \boldsymbol{x}}$	partial derivative with respect to vector $\boldsymbol{x}$
$\mathbf{i}$	$\sqrt{-1}$
$\vert z\vert$	absolute value of real or complex number z
z^*	complex conjugate of complex number z
Ω	sample space in probability theory
$\Pr(A \mid B)$	probability of event A conditional on event B
$P(g)$	distribution function for random variable G
$P(\boldsymbol{g})$	joint distribution function for random vector $\boldsymbol{G}$
$p(g)$	density function for G
$p(\boldsymbol{g})$	joint density function for $\boldsymbol{G}$
$\langle G\rangle$, μ	mean (or expected) value
$\mathrm{var}(G)$, σ^2	variance
$\langle\boldsymbol{G}\rangle$, $\boldsymbol{\mu}$	mean vector
$\boldsymbol{\Sigma}$	variance–covariance matrix

$\mathcal{C}$	variance–covariance matrix estimate
$\mathcal{K}$	kernel matrix with elements $k(g,g')$
$\Phi(g)$	standard normal distribution function
$\phi(g)$	standard normal density function
$p_{\chi^2;n}(z)$	chi-square density function with n degrees of freedom
$p_{t;n}(t)$	Student-t density function with n degrees of freedom
$p_{f;m,n}(f)$	F-density function with m and n degrees of freedom
$f(x)\vert_{x=x^*}$	$f(x)$ evaluated at $x = x^*$
$(\hat{g}(0), \hat{g}(1)\ldots)$	discrete Fourier transform of array $(g(0), g(1)\ldots)$
$\boldsymbol{f} * \boldsymbol{g}$	discrete convolution of vectors $\boldsymbol{f}$ and $\boldsymbol{g}$
$\langle\phi, \psi\rangle$	inner product $\int \phi(x)\psi(x)dx$ of functions ϕ and ψ
$\mathcal{N}_i$	neighborhood of ith pixel
$\{x \mid c(x)\}$	set of elements x that satisfy condition $c(x)$
$\mathcal{K}$	set of class labels $\{1\ldots K\}$
ℓ	vector representation of a class label
$u \in U$	u is an element of the set U
$U \otimes V$	Cartesian product set
$V \subset U$	V is a (proper) subset of the set U
$\arg\max_x f(x)$	the set of x that maximizes $f(x)$
$f : A \mapsto B$	the function f that maps the set A to the set B
$\mathbb{R}$	set of real numbers
$\mathbb{Z}$	set of integers
$=:$	equal by definition
$\Box$	end of a proof

References

Aboufadel, E. and Schlicker, S. (1999). *Discovering Wavelets*. John Wiley & Sons, New York.

Abrams, M., Hook, S., and Ramachandran, B. (1999). ASTER user handbook. Technical Report, Jet Propulsion Laboratory and California Institute of Technology, Pasadena, CA.

Aiazzi, B., Alparone, L., Baronti, S., and Garzelli, A. (2002). Context-driven fusion of high spatial and spectral resolution images based on oversampled multiresolution analysis. *IEEE Transactions on Geoscience and Remote Sensing*, **40**(10), 2300–2312.

Anderson, T. W. (2003). *An Introduction to Multivariate Statistical Analysis. Wiley Series in Probability and Statistics*, 3rd edition. John Wiley & Sons, Hoboken, NJ.

Beisl, U. (2001). Correction of bidirectional effects in imaging spectrometer data. PhD thesis, Remote Sensing Laboratories, Department of Geography, University of Zurich, Zurich, Switzerland.

Bellman, R. (1961). *Adaptive Control Processes, A Guided Tour*. Princeton University Press, Princeton, NJ.

Belousov, A. I., Verzakov, S. A., and von Frese, J. (2002). A flexible classification approach with optimal generalization performance: Support vector machines. *Chemometrics and Intelligent Laboratory Systems*, **64**, 15–25.

Benz, U. C., Hoffmann, P., Willhauck, G., Lingfelder, I., and Heynen, M. (2004). Multi-resolution, object-oriented fuzzy analysis of remote sensing data for GIS-ready information. *ISPRS Journal of Photogrammetry & Remote Sensing*, **258**, 239–258.

Bilbo, C. M. (1989). Statistisk analyse af relationer mellem alternative antistoftracere. Master's thesis, Informatics and Mathematical Modeling, Technical University of Denmark, Lyngby, Denmark (in Danish).

Bishop, C. M. (1995). *Neural Networks for Pattern Recognition*. Oxford University Press, Oxford, U.K.

Bishop, C. M. (2006). *Pattern Recognition and Machine Learning*. Springer, New York.

Bishop, Y. M. M., Feinberg, E. E., and Holland, P. W. (1975). *Discrete Multivariate Analysis, Theory and Practice*. MIT Press, Cambridge, MA.

Bradley, A. P. (2003). Shift-invariance in the discrete wavelet transform. In C. Sun, H. Talbot, S. Ourselin, and T. Adriaansen, editors, *Proceedings of the VIIth Digital Image Computing: Techniques and Applications*, Sydney, Australia, pp. 29–38.

Breiman, L. (1996). Bagging predictors. *Machine Learning*, **24**(2), 123–140.

Bridle, J. S. (1990). Probabilistic interpretation of feedforward classification outputs, with relationships to statistical pattern recognition. In F. F. Soulié and J. Hérault, editors, *Neurocomputing: Algorithms, Architectures and Applications*. Springer-Verlag, Berlin, Germany, pp. 227–236.

Bruzzone, L. and Prieto, D. F. (2000). Automatic analysis of the difference image for unsupervised change detection. *IEEE Transactions on Pattern Analysis and Machine Intelligence*, **11**(4), 1171–1182.

Canty, M. J. (2009). Boosting a fast neural network for supervised land cover classification. *Computers and Geosciences*, **35**(6), 1280–1295.

Canty, M. J. and Nielsen, A. A. (2006). Visualization and unsupervised classification of changes in multispectral satellite imagery. *International Journal of Remote Sensing*, **27**(18), 3961–3975. http://www.imm.dtu.dk/pubdb/p.php?3389.

Canty, M. J. and Nielsen, A. A. (2007). Automatic radiometric normalization of multitemporal satellite imagery with the iteratively re-weighted MAD transformation. *Remote Sensing of Environment*, **112**(3), 1025–1036. http://www.imm.dtu.dk/pubdb/p.php?5362.

Canty, M. J., Nielsen, A. A., and Schmidt, M. (2004). Automatic radiometric normalization of multitemporal satellite imagery. *Remote Sensing of Environment*, **91**(3–4), 441–451. http://www.imm.dtu.dk/pubdb/p.php?2815.

Cohen, J. (1960). A coefficient of agreement for nominal scales. *Educational and Psychological Measurement*, **20**, 37–46.

Comaniciu, D. and Meer, P. (2002). Mean shift: A robust approach toward feature space analysis. *IEEE Transactions on Pattern Analysis and Machine Intelligence*, **24**(5), 603–619.

Congalton, R. G. and Green, K. (1999). *Assessing the Accuracy of Remotely Sensed Data: Principles and Practices*. Lewis Publishers, Boca Raton, FL.

Coppin, P., Jonckheere, I., Nackaerts, K., and Muys, B. (2004). Digital change detection methods in ecosystem monitoring: A review. *International Journal of Remote Sensing*, **25**(9), 1565–1596.

Cristianini, N. and Shawe-Taylor, J. (2000). *Support Vector Machines and Other Kernel-based Learning Methods*. Cambridge University Press, Cambridge, U.K.

Daubechies, I. (1988). Orthonormal bases of compactly supported wavelets. *Communications on Pure and Applied Mathematics*, **41**, 909–996.

Dhillon, I., Guan, Y., and Kulis, B. (2005). A unified view of kernel K-means, spectral clustering and graph partitioning. Technical Report UTCS TR-04-25, University of Texas at Austin, Austin, TX.

Dietterich, T. G. (1998). Approximate statistical tests for comparing supervised classification learning algorithms. *Neural Computation*, **10**(7), 1895–1923.

Du, Y., Teillet, P. M., and Cihlar, J. (2002). Radiometric normalization of multitemporal high-resolution images with quality control for land cover change detection. *Remote Sensing of Environment*, **82**, 123–134.

Duda, R. O. and Hart, P. E. (1973). *Pattern Classification and Scene Analysis*. John Wiley & Sons, New York.

Duda, T. and Canty, M. J. (2002). Unsupervised classification of satellite imagery: Choosing a good algorithm. *International Journal of Remote Sensing*, **23**(11), 2193–2212.

Duda, R. O., Hart, P. E., and Stork, D. G. (2001). *Pattern Classification*, 2nd edition. Wiley Interscience, New York.

Dunn, J. C. (1973). A fuzzy relative of the isodata process and its use in detecting compact well-separated clusters. *Journal of Cybernetics*, **3**(3), 32–57.

Fahlman, S. E. and LeBiere, C. (1990). The cascade correlation learning architecture. In D. S. Touertzky, editor, *Advances in Neural Information Processing Systems 2*. Morgan Kaufmann, Los Altos, CA, pp. 524–532.

Fanning, D. W. (2000). *IDL Programming Techniques*. Fanning Software Consulting, Fort Collins, CO.

Fraley, C. (1996). Algorithms for model-based Gaussian hierarchical clustering. Technical Report 311, Department of Statistics, University of Washington, Seattle, WA.

Freund, J. E. (1992). *Mathematical Statistics*, 5th edition. Prentice-Hall, Englewood Cliffs, NJ.

Freund, Y. and Shapire, R. E. (1996). Experiments with a new boosting algorithm. In *Proceedings of the 13th International Conference on Machine Learning*. Bari, Italy. Morgan Kaufmann, San Francisco, CA.

Freund, Y. and Shapire, R. E. (1997). A decision-theoretic generalization of on-line learning and an application to boosting. *Journal of Computer and System Sciences*, **55**, 119–139.

Fukunaga, K. (1990). *Introduction to Statistical Pattern Recognition*, 2nd edition. Academic Press, San Diego, CA.

Fukunaga, K. and Hostetler, L. D. (1975). The estimation of the gradient of a density function, with applications to pattern recognition. *IEEE Transactions on Information Theory*, **IT-21**, 32–40.

Gath, I. and Geva, A. B. (1989). Unsupervised optimal fuzzy clustering. *IEEE Transactions on Pattern Analysis and Machine Intelligence*, **3**(3), 773–781.

Gonzalez, R. C. and Woods, R. E. (2002). *Digital Image Processing*. Addison-Wesley, Reading, MA.

Green, A. A., Berman, M., Switzer, P., and Craig, M. D. (1988). A transformation for ordering multispectral data in terms of image quality with implications for noise removal. *IEEE Transactions on Geoscience and Remote Sensing*, **26**(1), 65–74.

Groß, M. H. and Seibert, F. (1993). Visualization of multidimensional image data sets using a neural network. *The Visual Computer*, **10**, 145–159.

Gumley, L. E. (2002). *Practical IDL Programming*. Morgan Kaufmann, San Francisco, CA.

Haberächer, P. (1995). *Praxis der Digitalen Bildverarbeitungen und Mustererkennung*. Carl Hanser Verlag, Munich, Germany.

Harsanyi, J. C. (1993). Detection and classification of subpixel spectral signatures in hyperspectral image sequences. PhD thesis, University of Maryland, Baltimore, MD, 116 pp.

Harsanyi, J. C. and Chang, C.-I. (1994). Hyperspectral image classification and dimensionality reduction: An orthogonal subspace projection approach. *IEEE Transactions on Geoscience and Remote Sensing*, **32**(4), 779–785.

Hayken, S. (1994). *Neural Networks: A Comprehensive Foundation*. Macmillan, New York.

Hertz, J., Krogh, A., and Palmer, R. G. (1991). *Introduction to the Theory of Neural Computation*. Addison-Wesley, Redwood City, CA.

Hilger, K. B. (2001). Exploratory analysis of multivariate data. PhD thesis, IMM-PHD-2001-89, Technical University of Denmark, Lyngby, Denmark.

Hilger, K. B. and Nielsen, A. A. (2000). Targeting input data for change detection studies by suppression of undesired spectra. In *Proceedings of a Seminar on Remote Sensing and Image Analysis Techniques for Revision of Topographic Databases*, KMS, The National Survey and Cadastre, Copenhagen, Denmark, February 2000.

Hotelling, H. (1936). Relations between two sets of variates. *Biometrika*, **28**, 321–377.

Hu, M. K. (1962). Visual pattern recognition by moment invariants. *IRE Transactions on Information Theory*, **IT-8**, 179–187.

Jensen, J. R. (2005). *Introductory Digital Image Analysis: A Remote Sensing Perspective*. Prentice Hall, Upper Saddle River, NJ.

Kendall, M. and Stuart, A. (1979). *The Advanced Theory of Statistics*, volume 2, 4th edition. Charles Griffen & Company Limited, London, U.K.

Kohonen, T. (1989). *Self-Organization and Associative Memory*. Springer-Verlag, New York.

Kruse, F. A., Lefkoff, A. B., Boardman, J. B., Heidebrecht, K. B., Shapiro, A. T., Barloon, P. J., and Goetz, A. F. H. (1993). The spectral image processing system (SIPS), interactive visualization and analysis of imaging spectrometer data. *Remote Sensing of Environment*, **44**, 145–163.

Kurz, F., Charmette, B., Suri, S., Rosenbaum, D., Spangler, M., Leonhardt, A., Bachleitner, M., Stätter, R., and Reinartz, P. (2007). Automatic traffic monitoring with an airborne wide-angle digital camera system for estimation of travel times. In U. Stilla, H. Mayer, F. Rottensteiner, C. Heipke, and S. Hinz, editors, *Photogrammetric Image Analysis*. International Archives of the Photogrammetry, Remote Sensing and Spatial Information Service PIA07, Munich, Germany.

Lang, H. R. and Welch, R. (1999). Algorithm theoretical basis document for ASTER digital elevation models. Technical Report, Jet Propulsion Laboratory, Pasadena, CA and University of Georgia, Athens, GA.

Li, S. Z. (2001). *Markov Random Field Modeling in Image Analysis*, 2nd edition. *Computer Science Workbench*. Springer, Tokyo, Japan.

Li, H., Manjunath, B. S., and Mitra, S. K. (1995). A contour-based approach to multisensor image registration. *IEEE Transactions on Image Processing*, **4**(3), 320–334.

Liao, X. and Pawlak, M. (1996). On image analysis by moments. *IEEE Transactions on Pattern Analysis and Machine Intelligence*, **18**(3), 254–266.

Mallat, S. G. (1989). A theory for multiresolution signal decomposition: The wavelet representation. *IEEE Transactions on Pattern Analysis and Machine Intelligence*, **11**(7), 674–693.

Masters, T. (1995). *Advanced Algorithms for Neural Networks, A C++ Sourcebook*. John Wiley & Sons, New York.

Milman, A. S. (1999). *Mathematical Principles of Remote Sensing*. Sleeping Bear Press, Chelsea, MI.

Moeller, M. F. (1993). A scaled conjugate gradient algorithm for fast supervised learning. *Neural Networks*, **6**, 525–533.

Müller, K.-R., Mika, S., Rätsch, G., Tsuda, K., and Schölkopf, B. (2001). An introduction to kernel-based learning algorithms. *IEEE Transactions on Neural Networks*, **12**(2), 181–202.

Murphey, Y. L., Chen, Z., and Guo, H. (2001). Neural learning using AdaBoost. In *Proceedings of the IJCNN'01, International Joint Conference on Neural Networks*, Washington, DC, volume 2, pp. 1037–1042.

Mustard, J. F. and Sunshine, J. M. (1999). Spectral analysis for Earth science: Investigations using remote sensing data. In A. Rencz, editor, *Manual of Remote Sensing*, 2nd edition. John Wiley & Sons, New York.

Nielsen, A. A. (2001). Spectral mixture analysis: Linear and semi-parametric full and iterated partial unmixing in multi- and hyperspectral data. *Journal of Mathematical Imaging and Vision*, **15**, 17–37.

Nielsen, A. A. (2007). The regularized iteratively reweighted MAD method for change detection in multi- and hyperspectral data. *IEEE Transactions on Image Processing*, **16**(2), 463–478. http://www.imm.dtu.dk/pubdb/p.php?4695.

Nielsen, A. A. and Canty, M. J. (2008). Kernel principal component analysis for change detections. In *SPIE Europe Remote Sensing Conference*, Cardiff, U.K., September 15–18, volume 7109.

Nielsen, A. A., Conradsen, K., and Simpson, J. J. (1998). Multivariate alteration detection (MAD) and MAF post-processing in multispectral, bitemporal image data: New approaches to change detection studies. *Remote Sensing of Environment*, **64**, 1–19. http://www.imm.dtu.dk/pubdb/p.php?1220.

Núnez, J., Otazu, X., Fors, O., Prades, A., Palà, V., and Arbiol, R. (1999). Multiresolution-based image fusion with additive wavelet decomposition. *IEEE Transactions on Geoscience and Remote Sensing*, **37**(3), 1204–1211.

Palubinskas, G. (1998). K-means clustering algorithm using the entropy. *SPIE* (*European Symposium on Remote Sensing, Conference on Image and Signal Processing for Remote Sensing*, September, Barcelona), **3500**, 63–71.

Patefield, W. M. (1977). On the information matrix in the linear functional problem. *Journal of the Royal Statistical Society, Series C*, **26**, 69–70.

Polikar, R. (2006). Ensemble-based systems in decision making. *IEEE Circuits and Systems Magazine*, Third Quarter 2006, 21–45.

Press, W. H., Teukolsky, S. A., Vetterling, W. T., and Flannery, B. P. (2002). *Numerical Recipes in C++*, 2nd edition. Cambridge University Press, New York.

Price, D., Knerr, S., Personnaz, L., and Dreyfus, G. (1995). Pairwise neural network classifiers with probabilistic outputs. In M. A. Fischler and O. Firschein, editors, *Neural Information Processing Systems*. MIT Press, Cambridge, MA, pp. 1109–1116.

Prokop, R. J. and Reeves, A. P. (1992). A survey of moment-based techniques for unoccluded object representation and recognition. *Graphical Models and Image Processing*, **54**(5), 438–460.

Quam, L. H. (1987). Hierarchical warp stereo. In M. A. Fischler and O. Firschein, editors, *Readings in Computer Vision*. Morgan Kaufmann, Los Altos, CA, pp. 80–86.

Radke, R. J., Andra, S., Al-Kofahi, O., and Roysam, B. (2005). Image change detection algorithms: A systematic survey. *IEEE Transactions on Image Processing*, **14**(4), 294–307.

Ranchin, T. and Wald, L. (2000). Fusion of high spatial and spectral resolution images: The ARSIS concept and its implementation. *Photogrammetric Engineering and Remote Sensing*, **66**(1), 49–61.

Reddy, B. S. and Chatterji, B. N. (1996). An FFT-based technique for translation, rotation and scale-invariant image registration. *IEEE Transactions on Image Processing*, **5**(8), 1266–1271.

Redner, R. A. and Walker, H. F. (1984). Mixture densities, maximum likelihood and the EM algorithm. *SIAM Review*, **26**(2), 195–239.

Riano, D., Chuvieco, E., Salas, J., and Aguado, I. (2003). Assessment of different topographic corrections in LANDSAT-TM data for mapping vegetation types. *IEEE Transactions on Geoscience and Remote Sensing*, **41**(5), 1056–1061.

Richards, J. A. and Jia, X. (2006). *Remote Sensing Digital Image Analysis*, 4th edition. Springer, Berlin, Germany.

Ripley, B. D. (1996). *Pattern Recognition and Neural Networks*. Cambridge University Press, Cambridge, U.K.

Schölkopf, B., Smola, A., and Müller, K.-R. (1998). Nonlinear component analysis as a kernel eigenvalue problem. *Neural Computation*, **10**(5), 1299–1319.

Schott, J. R., Salvaggio, C., and Volchok, W. J. (1988). Radiometric scene normalization using pseudo-invariant features. *Remote Sensing of Environment*, **26**, 1–16.

Schowengerdt, R. A. (1997). *Remote Sensing, Models and Methods for Image Processing*, 2nd edition. Academic Press, London, U.K.

Schroeder, T. A., Cohen, W. B., Song, C., Canty, M. J., and Zhiqiang, Y. (2006). Radiometric calibration of Landsat data for characterization of early successional forest patterns in western Oregon. *Remote Sensing of Environment*, **103**(1), 16–26.

Schwenk, H. and Bengio, Y. (2000). Boosting neural networks. *Neural Computation*, **12**(8), 1869–1887.

Settle, J. J. (1996). On the relation between spectral unmixing and subspace projection. *IEEE Transactions on Geoscience and Remote Sensing*, **34**(4), 1045–1046.

Shah, S. and Palmieri, F. (1990). Meka—A fast, local algorithm for training feed forward neural networks. In *Proceedings of the International Joint Conference on Neural Networks*, San Diego, CA, **I**(3), pp. 41–46.

Shawe-Taylor, J. and Cristianini, N. (2004). *Kernel Methods for Pattern Analysis*. Cambridge University Press, Cambridge, U.K.

Shekarforoush, H., Berthod, M., and Zerubia, J. (1995). Subpixel image registration by estimating the polyphase decomposition of the cross power spectrum. Technical Report 2707, Institut National de Recherche en Informatique et en Automatique (INRIA), Grenoble, France.

Singh, A. (1989). Digital change detection techniques using remotely-sensed data. *International Journal of Remote Sensing*, **10**(6), 989–1002.

Smith, S. M. and Brady, J. M. (1997). SUSAN—A new approach to low level image processing. *International Journal of Computer Vision*, **23**(1), 45–78.

Strang, G. (1989). Wavelets and dilation equations: A brief introduction. *SIAM Review*, **31**(4), 614–627.

Strang, G. and Nguyen, T. (1997). *Wavelets and Filter Banks*, 2nd edition. Wellesley-Cambridge Press, Wellesley, MA.

Stuckens, J., Coppin, P. R., and Bauer, M. E. (2000). Integrating contextual information with per-pixel classification for improved land cover classification. *Remote Sensing of Environment*, **71**(2), 82–96.

Sulsoft (2003). AsterDTM 2.0 installation and user's guide. Technical Report, SulSoft Ltd., Porto Alegre, Brazil.

Tao, C. V. and Hu, Y. (2001). A comprehensive study of the rational function model for photogrammetric processing. *Photogrammetric Engineering and Remote Sensing*, **67**(12), 1347–1357.

Teague, M. (1980). Image analysis by the general theory of moments. *Journal of the Optical Society of America*, **70**(8), 920–930.

Teillet, P. M., Guindon, B., and Goodenough, D. G. (1982). On the slope-aspect correction of multispectral scanner data. *Canadian Journal of Remote Sensing*, **8**(2), 84–106.

Tran, T. N., Wehrens, R., and Buydens, L. M. C. (2005). Clustering multispectral images: A tutorial. *Chemometrics and Intelligent Laboratory Systems*, **77**, 3–17.

Tsai, V. J. D. (2004). Evaluation of multiresolution image fusion algorithms. In *Proceedings of the Geoscience and Remote Sensing Symposium, IGARSS 04*, Anchorage, AK, volume 1, pp. 20–24.

van Niel, T. G., McVicar, T. R., and Datt, B. (2005). On the relationship between training sample size and data dimensionality: Monte Carlo analysis of broadband multi-temporal classification. *Remote Sensing of Environment*, **98**(4), 468–480.

von Luxburg, U. (2006). A tutorial on spectral clustering. Technical Report TR-149, Max-Planck-Institut für biologische Kybernetik, Tübingen, Germany.

Vrabel, J. (1996). Multispectral imagery band sharpening study. *Photogrammetric Engineering and Remote Sensing*, **62**(9), 1075–1083.

Wang, Z. and Bovik, A. C. (2002). A universal image quality index. *IEEE Signal Processing Letters*, **9**(3), 81–84.

Weiss, S. M. and Kulikowski, C. A. (1991). *Computer Systems That Learn*. Morgan Kaufmann, San Mateo, CA.

Welch, R. and Ahlers, W. (1987). Merging multiresolution SPOT HRV and LANDSAT TM data. *Photogrammetric Engineering and Remote Sensing*, **53**(3), 301–303.

Wiemker, R. (1997). An iterative spectral-spatial Bayesian labeling approach for unsupervised robust change detection on remotely sensed multispectral imagery. In *Proceedings of the Seventh International Conference on Computer Analysis of Images and Patterns*, Kiel, Germany, *Lecture Notes in Computer Science*, **1296**, 263–370.

Winkler, G. (1995). *Image Analysis, Random Fields and Dynamic Monte Carlo Methods. Applications of Mathematics*, volume 27. Springer, Berlin, Germany.

Wu, T.-F., Lin, C.-J., and Weng, R. C. (2004). Probability estimates for multi-class classification by pairwise coupling. *Journal of Machine Learning Research*, **5**, 975–1005.

Yang, X. and Lo, C. P. (2000). Relative radiometric normalization performance for change detection from multi-date satellite images. *Photogrammetric Engineering and Remote Sensing*, **66**, 967–980.

Yocky, D. A. (1996). Artifacts in wavelet image merging. *Optical Engineering*, **35**(7), 2094–2101.

Index

N

Q

R

S